Discovering
Alvarez

Discovering Alvarez

Selected Works of Luis W. Alvarez, with Commentary by His Students and Colleagues

EDITED BY

W. Peter Trower

THE UNIVERSITY OF CHICAGO PRESS

CHICAGO AND LONDON

The University of Chicago Press, Chicago 60637
The University of Chicago Press, Ltd., London

96 95 94 93 92 91 90 89 88 87   5 4 3 2 1

Library of Congress Cataloging in Publication Data

Alvarez, Luis W., 1911–
    Discovering Alvarez.

    Includes bibliographies and index.
    1. Physics.   2. Nuclear physics.   I. Trower,
W. Peter.   II. Title.
QC71.A462   1987   530   87-12766
ISBN 0-226-81304-5

LUIS W. ALVAREZ is senior research scientist at the
Lawrence Berkeley National Laboratory and pro-
fessor emeritus of physics at the University of Califor-
nia, Berkeley. His autobiography, *Alvarez: Adventures
of a Physicist*, has just been published. W. Peter
Trower, professor of physics at Virginia Polytechnic
Institute and State University, also collaborated on
Alvarez's autobiography.

Contents

Galleries of photographs follow pages 54 and 202.

Preface

In the early 1970s, as Luis W. Alvarez's sixtieth birthday approached, his friends began discussing a traditional *Festschrift* to commemorate his many contributions to physics and to honor him. But if anything Luie is not traditional, so he discouraged the proposed meeting and squashed a celebratory party, saying, "There is nothing interesting in getting a year older."

Not to be dissuaded, Art Rosenfeld and Frank Solmitz discussed the possibility of motivating a set of essays based on some of Luie's papers. Because of Frank's unfortunate death and Art's changing research, this didn't materialize. In 1982, anticipating a lull in my own research, I revived, with Art, the project, selected the papers, and contracted the commentators.

Many have contributed. First and foremost are the authors of the commentary, all extraordinarily busy men, who out of respect and affection found the time to provide these reminiscences. Barbara Barnett, Robin Craven, Jerry Silvious, and Alex Trower were skillful text processors. And finally we had the valuable bibliographic services of Richard Robinson (Lawrence Berkeley Laboratory), May West (Fermilab), and John Crissinger (Virginia Tech).

My first goal for this collection of papers was to provide a feeling for the context and consequences of discovery and invention in physics. That these are the product of a single man's lifework is impressive, but it is their importance and variety that is breathtaking. So my second goal was to allow the reader to know the manner of this man who has created so much science. Finally, by this work we all say Thanks to Luie for so enriching our lives by his example, inspiration, and often intervention.

This book is dedicated to those students of our science who may never have the good fortune of knowing Luie personally.

W. Peter Trower
Blacksburg, Virginia

The Diffraction Grating at Grazing Incidence

Franklin Miller, Jr.

When I started to teach physics and astronomy at Rutgers University in 1937, using a broken phonograph record to measure the wavelength of light was part of our heritage. Now I see that I must have picked up the method from Luis W. Alvarez, who published it while he was an undergraduate at the University of Chicago in 1932. Luie was a year ahead of me in graduate work there.

In 1932 no United States journal was devoted explicitly to the teaching of physics. *The American Physics Teacher,* later renamed *American Journal of Physics* (AJP), did not begin publication until February, 1933. *School Science and Mathematics*, where my own first published paper also appeared, was a fixture at my high school in St. Louis, Missouri. The magazine had (and still has) mathematics problems to challenge high-school students. I vividly recall in detail one simultaneous-equations problem of 1928 that I fully intend to solve, someday.

In the accompanying article Luie gives a derivation of his working equation for the reflection grating at small angles. Perhaps because of the level of sophistication of his audience, he chose to get around equating sin θ to tan θ by using straightforward algebra and the binomial theorem.

In a 1965 lecture-demonstration, Art Schawlow measures the wavelength of the red helium-neon laser line by grazing reflection from a coarse grating formed by the grooves of a metal ruler. He writes, "It is well known that a reflection grating with widely spaced grooves gives good diffraction spectra if the light is incident nearly parallel to the surface of the grating. . . [This was] pointed out to me some years ago by Professor R. R. Richmond, University of Toronto."[1] The method was published in 1942 by R. W. Ditchburn[2] of Trinity College, Dublin, who used about 2 inches of a millimeter-ruled steel scale (grating space of one-half millimeter). Ditchburn obtained spectra by bending the scale into an arc of a circle, thus demonstrating the focusing properties of curved gratings. Grazing angles relative to the surface were 3°–10°. A search through the pages of AJP shows no other reinvention of this particular wheel.

Luie's first paper was no doubt a significant input to his 1936 Ph.D. thesis, published as "The Diffraction Grating at Grazing Incidence."[3] To see the motivation for this work and to assess its importance we must look back to the 1930s, when there was concern about the universality of two of the fundamental atomic constants, the electric charge of the electron (e) and its specific charge, its charge divided by its mass (e/m).

Deflection measurements give e/m for a free electron. However, e/m can also be deduced for electrons bound to an atom by measuring the frequency separation of the H_α red Balmer series lines for the hydrogen isotopes[4] and from the Zeeman effect.[5] In the early 1930s there was enough discrepancy between the deflection and spectroscopic e/m values to support speculation that a bound electron might in some way be slightly different from a free electron traveling in a magnetic field. In his article on the physical constants—the first in the initial issue of *Reviews of Modern Physics*—R. T. Birge gave two e/m values, suggesting that the researcher use whichever value was appropriate for his particular experiment.[6] F. G. Dunnington's 1937 all-out precision deflection measurement of e/m was consistent with the spectroscopic value.[7]

A similar dilemma existed regarding the value of e. Millikan's famous oil-drop experiments, critically evaluated by Birge,[8] gave 1.591×10^{-19} coulomb $\pm 0.1\%$, a value attributed to a free electron. For a bound electron, e was inferred from x-ray experiments as follows: Use a ruled grating at grazing incidence to measure the wavelength of an x-ray line. Measure the Bragg diffraction angle of this x-ray line with a calcite crystal, thus finding the grating space of the calcite cleavage planes. Using the grating space and the known structure and density of calcite, calculate Avogadro's number N_A. Finally, with the well-measured value of the faraday Q_F, determine e from $Q_F = eN_A$. Various corrections need to be applied, such as that for the calcite

1. A. L. Schawlow, Am. J. Phys. **33,** 922 (1965).
2. R. W. Ditchburn, Am. J. Phys. **10,** 195 (1942).
3. L. W. Alvarez, J. Opt. Soc. Am. **26,** 343 (1936).
4. C. D. Shane and F. H. Spedding, Phys. Rev. **47,** 33 (1935); and W. V. Houston, Phys. Rev. **51,** 446 (1937).
5. L. E. Kinsler and W. V. Houston, Phys. Rev. **45,** 104 (1934); and **46,** 533 (1934).
6. R. T. Birge, Rev. Mod. Phys. **1,** 1 (1929).
7. F. G. Dunnington, Phys. Rev. **52,** 475 (1937).
8. R. T. Birge, *loc. cit.*

x-ray index of refraction; and various assumptions need to be explicated, such as the absence of crystal voids and surface defects. In 1926 Arthur Compton and Richard Doan first measured *e* using a ruled grating.[9] Their work was refined by others, and in 1931 J. A. Bearden measured four different x-ray wavelengths to be 0.20%–0.22% greater than the corresponding crystal values.[10] Since the crystal-determined wavelength depends on the cube root of *e*, this implied that Millikan's oil-drop *e*-value was 0.6% low, an apparently untenable conclusion. Could the free electron's charge be 0.6% less than that of one bound in a crystal lattice? Against this background of confusion about the true value or values of *e*, Luie conceived an experiment to test the grating formula at grazing incidence. In 1933 C. Eckart made an exhaustive theoretical analysis of this problem and concluded that the elementary grating law was valid.[11] However, it had not been experimentally tested since 1888, when L. Bell[12] used Rowland's high-quality gratings to make precision absolute wavelength measurements. Bell used glancing angles of about 45° and found only a slight discrepancy between his values and those obtained with the newly developed interferometer. But perhaps there *was* some unsuspected systematic error in the grating formula when applied at grazing incidence, as in x-ray experiments. So Luie replicated Bearden's work on a large scale, measuring the wavelength of the mercury blue line. His reflection grating was ruled on an iron surface plate with grating space of 4.0000 millimeters, with the groove width one-half the grating space. Thus the wavelength and grating space were each increased 2400 times. The glancing angle was 38 minutes 52.39 seconds in a typical run.

Although I will not recount Luie's thesis in detail, I remember well the perseverance and ingenuity which became the hallmark of his later work. I can still see the fans he set up in a long basement corridor of the University of Chicago's Ryerson and Eckhart Halls to impose a regular air flow and so eliminate random turbulence along his 20-meter optical path. A nostalgic sentence in Luie's precomputer thesis stated that "An-

doyer's 15 place trigonometric tables were used throughout and all computations were made on a machine." His thesis concluded ". . . that the wave-length of the blue mercury line, as measured in this way is 4358.71 ± 0.8 A. The accepted value, as given in Kayser's tables, is 4358.30 A. This agreement upholds the view that the diffraction law is valid at grating incidence."

It turned out that Millikan's oil-drop experiment was, in fact, in error because of an erroneous value for the viscosity of air, which enters into Stokes' Law for the moving drop. By 1932 it was becoming apparent that the viscosity of air measured by capillary flow differed from Millikan's value, which was obtained by his student Harrington using the classical cylinder-drag method. The systematic error was eventually traced to a tiny necking-down of the torsion wire to a slightly smaller radius when a standard calibration ring was added during the measurement of the moment of inertia of the suspended cylinder. The inconstancy of the torsion constant, which depends on the fourth power of the wire's radius, translated into a viscosity error and thus into an error in *e*.

Millikan's reputation loomed large in all this. What graduate student or postdoc would dare challenge Millikan on his own turf? Even Bearden did not draw a sufficiently strong conclusion from his own work. At Chicago, prior to his precision ruled-grating wavelength measurements, his thesis advisor was Sam Allison, who later was my advisor. Allison once said that Bearden was the first person to measure the charge of the electron correctly without knowing that he had done so! In *X-Rays in Theory and Experiment*,[13] Allison devotes ten pages to a detailed discussion of the "x-ray" value of *e* and the controversy that was to be resolved within a year or two, partly with input from Luie's Ph.D. thesis.

This first published paper of Luie's concerns a teaching device. He would have been one of the great teachers of undergraduate physics had he chosen that path. Of course, over the years Luie has not been aloof from teaching, which he thoroughly enjoys. Through his writings and his wide-ranging interests in important but offbeat physics, we have all been placed in his debt.

9. A. H. Compton and R. L. Doan, Proc. Nat. Acad. Sci. **11**, 598 (1926).

10. J. A. Bearden, Phys. Rev. **37**, 1210 (1931).

11. C. Eckart, Phys. Rev. **44**, 12 (1933).

12. L. Bell, Phil. Mag. **25**, 350 (1888).

13. A. H. Compton and S. K. Allison, *X-Rays in Theory and Experiment* (Van Nostrand, New York, 1935), pp. 694–704.

A SIMPLIFIED METHOD FOR THE DETERMINATION OF THE WAVE LENGTH OF LIGHT.

By Luis Alvarez,

Chicago, Ill.

The wave length of light may be measured with the aid of a phonograph record, an electric light, and a meter stick. If the reflection of an electric light in the surface of a phonograph record is viewed at a large angle of incidence, it will be seen to be accompanied by several spectra. From measurements of the spectrum closest to the reflected image, the wave lengths of the various colors of the rainbow may be calculated. The record is used as a coarse diffraction grating, and a spectroscope is not needed.

A piece about two inches wide and as long as the grooved portion of the record is broken from the disk. On this piece, the grooves will be almost straight and parallel, as on a reflection grating. The only other piece of apparatus needed in the experiment is an electric bulb with some sort of reflector or shade attached to it. The reflector should be arranged to throw the beam horizontally, and with equal portions above and below the center of the bulb. The lamp is placed against a wall at a distance of five or ten meters from the observer. The latter should be seated directly behind some support for the record, such as the back of a chair. The record is held in the hand in a nearly horizontal position, and the eye is placed as close to it as possible, and directed downward. The record is held so that it may be rotated about a horizontal axis, which is perpendicular to the line joining the lamp with the observer. It is placed before only one eye; the other is used to determine the position of the spectrum on the wall in relation to the lamp. On looking into the record, a reflection of the lamp and its reflector will be seen. If a line joining the lamp and the record is in the surface of the latter, it will be impossible to see the image of the former. But in this case, the image of the upper edge of the shade will coincide with its lower edge, as seen with the other eye. The spectrum will still be seen plainly enough for the purposes of the experiment. Its position on the wall

Reprinted by permission from *School Science and Mathematics*.

below the lamp should be noted roughly, and a sheet of paper crossed with lines about an inch apart should be pinned on the wall so that the apparent position of the spectrum is on the paper. The lines should be numbered in large numbers for convenience in identification at a distance. Now, any color such as red, will be chosen on which to make the measurements. The record is held in the hand as described before, and steadied on the rest to prevent vibration. It is adjusted so that the image of the shade coincides with the shade as seen in the other eye. Now, the first spectrum will appear to be located on the wall behind the lamp, and the red portion of it will fall on one of the lines drawn on the paper, or between two of them. The particular line is noted by its number, and the distance from this line to the point directly in back of the center of the lamp is measured with a meter stick. The distance from the wall to the position occupied by the record is also measured. The number of grooves per centimeter is determined by using a low powered microscope, or by counting the number of revolutions necessary to move a needle a distance of one centimeter across the disk. Representing these three figures by d, D, and n, respectively, the wavelength of

red light is given by the relation: wavelength$=\dfrac{d^2}{2nD^2}$ cm.

In this manner, the wavelength of any color may be determined.

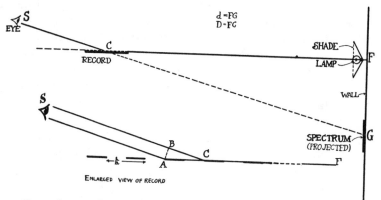

ENLARGED VIEW OF RECORD

Sample results obtained by using a small Victor Record, and a frosted electric light globe.

D = 762 cm.
n = 34.9 per cm.
d (green) = 46.5 cm.
d (red) = 52.5 cm.

Wavelengths	Calculated	Accepted
green	0.00005340 cm.	0.00005200 cm.
red	0.00006810 cm.	0.00006800 cm.

The derivation of the formula is given below. It is based on the wave theory of light, and Huygen's Principle. (See diagram.)

Distance between grooves = 1/n cm.

Path difference between SAF and SCF = AC − BC = λ, since first spectrum only is measured. (λ represents wavelength).

$AB/AC = d/D$

$AB = d/Dn$

$BC^2 = AC^2 - AB^2 = 1/n^2 - \dfrac{d^2/D^2}{n^2} = \dfrac{1-d^2/D^2}{n^2}$

$BC = 1/n \cdot \sqrt{1 - d^2/D^2}$

$AC - BC = \lambda = \dfrac{1}{n} - \dfrac{1}{n}\sqrt{1 - \dfrac{d^2}{D^2}} = \dfrac{1 - \sqrt{1 - \dfrac{d^2}{D^2}}}{n} = \dfrac{1 - 1 + \dfrac{1}{2}\dfrac{d^2}{D^2}}{n} + \text{-------}$ (neglect)

$\lambda = \dfrac{d^2}{2nD^2}$

2

Some Early Days of Cosmic Rays

H. Victor Neher

There are times in the history of science when our understanding of nature's behavior builds in a crescendo. Before, progress is usually slow, but essential. The activity following such a peak, the diminuendo, is also important. But activities before and after lack the excitement that characterizes the crescendo. Thus it was with the search for the nature of cosmic rays and how they interact with matter. The crescendo occurred in the 1930s, and following the hiatus of the 1940s work has continued at a slower pace.

Every era is seen by its denizens as unique, but some eras truly are. We entered the 1930s with much of the background necessary to study cosmic rays. Relativity had been around for some time, quantum mechanics had come into being, and the consequences of applying the one to the other were being worked out. The experiments that prove these theories formed the foundation of progress in the 1930s.

There is always the personal element, especially for younger scientists who find themselves in the midst of new discoveries and are getting to know the people responsible for them. For those of us who were beginning to try our wings, visitors, especially from Europe, had a profound effect. There were numerous distinguished visiting physicists at the University of Chicago, where Luis W. Alvarez started his career; at the University of California at Berkeley, where he has been since; and at the California Institute of Technology, where I spent 42 years.

Robert Oppenheimer served to connect Luie's and my worlds, as he spent parts of most years at Berkeley and at Caltech. He would lecture in a hushed whisper and was difficult to follow even in a small room. Carl Anderson, the discoverer of the positron, was among those signed up as a graduate student to take Oppie's quantum mechanics course. The going was not easy, so just before the deadline to drop the course without penalty, Carl told Oppie that he wasn't getting much out of it and wanted to drop it. Oppie protested that he couldn't do that, since he was the only one still registered!

Oppie also had difficulty making his lectures intelligible. One of the great teachers at Caltech was Richard Tolman, who was able to sense his students' difficulties and would go out of his way to clear up obscure points. Tolman felt that you had to be comfortable about an idea before you really understood it. Oppie's students weren't comfortable, so Tolman proposed that Oppie slow down. He suggested that to accomplish this, in preparing lectures Oppie include a second subject on which his mind could focus during his lecture. In that way his pace might be reduced, to his students' benefit.

In the summer of 1930 I made a short visit to the Berkeley Physics Department to get acquainted with the staff and learn about their research. It was nice to see my former Pomona College roommate, Stan Livingston, who was then working with Ernest Lawrence on the cyclotron. New generations of cyclotrons were appearing at a rapid rate, and Lawrence had heard of a big magnet at Palo Alto which he wanted. So the three of us drove down to see it in the yards of the Federal Telephone and Telegraph Company, who had built it during World War I for a radio voice transmitter. It ended up as the magnet for the 37-inch cyclotron.

The search for the nature of cosmic rays was begun in 1910 with the manned balloon flights of Gockel, continued by Hess, and extended by Kohlhorester.[1] From these studies the extraterrestrial origin of this radiation was inferred, since their intensity increased with altitude once the balloon rose beyond the range of Earth's radioactivity.

After World War I several physicists, including Swann at the Bartol Institute and Millikan at Caltech, became interested in cosmic rays. Some suggested that the Hess-Kohlhorester results could be explained by radiation originating in the atmosphere, making a source in outer space unnecessary. To gain more information, Millikan went into the California mountains and sank instruments into two lakes at different altitudes. He found that the absorption of the radiation in the air at different elevations was the same as that of an equivalent amount of water. Millikan concluded that this could only happen if the radiation was coming from outside Earth and so named them cosmic rays.

Millikan reasoned that, in passing through space, such radiation would interact with matter and liberate

---

1. Essays by participants in the early cosmic-ray and particle physics are included in (a) *The Birth of Particle Physics*, edited by Laurie Brown and Lillian Hoddeson (Cambridge University Press, Cambridge, 1983) and (b) *Early History of Cosmic-Ray Studies*, edited by Y. Sekido and H. Elliot (Reidel, Boston, 1982).

electrons. Using the theories of gamma-ray absorption, the cosmic-ray photon energies were estimated, as were those of the accompanying electrons. Paul Epstein at Caltech then calculated how Earth's magnetic field deflects such electrons. From these results Millikan was led to expect fewer cosmic-ray electrons as he approached the equator. To test his hypothesis, Millikan and Cameron in 1926 took three instruments by ship to Peru, making measurements on the way.

An unfortunate series of circumstances prevented them from finding any latitude effect at sea level. In Bolivia they did observe less radiation than at the same air pressure in California, but possible leaks in their air chambers and other experimental uncertainties made them decide that there was no latitude effect at these elevations either. From their result and the way the radiation was absorbed in the atmosphere, Millikan concluded that the primary cosmic radiation consisted only of photons.

Thus, when Clay, and later Compton, found a latitude effect, Millikan decided that further experimental work was necessary. In the spring of 1932, he and Ira Bowen invited me to participate in their ambitious program to measure cosmic rays at different latitudes from sea level to airplane altitudes. My job was to make a more accurate self-recording instrument that would be insensitive to vibration, tilt, humidity, and temperature.

By autumn, I was on a ship from Los Angeles to Peru and also had instrument troubles. But going from Peru to New York and back to Los Angeles through the Panama Canal, I saw very clear evidence of the latitude effect. Airplane flights in that same year in Peru, Panama, and other latitudes extending into northern Canada also showed the latitude effect up to 22,000 feet.

Now, since some primary cosmic rays were electrically charged, it was possible to determine their kind—positive or negative. If they carried a negative charge, Earth's magnetic field would bend them, so that more would come from the east than from the west. On the other hand, if the primaries carried a positive charge, more particles would come from the west. Thus, the latitude and the east-west cosmic-ray effects are intimately related. If there were no latitude effect, there could be no east-west effect; and where the latitude effect is greatest, there too would be the greatest east-west effect. At sea level there is no latitude effect from about 40° to the poles, since the primaries get deflected away before they reach Earth's surface. With increasing altitude, the onset of the latitude effect moves toward the poles.

In 1931, Bruno Rossi realized the importance of a possible east-west effect and set up an experiment near Florence, Italy. He used a directional cosmic-ray telescope of Geiger-Muller tubes to detect individual charged particles. Rossi also invented an electrical circuit that only responded when all his tubes discharged at once. The circuit Luie used in his east-west telescope worked well with his two counters, but Rossi's method could be used with any number of counters.

The study of charged-particle paths in Earth's magnetic field has been interesting since Stormer showed that Earth's aurora was caused by charged particles from the Sun. Later calculations were made for cosmic rays by Lemaître and Vallarta, Epstein and others. The presence of the earth complicates the situation but at the equator and the poles the calculations are simplified. Fermi pointed out that a general mechanics theorem, Liouville's, applied to cosmic rays predicts that cosmic rays far from the earth going uniformly in all directions will arrive at the earth with their intensity undiminished.[2] In the 1930s, Vallarta with the MIT electronic computer, a room of vacuum tubes and circuits, kept at this problem for many years. But by 1931 Rossi had mastered enough of the theory to guide his own experiments. He found no east-west effect in Florence because there was no latitude effect there. In 1933 he went to Asmara, Eritrea (11° north geomagnetic latitude and elevation 7400 feet) and measured an east-west effect of 26% at an angle of 45° from the vertical, with more particles coming from the west. In the United States Tom Johnson of Bartol measured no east-west effect in Swarthmore, Pennsylvania, or on top of Mount Washington in New Hampshire.

Luie was invited by Compton to an informal meeting in November, 1932. Lemaître and Vallarta had concluded that a latitude effect implied an east-west effect if the responsible particles were predominantly either positively or negatively charged. Vallarta lived in Mexico City, where the low latitude and high altitude made the effect quite significant. Johnson and Luie accepted Vallarta's invitation to host any physicists who would perform an east-west experiment in Mexico City. Johnson was ready, while Luie took some time to prepare. They both, however, were met by Vallarta as they got off the same train in Mexico City.

Luie and Johnson set up their equipment on the roof of the Genéve Hotel where they both were also staying. Luie's setup was primitive, and to eliminate errors he bought a wheelbarrow in which he installed his equipment. To orient his telescope east or west he simply turned the wheelbarrow around. Johnson's counters were mounted so that he could rotate his telescope through 180° to change its orientation from east-pointing to west-pointing. After a few days, each found more particles coming in from the west than from the east, ~12% at 45° from the vertical. Thus, the primaries causing this difference had a positive electrical charge. Luie published his results in the accompanying article.

2. See Bruno Rossi, *Cosmic Rays* (McGraw-Hill, New York, 1964), p. 65.

Rossi's experiment in Eritrea confirmed Luie's results several months later.

Missetting the telescope angle can introduce important systematic errors. For Johnson, an error could result if his telescope axis was not vertical. In the lower atmosphere, cosmic rays become rarer toward the horizon because they pass through more air, which absorbs them. The count rate n is approximately $n_0\cos^2\theta$, where n_0 is the count rate at the vertical and θ the zenith angle. For Johnson's telescope, which rotated about a vertical axis, an error in θ would increase the count rate in one direction and decrease it in the other, giving a spurious east-west effect, $\Delta n/n = 4\tan\theta\,\Delta\theta$. Thus a 1° error in setting the vertical axis with the telescope at 45° would give a 7% east-west counting error.

Although nothing was said in their 1933 publications about errors from misalignment, both Luie and Johnson must have been aware of the consequences of such misadjustments. Luie's wheelbarrow technique to get things lined up was crude, but he must have succeeded.

Although the east-west measurements of Johnson, Alvarez, and Rossi each established that more primary cosmic rays carried a positive charge, they said nothing about the nature of the negative particles. In the following years, east-west measurements were made in different parts of the world and at various altitudes.

Above a primary energy of 25 billion electron volts (GeV), Earth's magnetic field no longer acts as a magnetic analyzer, and the east-west effect is unobservable. But up to that value, all the measured geomagnetic effects are consistent with the primaries having no negatively charged particles. Cloud chambers carried in high-altitude balloons showed that high-energy electrons constitute less than 1% of the positives. By the 1950s, the cosmic rays were identified, using photographic emulsions, as nuclei with roughly the same relative abundance as found on Earth.

By what mechanisms are cosmic rays absorbed in the atmosphere? The history of this search is the history of particle physics, for it was in cosmic rays that the positron, the muon, and other important new particles were first found. But these events occur too infrequently in nature, so machines with many times nature's intensity were built. It was at the center of this activity that Luie found himself in 1936. I agree with Rossi that the early thirties were "the age of innocence of experimental particle physics."[3] There were challenges on every side, and no one sensed them more clearly or devised ways of meeting them better than Luie.

3. B. Rossi, ref. 1a, p. 204.

Reprinted from THE PHYSICAL REVIEW, Vol. 43, No. 10, May 15, 1933

A Positively Charged Component of Cosmic Rays

The relatively low intensity of cosmic rays at low geomagnetic latitudes, as recently found by our associated expeditions[1] and others,[2] indicates that a part of the cosmic rays consists of electrified particles. When interpreted in terms of Lemaitre and Vallarta's theory[3] of the deflection of electrified particles by the earth's magnetic field, these results indicate that at geomagnetic latitudes higher than about 45° the earth's magnetic field should not alter the direction of the incoming rays as observed at sea level. This is in accord with the sea-level observations of Johnson and Street,[4] which show a symmetrical East-West distribution. At the geomagnetic equator an analysis of our intensity-latitude curves suggests that most of the cosmic rays which are affected by the earth's magnetic field are too strongly deflected to reach the earth's surface. If this is correct, there should be but little asymmetry in the direction of approach of the cosmic rays near the equator. In an intermediate zone, however, where the intensity *vs.* latitude curve is steep, the rays that are being affected by the earth's magnetic field should strike the earth from certain directions but not from others. If the rays are positively charged, they should come mostly from the west, if negatively, predominantly from the east, due to deflection by the earth's magnetic field. From such considerations Vallarta has suggested that Mexico City should be a good place to search for the predicted asymmetry in the direction of the incoming cosmic rays. Besides being in the favorable zone of geomagnetic latitude (29°N), its elevation (2310 meters) is sufficient to avoid some of the disturbing effects of the atmosphere.

In order to observe the direction of the incoming particles we have used a double coincidence counter, as shown diagrammatically in Figs. 1 and 2. Tests made by separating the tubes indicate that chance coincidences occur at the rate of only about 1.5 per hour, so that with a normal

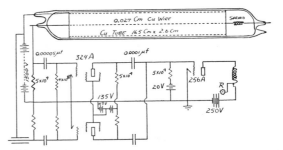

FIG. 2. Circuit used with double coincidence counter.

counting rate of about 5 per minute these were of negligible importance. The zenith angle θ of the line joining the axes of the tubes with the vertical was measured with the help of a protractor and spirit level. In order to avoid any possible change of conditions, the whole apparatus was mounted on a platform, which was rotated through 180 degrees when the changes between east and west were made. Readings of about a half hour's duration were taken alternately between east and west at the same zenith angle θ. For each angle the final series of readings totaled about fourteen hours on either side. By changing thus back and forth, enough readings were obtained to make a good estimate of the probable errors of the observed counting rates. The errors thus estimated from the consistency of successive readings under similar conditions were but little greater than those calculated as the statistical error from the total number of coincidences in the series. This means that no serious disturbing factor was affecting the readings. Table I summarizes our results.

TABLE I. *East-west measurements at Mexico City, April, 1933.*

Geomagnetic latitude 29°N, elevation 2310 m, barometer, 56.5 cm.

Zenith angle		West	East	West/East
15°	Counts	5370	4856	
	Rate	6.83±0.07	6.64±0.07	1.03±0.02
30°	Counts	4897	4869	
	Rate	5.79±0.06	5.49±0.06	1.055±0.015
45°	Counts	2691	2693	
	Rate	3.70±0.05	3.30±0.05	1.12±0.02

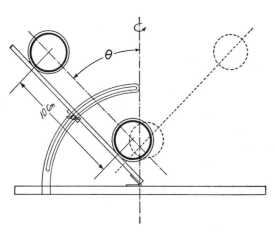

FIG. 1. Arrangement of coincidence counting tubes for studying East-West asymmetry of cosmic rays.

[1] A. H. Compton, Phys. Rev. **41**, 111 (1932); **43**, 387 (1933).

[2] J. Clay and H. P. Berlage, Naturwiss. **37**, 687 (1932).

[3] G. Lemaitre and M. S. Vallarta, Phys. Rev. **43**, 87 (1933).

[4] T. H. Johnson, J. Frank. Inst. **214**, 689 (1932).

It will be noted that at the larger zenith angles the rate at which the rays come from the west is greater than that from the east by several times the probable error of the measurements. It would appear that the asymmetry thus observed at Mexico City is considerably larger and more definite than that found by Johnson and Street[5] on Mt. Washington, geomagnetic latitude 55°N, elevation 1920 meters. This preponderance of rays from the west seems necessarily to imply the existence of a positively charged component of the cosmic rays.

Since our earlier measurements have shown that the cosmic rays at geomagnetic latitude 29° differ by only about 14 percent from those in high latitudes for this elevation, the difference in counts in the east and west directions is of the order of magnitude to be expected due to the deflection of the particles by the earth's magnetic field. The smallness of the effect confirms our earlier conclusion that most of the rays capable of penetrating the earth's atmosphere are not sufficiently bent by the earth's magnetic field to prevent them from reaching the earth. We may add that these data are consistent with the view that the positively charged component here found consists of Anderson's newly discovered[6] positrons.

We wish to thank Professor R. D. Bennett for suggesting the circuit used, Dr. M. S. Vallarta and Dr. T. H. Johnson for valuable suggestions in carrying on the experiment, and the Carnegie Institution of Washington for financial aid.

Luis Alvarez
Arthur H. Compton
University of Chicago,
April 22, 1933.

[5] T. H. Johnson and J. C. Street, Phys. Rev. **43**, 381 (1933).

[6] C. D. Anderson, Science **76**, 238 (1932); Phys. Rev. **43**, 491 (1933).

3

K-Electron Capture by Nuclei

Emilio Segrè

The theory of positron emission and orbital electron capture was first considered by G. C. Wick in a paper presented to the Accademia dei Lincei on March 3, 1934 which begins:

This paper contains an application of beta ray theory as recently proposed by E. Fermi, to the phenomena of artificially provoked radioactivity observed by F. Joliot and I. Curie. . . .

The emission of a positive electron is not the only way in which the nuclei considered may decay. Instead of an electron of negative kinetic energy, one may annihilate one of the electrons of the *K, L, M* atomic shells. In this case the transmutation occurs without positron emission; one will have, however, an emission of x-rays or of Auger electrons. These phenomena will be difficult to observe in our case.

One begins to realize that the number of radioactive atoms formed may be much greater than that found by counting the positrons emitted . . . Fortunately the theory . . . permits a fairly precise evaluation of the branching ratio between positron emission and *K*-capture. . . .[1]

Wick then proceeds to calculate these two nuclear decay probabilities. He ends by thanking Fermi for showing him his beta-decay manuscript[2] before publication. Fermi had explained his work to friends while skiing in the Alps during the 1933 Christmas vacation. Wick learned of it at about the same time.

Curie and Joliot had announced their discovery of artificial radioactivity by positron emission to the Paris Academy of Sciences only six weeks before Wick's communication.[3] Thus there is little doubt that Wick was the first to mention *K*-capture and calculate its probability.

Bethe and Peierls also pointed out the possibility of *K*-capture three weeks after Wick's paper, when they wrote, "In addition to the nuclear processes mentioned in our previous communication, it may also be expected that a nucleus catches one of its orbital electrons, decreases by one its atomic number, and emits a neutrino."[4] The next year Yukawa and Sakata re-

peated and improved Wick's calculation,[5] showing that it had not escaped their attention, in spite of being written in Italian and published in an obscure journal.

The discovery of the precise form of the beta-interaction proved to be a difficult task.[6] Because of errors in the measurements of beta spectra, it was thought that the vector interaction originally proposed by Fermi, should be replaced by a more complicated one proposed by Konopinski and Uhlenbeck. Yukawa and Sakata showed that the two interactions give about the same results for *K*-capture.[7]

The theoreticians had emphasized that the observation of *K*-electron capture was a difficult problem. In fact, three years elapsed before Alvarez succeeded in giving indisputable proof of the effect. His first communication is dated June 21, 1937.

An atom with Z electrons is transformed by *K*-capture into one with $Z - 1$ electrons, emitting an undetectable neutrino and possibly nuclear gamma rays. The clearest indication of *K*-capture comes from the resulting atomic x-rays and associated Auger electrons. These originate in the rearrangement of the atomic electrons provoked by the hole left in the innermost atomic shell, the *K*-shell, when the *K*-electron is captured by the nucleus. In addition, there is annihilation radiation from emitted positrons that stop in the source. Alvarez tackled the problem of detecting and interpreting the radiations resulting from *K*-capture with great ingenuity, taking advantage of whatever techniques came to mind and the uniquely strong sources available at the cyclotron.

In the first of his two *Physical Review* letters on nuclear *K*-capture[8], Alvarez discusses the problems associated with identifying *K*-capture of vanadium-48 produced in deuteron bombardment of titanium. The accurate measurement of the absorption of titanium's characteristic x-rays in aluminum is the key feature. Vanadium-48 is also a positron emitter, and Alvarez

1. G. C. Wick, Rendiconti Accad. Lincei **19**, 319 (1934).
2. E. Fermi, Nuovo Cimento **11**, 1 (1934).
3. I. Curie and F. Joliot, C. R. Acad. Sci. **198**, 254 (1934).
4. H. A. Bethe and R. Peierls, Nature (London) **133**, 689 (1934).

5. H. Yukawa and S. Sakata, Phys. Math. Soc. Japan, Proc. **17**, 467 (1935).
6. C. Wu and S. Moszkowski, *Beta Decay* (Interscience, New York, 1966), pp. 3ff.
7. A concise review of the subject, with references, is contained in F. Rasetti, *Elements of Nuclear Physics* (Prentice Hall, New York, 1936), which was widely used at the Radiation Laboratory.
8. L. W. Alvarez, Phys. Rev. **52**, 134 (1937).

concluded from his measurements that the ratio of electron capture to positron emission was 1.0 ± 0.4. The modern value, 50.4/49.6, agrees very well with his observation and shows Alvarez's ability for obtaining the right results by simple means as well as his luck with statistical errors.

The second letter is devoted to gallium-67.[9] To pin down its K-capture, Alvarez had to find characteristic x-rays from zinc-67 of such small intensity that they were undetectable in conventional x-ray spectrographs. Alvarez overcame this difficulty by measuring the x-ray absorption in nickel and copper foils, where the absorption edges bracket the characteristic x-rays from zinc. The method, described in a different context elsewhere,[10] had been used extensively in x-ray investigations. Alvarez used it repeatedly for nuclear problems and I learned it from him. Alvarez's conclusions on the nature of the radiations compare very well with present knowledge.[11]

In the accompanying paper, Alvarez recapitulated and extended his work on K-capture. He gives details on the experiments described in his two Letters and discusses other possible K-capturers. Using his own experiments and published data, he considers tantalum-180, potassium-40, mercury-197 (an erroneous assignment—the activity is due to mercury-199m), silver-106, cadmium-107, beryllium-7 and manganese 54.

Alvarez's K-capture papers show the way in which the Radiation Laboratory then operated. We see Alvarez freely borrowing beta-spectrographs and other instruments from colleagues and students: Lyman, Kalbfell, Van Voorhis, Abelson, and others. He is also able to immediately use results obtained by McMillan, Kamen, Ruben, Livingood, and Seaborg. These exchanges went both ways, fostered by Monday Meetings in the Physics Department, encounters and conversations at the Faculty Club, and the many hours spent by everyone tending the cyclotron. The papers also show Alvarez's devotion to and admiration for Lawrence and the interaction with Oppenheimer and his students for theoretical consultations.

Alvarez's work on K-capture was important for the progress of nuclear physics, as it confirmed the theory of beta-decay. It also connected with other problems, such as that of internal conversion, which were of central interest at the time.

9. L. W. Alvarez, Phys. Rev. **53**, 606 (1937).

10. A. H. Compton and S. K. Allison, *X-Rays in Theory and Experiment* (Van Nostrand, New York, 1935).

11. *Table of Isotopes,* 7th edition, edited by C. M. Lederer and V. S. Shirley (Wiley, New York, 1978).

OCTOBER 1, 1938   PHYSICAL REVIEW   VOLUME 54
Printed in U. S. A.

The Capture of Orbital Electrons by Nuclei

Luis W. Alvarez

Radiation Laboratory, Physics Department, University of California, Berkeley, California

(Received July 26, 1938)

The simple theory of electron capture is outlined and three general methods for its detection are suggested. The first experimental evidence for the process (in activated titanium) is described. A rigorous experimental proof of the hypothesis is given for the case of Ga^{67}. A summary of several isotopes whose properties are best explained on this hypothesis is appended. The properties of Ga^{67} are described in considerable detail, and include the first evidence for internal conversion in artificially radioactive atoms.

Introduction

THE suggestion that positron emitters might decay by the alternate process of electron capture was first advanced by Yukawa[1] from considerations based on the Fermi theory of beta-ray emission. In this theory, the electrons and positrons are pictured as being created at the moment they are ejected, during neutron-proton transitions. The continuous beta-ray spectrum and the conservation of spin are explained by the simultaneous emission of a neutrino and electron. One may represent the transition involved in electron and positron decay by the following equations:

$$N \to P + e^- + \nu \tag{1}$$

$$P \to N + e^+ + \nu. \tag{2}$$

On the basis of Dirac's theory, however, the positron is merely the "hole" left in the con-

[1] Yukawa and Sakata, Proc. Phys. Math. Soc. Japan **17**, 467 (1935); **18**, 128 (1936).

tinuum of negative energy electrons when one of these electrons is given a positive energy by the addition of at least $2mc^2$. The proton in (2) does not transform into a neutron and positron, but rather captures a negative energy electron, and turns into a neutron, leaving the hole in the negative energy sea, or positron. Eq. (2) may then be written

$$e^- + P \to N + \nu. \tag{3}$$

The experimental observation that positrons may be annihilated (a positive energy electron falling into the hole), shows that there is no essential difference between electrons in the two energy states. Therefore, there is no *a priori* reason why Eq. (3) demands the use of a negative energy electron. In fact, when the energy difference between parent and daughter nucleus is less than $2mc^2$, it would be impossible for the relation to be satisfied unless a proton could capture an ordinary electron. Since there are many cases of negative beta-ray decay with an energy release of less than this value, it is natural to suppose that there would be excited nuclei whose desire to emit positrons could not be allowed on energetic grounds. Yukawa suggested that in these cases, the decay would proceed by the capture of an orbital electron. In addition, he calculated, and others[2, 3, 4, 5] extended the calculations, that even when there was enough energy to create a pair, a certain fraction of the excited nuclei would decay by electron capture. The branching ratio of the two processes was found to depend on the energy available, the spin change involved, and the nuclear charge (density of electrons at the nucleus). Electron capture should become more probable as the energy decreases, and as the spin change, atomic number and half-life increase.

Experimental Methods

Alpha-decay and the two well-established methods of beta-decay are easy to observe, since ionizing radiations are emitted in these processes. Eq. (3) shows that only the undetectable neutrino is given off by the nucleus in this new type of transition, so that more refined experimental methods must be employed if the effect is to be demonstrated.

(1) It is possible in theory at least to count the number of positron-radioactive nuclei formed in a given reaction, and to compare this with the number of positrons emitted in the subsequent decay. Unfortunately, most reactions leading to e^+ emitters are formed in (α, n), (d, n), or (p, n) reactions, which makes the accurate counting of the disintegrations very difficult. In the one case where the number of neutrons has been compared with the number of positrons—in the reaction $C^{12}+d \rightarrow N^{13}+n$; $N^{13} \rightarrow C^{13}+e^+$—a discrepancy was found,[6] which might be interpreted as evidence for electron capture.[7] But the data are not sufficiently precise to make that conclusion necessary. In addition, from theoretical

considerations, it is unlikely that N^{13}, which is energetic, light, and involves no spin change, should have a capture branching-ratio large enough to observe by this method.[7a]

(2) If it were possible to count the number of atoms formed when a positron-active substance decayed, this number could be compared with the total number of positrons given off, to give a measure of the branching ratio. This might be applicable if there were any known cases of successive positron activities, i.e., radioactive series, where one could compare the activities of the parent and daughter substances. Since no such cases have as yet been discovered, the alternate, but much more difficult, method of counting the number of stable product nuclei is worth investigation. A radioactive sample with an initial strength of 1 millicurie and a half-life of t days contains $1.6 \times 10^{12}t$ active atoms, which will give rise to $2.6 \times 10^{-12}tA$ gram of decay product. ($A=$ atomic weight.) For anything but a noble gas, this is beyond the limit of chemical or spectroscopic detection. It is very fortunate that Na^{22},[8] the longest lived positron-active substance, decays into Ne, the most easily detectable rare gas.[9] Calculation shows that a one day bombardment of Mg in the cyclotron will yield enough Na^{22} to allow the easy measurement of the quantity of Ne produced each month, if no electrons are captured. Theory suggests that 30 times as much Ne will be formed[10] by electron capture as is expected from the positron emission. The small bulk of Na containing the activity can be freed of all gases by prolonged heating in a vacuum, and then the Ne can be allowed to grow in the cold, an ideal case for measurement of small quantities of noble gases by the Paneth method.[11]

(3) In the two previous methods suggested for detecting the capture of electrons, no advantage was taken of the fact that the electron was originally part of the stable electronic system of the parent atom, and that this system is dis-

[2] Mercier, Nature 139, 797 (1937); Comptes rendus 204, 1117 (1937).
[3] Hoyle, Nature 140, 235 (1937); Proc. Camb. Phil. Soc. 33, 286 (1937).
[4] Møller, Physik. Zeits. d. Sowjetunion 11, 9 (1937); Phys. Rev. 51, 84 (1937).
[5] Uhlenbeck and Kuiper, Physica 4, 601 (1937).
[6] Alvarez, Phys. Rev. 53, 326 (1937).
[7] Roberts and Heydenburg, Phys. Rev. 53, 374 (1937).

[7a] Note added in proof. Crane and Halpern [Phys. Rev. 54, 306 (1938)] have recently shown that electron capture plays a negligible role in the decay of N^{13}.
[8] Laslett, Phys. Rev. 52, 529 (1937).
[9] Günther and Paneth, Zeits. f. physik. Chemie 173, 401 (1935).
[10] Lamb, Phys. Rev. 50, 388 (1936).
[11] Professor H. E. White and Mr. H. Weltin are working on this problem at present, in collaboration with the author.

turbed by the loss of one of its component parts. It is well known that x-rays are given off by an atom which has lost one of its inner electrons by photo-ionization, and the same will be true of one which has lost an inner electron to the nucleus. The vacant place in the inner shell will be immediately filled, and a quantum of x-radiation (or an Auger electron) will be emitted in the process. It is this phenomenon which is the basis of the third method of detection. One has again a technical difficulty that the x-rays from the light elements (and no positron emitters are known among the heavy ones[12]) are soft, and difficult to observe in the presence of positrons and gamma-rays. Jacobsen[13] was the first to try this method, in the case of Sc^{43}, but he found no trace of the expected x-rays in a cloud chamber.

CHARACTERISTIC X-RAYS FROM ACTIVE TITANIUM[14]

Walke[15] has shown that a strong positron activity of 16 days half-life is induced in titanium when it is bombarded with high energy deuterons. The energy of the positrons shows that the transition is not an allowed one (second Sargent curve), and this fact, together with the long life and relatively high atomic number for a positron emitter, suggested that it would be an ideal starting point in a search for the x-rays following electron capture. The isotope responsible for the activity has been identified chemically by Walke as vanadium, so any x-rays would have the wave-length characteristic of the daughter element, titanium. The fact that this wave-length, 2.7A, is just below the "vacuum region" had much to do with the choice. From the arguments given above, Na^{22} would have been a far better choice, but Ne $K\alpha$ has a wave-length of about 14A.

The detection of the soft x-ray quanta presented several difficulties. It was, of course, necessary to eliminate the positrons, and this was easily accomplished with the aid of an electromagnet. Activated Ti has an abnormally high ratio of gamma-rays to positrons, so the

problem resolved itself into the detection of a weak component of very soft radiation superposed on a strong, hard one. A counter has an obvious advantage over an electroscope here, but the problem of getting the soft quanta into the counting volume is a serious one. This was solved by constructing the envelope of thin Cellophane, and filling it with argon at atmospheric pressure. This last feature greatly increases its relative sensitivity to 2.7A x-rays, as almost all the quanta are stopped in the gas, while very few gamma-rays are absorbed in the thin walls or gas.

The experimental arrangement is shown in Fig. 1. The Ti sample was placed in a five-sided aluminum box between the poles of an electromagnet, in a field of 2000 oersteds, so that all positrons were kept from reaching the counter. The sample was aged for two weeks, so that all short periods were of negligible intensity. The tube between source and counter was filled with He at atmospheric pressure and capped with Cellophane ends. This reduced the solid angle (counting rate) without discriminating against the soft x-rays. The counter cathode was made of copper foil in the shape of a C, so that x-rays could enter the active volume without passing through too much absorber. The copper foil was only 0.00025 cm thick, to prevent the absorption and counting of gamma-rays. Lead blocks screened the glass tubes which supported the Cellophane wall and copper cathode, for the latter reason also. The anode wire was of 0.0025 cm tungsten wire to keep the counter voltage low—the working potential was about 2000 volts. Alcohol in a side tube made the counter more reliable, and to keep the sensitivity constant, it was found necessary to counteract leaks in the Cellophane system by flowing argon through the counter whenever it was in operation. The background was high, but the counting rate was so great that this caused no inconvenience. Absorbers of thin aluminum foil were arranged on sliding frames so that it was a simple matter to take absorption data.

The experimental procedure consisted in taking 2000 counts at each of seven values of absorber thickness, and in repeating each setting eight times. Each point then had a probable error of less than one percent. The background

[12] Alvarez, Phys. Rev. **53**, 213 (1938).
[13] Jacobsen, Nature **139**, 879 (1937).
[14] Alvarez, Phys. Rev. **52**, 134 (1937).
[15] Walke, Phys. Rev. **51**, 1011 (1937).

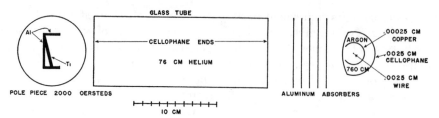

F<small>IG</small>. 1. Experimental arrangement.

count and the linearity of the counting circuit were checked after each series. The counting rate dropped from 21.3 to 15.4 per second as the absorbers were introduced into the beam. The rate remained at the latter value from 0.0025 cm to 0.0075 cm of Al, indicating that the decrease was due to a soft component in addition to the hard gamma-rays.

Figure 2 shows the actual absorption curve, and Fig. 3 is the absorption curve of the soft component. The three lines are drawn through the upper, most accurately known point, with slopes calculated from the measured absorption coefficients in Al of K-radiation from Sc, Ti and V. The agreement of the data with the Ti line is very striking, and indicates that the radiation is homogeneous, and of the expected wavelength, giving support to the electron capture hypothesis.

On the remote chance that this radiation could have been excited by the positrons in passing through the Ti sample, the sample was covered with a thin layer of chromium (actually a 0.00025 cm Ni foil plated on each side with 0.0012 cm of Cr). This foil was thick enough to absorb all the observed Ti $K\alpha$ and if these x-rays were of secondary origin, the positrons should have excited about equal amounts of Cr $K\alpha$ in the Cr foil. Absorption data showed that no soft component was present under these circumstances, so it seemed safe to assume that the x-rays arose from Ti atoms which had been formed by electron capture from the 16-day vanadium period. The assumption as to the period was based on Walke's finding that all but a negligible fraction of the activity in a two-week-old sample of Ti bombarded with deuterons was due to this period. This fact, plus the theory, which suggested that K-radiation should be found in this period made it seem

unnecessary to repeat the expensive (because of the argon waste) experiment two weeks later just to check the period.

However, Walke carried this Ti sample to Liverpool shortly after, and Williams and Pickup[16] have followed its activity in a cloud chamber since then. They improved Jacobsen's technique by bending the positrons out of the chamber by a magnetic field, and were thus able to see x-ray photoelectrons much closer to the source, where they are more plentiful. A count of these short tracks against distance verified that they had approximately the correct absorption coefficient to be Ti $K\alpha$. The surprising observation was made that there were about 500 photoelectrons per positron, and that while the positrons decayed with the 16-day period, the number of photoelectrons remained constant for several months, within the statistical error. This indicated that some of the x-rays observed in this laboratory were due to a new, long period in Ti, which had escaped Walke's attention because of its soft radiation. To find what percentage, if any, of the x-rays were due to the 16-day period, the author prepared and sent a fresh sample of Ti to Liverpool, where Williams and Pickup showed that no appreciable fraction was emitted by the 16-day isotope. While this changes the identification of the responsible isotope, the original interpretation that electron capture was probably the cause of the radiation, was still the most plausible guess.

But this interpretation is not the only one which will fit the facts, and if one is to establish the existence of a new phenomenon, there must be no alternative explanations. It is well known that x-rays appear when gamma-rays are internally converted. Here the shell is vacated by an expelled electron, instead of one lost to the

[16] Williams and Pickup, Nature **141**, 199 (1938).

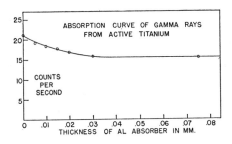

FIG. 2.

nucleus. It is entirely possible that the long-lived x-ray emitting isotope is a metastable state of a stable or radioactive Ti or V isotope. Low lying states of this kind are usually assumed to be responsible for the existence of nuclear isomers, of which this long period would then be an example. It is assumed that the higher level may radiate a gamma-ray to the ground state, and if these states are close, as they must be for a long life to be possible, the soft gamma-rays could be largely internally converted. The soft conversion electrons would not be seen in Williams and Pickup's chamber. While this is not suggested as the probable explanation, it fits all the observed data as well as the electron capture hypothesis. It serves best to emphasize that a more decisive test must be made if electron capture is to be accepted as an established fact.

PROOF OF THE ELECTRON CAPTURE HYPOTHESIS[17]

A glance at the table of isotopes showed that several blank spaces existed on the positron side of the stability band, in the region between Mn and Ge, although this section of the table has been very carefully studied by several workers. It seemed quite probable that some of these missing isotopes might capture electrons, and that they had not been discovered because their soft x-rays had been masked by the strong beta- and gamma-rays of the other periods.

An exploratory investigation was therefore made of the soft quantum radiations emitted by several of these elements after bombardment with deuterons. Of the four elements tried, Fe, Ni, Cu, and Zn, all were found to give off various

[17] Alvarez, Phys. Rev. **53**, 606 (1938).

amounts of soft x-rays. Since the wave-lengths are shorter in this region, it was possible to use an air-filled Lauritsen type electroscope as the detector; the magnet was still necessary to suppress the beta-rays. The absorption curve on the iron radiation showed that about half of gamma-ray ionization was due to a component with an absorption coefficient in Al about equal to that of Cr $K\alpha$. There are several long periods induced in iron, so it did not lend itself easily to investigation. Ni showed about 8 percent K-radiation following the well-known 3.5 hr. period of Cu^{61}. Copper showed about 40 percent x-ray ionization following the 12 hr. Cu^{64}. This is interesting, as it is what one would predict on theory.[5] The activity with the shorter life has a smaller capture branching ratio, and in fact the numbers quoted above fit very well on the theoretical curves. The ratio of these two percentages is a number which does not depend on geometry or the relative sensitivity of the electroscope to various types of radiation, or on atomic number or spin change; so it is well adapted to a comparison with theory. It is encouraging that it fits so well, but it cannot be taken too seriously at present.

Activated zinc was the most interesting of the four elements bombarded, and the rest of this section will be devoted to it. Four of the activities induced in zinc by fast deuterons were found to be accompanied by x-rays with the correct absorption coefficient in Al to be Zn $K\alpha$. Electron periods as well as the positron periods were among these four, so it did not seem possible to

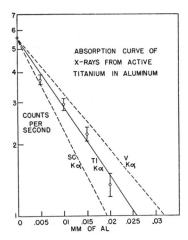

FIG. 3.

explain them all as due to electron capture. And if there was an alternate explanation, why could it not explain all the cases?

Most of the work was done on the 83-hour period. This activity has had the following interesting history: It was first found by Livingood,[18] who gave its half-life, without chemistry, as 97 hours. Later, Thornton[19] identified it as a zinc isotope with the same period, and assigned it to Zn^{71}. Next, Du Bridge, Barnes, Buck, and Strain[20] quoted its period as 82 hours, but said it was surprising that although it was prepared from Zn with protons, it was not detected in either copper, zinc, or gallium precipitates. The author gave the half-life as 83 hours, and showed that it was chemically similar to Ga. Strain and Buck[21] have since confirmed this, and Mann,[22] who reported a Ga isotope with a period of about 55 hours, has also shown that his activity is identical with the 83-hour Ga. The isotope has thus been prepared from zinc by deuteron, alpha-particle, and proton bombardment, which pins it down at Ga^{67}.

The Ga^{67} could be separated from the rest of the activities induced in Zn by means of the ether extraction process.[23] This procedure depends upon the fact that $GaCl_3$ is soluble in ether, while the chlorides of the neighboring elements are not. If one then shakes an HCl solution of these metals with ether in a separatory funnel, most of the $GaCl_3$ will be found in the ether layer. The ether may be then evaporated on a metal sheet, leaving an invisible layer of $GaCl_3$ which contains almost all of the original activity. This is probably the simplest way of preparing a pure radioactive sample—no carrier is needed. In the case under discussion, there were no doubt small traces of Ga impurities in the activated Zn, but very recently, Grahame and Seaborg[24] have activated specially prepared, Ga-free Zn and found that the ratio of activity in the ether layer to activity in the aqueous layer

is the same for pure radioactive $GaCl_3$ as it is for macroscopic amounts of stable $GaCl_3$.

When a weak, separated sample of Ga^{67} was measured without a magnetic field on a thin walled electroscope, a very soft component was found, which was electronic in nature as shown by the sign of the magnetic deflection. Rough range measurements showed that the energy of these electrons was about 100 kev. This was quite surprising for a period as short as 83 hours. The Sargent relations show that for any given half-life, there is a minimum beta-ray energy, corresponding to an allowed transition of spin change zero. For greater spin changes, the energy released is considerably higher. There are no known beta-ray energies less than that called for by the first Sargent curve, and there is no theoretical reason for expecting them.[25] So it seemed probable that the electrons observed here were conversion electrons instead of primary disintegration particles. No previous example of internal conversion in the artificially radioactive elements has been reported, and it has generally been assumed that it should be very unlikely.

To check this possibility, the absorption curve of these electrons was carefully investigated with a separated sample of Ga^{67}. The data are shown in Fig. 4. It is at once evident that the curve is not exponential, nor a combination of exponentials as one finds for beta-rays, since it is concave toward the origin. Its shape is strong evidence that the electrons responsible have a line structure. The intensity available in the sample was enough to allow its measurement on a high resolution mass spectrograph. Mr. D. C. Kalbfell photographed the electron spectrum in an instrument of his design, and showed that there was indeed a line structure. His plates showed an intense line at 90 kev, and a fainter one at about 99 kev. This is precisely what one would expect if the electrons were due to the internal conversion of a gamma-ray of 100 kev. The K and L absorption edges of zinc are at 1.2 and 12A, respectively, corresponding to energies of 10 and 1 kev. The strong line is then due to K-conversion of the gamma-ray, and the weak one to L-conversion. Dr. E. M. Lyman examined the specimen in his high

[18] Livingood, Phys. Rev. **50**, 425 (1936).
[19] Thornton, Phys. Rev. **53**, 326 (1938).
[20] Du Bridge, Barnes, Buck, and Strain, Phys. Rev. **53**, 447 (1938).
[21] Strain and Buck, Phys. Rev. **53**, 943 (1938).
[22] Mann, Phys. Rev. **53**, 212 (1938).
[23] Noyes and Bray, *Qualitative Analysis for the Rare Elements* (Macmillan Company).
[24] Grahame and Seaborg, Phys. Rev. **54**, 240 (1938).

[25] See later section on Hg^{197}.

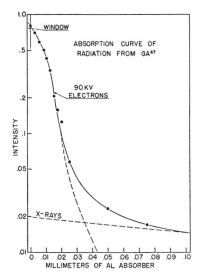

INTENSITY

WINDOW

ABSORPTION CURVE OF
RADIATION FROM GA⁶⁷

90 KV
ELECTRONS

X-RAYS

MILLIMETERS OF AL ABSORBER

Fig. 4.

resolution beta-ray spectrograph equipped with a Geiger counter, and obtained the spectrum shown in Fig. 5.[26] This electron group is interesting in that it is the purest radioactive line source known.

Attention was next directed to the x-rays. Although it would have been possible to determine the atomic number of the element whose characteristic radiation was emitted from active Ti, by careful absorption measurements in Al, this is not true of the radiation from Ga. The absorption coefficient of the K-radiation from Sc, Ti, and V is a very sensitive function of Z, as shown in Fig. 3, but it is a slowly varying function in the neighborhood of Zn. One must therefore resort to a more refined method to identify the wave-length of the Ga x-rays. There is enough intensity available to diffract the rays in a bent crystal spectrograph,[27] and this method was the first tried. If this were the only possibility, it is certain that results could have been obtained without too much difficulty. But the attempt was abandoned in favor of the simpler method described below.

It is well known that if one plots absorption coefficient in a given element against wave-length, sharp discontinuities appear at certain

[26] I wish to thank Dr. Lyman and Mr. Kalbfell for examining the active Zn in their spectrographs.
[27] Du Mond and Kirkpatrick, Rev. Sci. Inst. **1**, 88 (1930); Cauchois, J. de Physique **3**, 320 (1932).

wave-lengths which are known as "critical absorption limits," and which correspond to the binding energies of the K, L, etc. shells of the absorbing atom. If one now plots the same type of curves for the two neighboring elements, he finds that the curves are very similar, except that the discontinuities have shifted slightly to either side.

If one could find two neighboring elements which absorbed the Ga x-rays very differently, it would show that the wave-length of this radiation lay between the two critical limits. Experiments showed that Ni and Cu foils exhibited this property. The data are plotted in Fig. 6. The Ni and Cu absorption limits are at 1.48 and 1.38A, respectively, so most of the radiation is between these limits. Zn $K\alpha$ is the only strong line satisfying these conditions, so one may conclude that it is responsible for most of the effect. Zn $K\beta$ should accompany it, and since it has a shorter wave-length than either of the two absorption edges, it is strongly absorbed in both Ni and Cu. It is seen that the copper curve resolves into two components corresponding to $K\alpha$ and $K\beta$ of zinc, and the ratio of the intensities of the two lines is approximately correctly given by the intercepts of the resolved components on the vertical axis. That the absorption in Ni is not as great as one would expect from the tabulated absorption coefficients is due to the imperfection of the geometrical conditions. Ni $K\alpha$ is excited as fluorescence radiation when Zn $K\alpha$ is absorbed in Ni foil, and some of these x-rays are detected in the chamber, to give too high a reading. When the geometry was made still poorer, the apparent absorption coefficient fell further, showing that this explanation of

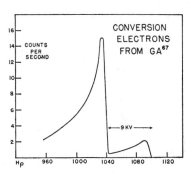

COUNTS
PER
SECOND

CONVERSION
ELECTRONS
FROM GA⁶⁷

9 KV

Hρ

Fig. 5.

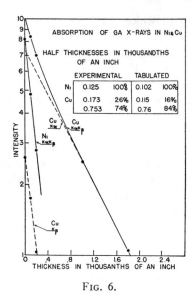

FIG. 6.

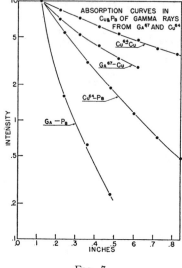

FIG. 7.

the discrepancy was correct. The evidence is therefore conclusive that the soft component of the undeflectable radiation is composed of zinc characteristic x-rays, and that therefore, the daughter substance is an isotope of zinc.

The gamma-ray spectrum was next investigated by two methods. Absorption data in copper and lead showed that there was a component in the neighborhood of 250 kev. It was particularly important to establish whether or not any annihilation radiation accompanied the decay. The absorption curve in lead was compared with one taken on Cu^{64} with identical geometry. The latter emits only annihilation radiation.[28] From the two curves (Fig. 7) one can see that if any of this type of gamma-rays is given off from Ga^{67}, it is a very small fraction indeed. Confirmatory evidence was obtained in a cloud chamber, where no positrons could be observed. No trace of the unconverted fraction of the 100 kev line was detectable in the absorption data, so a more sensitive method based on transition effects was used. The gamma-ray effects in an electroscope are due to Compton and photoelectrons ejected in the material between the source and the sensitive region of the chamber. As one piles lead over the source, the gamma-ray ionization at first increases, and then decreases. For lighter absorbers, the maximum will not be as high, and the rate of decrease will,

of course, be smaller. These initial rises are due to the production of electrons in the absorber (in heavier elements the equilibrium ratio of electrons to gamma-rays is higher). If one places enough lead over the source to bring the ionization current up to near its maximum value, and then absorbs out the electrons from the lead with aluminum sheets, the ionization current will fall to the low value for equilibrium in Al. The amount of Al necessary to accomplish this lowering is equal to the range of the electrons, and therefore is a measure of the energy of the gamma-ray which projected them from the lead.

This method was applied to the annihilation radiation from Cu^{64}, and gave a value near 500 kev. When Ga^{67} was substituted, the transition curve showed a large group of electrons from a gamma-ray at about 250 kev, in agreement with the absorption data. A careful search with very thin absorbers showed another group due to the unconverted 100 kev line. (See Fig. 8.) Thus the line is not totally internally converted, as had at first been suspected. An examination of the transition curve near the 500 kev portion showed no trace of a drop, so it is quite certain that positrons play no important part in this activity.

Summing up the evidence, we see that a radioactive Ga isotope has changed into a zinc isotope, and during the process, no positrons have been emitted. (The gamma-ray evidence

[28] Van Voorhis, Phys. Rev. **50**, 895 (1936).

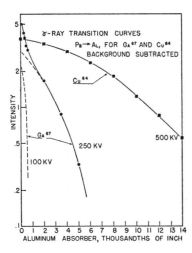

FIG. 8.

eliminates the possibility that the positrons might have been too soft to be detected.) To investigate the small chance that protons were emitted during the disintegration, a sample was examined in a linear amplifier, and also placed inside the electroscope chamber. No evidence was found for protons, and there are in addition good theoretical reasons for believing that extremely slow protons could not emerge in such numbers with a half-life as short as 83 hours. It must then be concluded that the transition from Ga to Zn has not been accompanied by the emission of any known particle of single positive charge. (The possibility that a heavy electron is responsible for the transition is ruled out on energetic grounds. There is not enough energy available for the transition to Zn^{67} and, if it went to Zn^{66}, the latter would be left too highly excited.)

We are therefore forced to the conclusion that the Ga^{67} nucleus has captured a negative electron from its orbital system, and been transformed into stable Zn^{67}.

OTHER DATA BEARING ON ELECTRON CAPTURE

Ta^{180}

The second artificially radioactive body shown to emit characteristic x-rays was Ta^{180}, reported by Oldenberg.[29] He irradiated tantalum with fast neutrons from the Li+D reaction, and separated an active form of tantalum by chemical

[29] Oldenberg, Phys. Rev. **53**, 35 (1938).

methods. Since this period could not be induced by slower neutrons, it was assumed to be formed by neutron loss, and therefore assigned to the isotope with atomic weight one less than the only known stable isotope of Ta, Ta^{181}. The beta-rays from this isotope were all negative, as determined by a trochoid analyzer. The gamma-rays had an absorption coefficient in Al equal to that of Hf K-radiation. No magnetic field was necessary, as the first Al absorber cut out all the electrons. The curve could then be followed for several half-thicknesses of Al, with the aid of a Geiger counter. Oldenberg interpreted the x-rays as due to electron capture, giving Hf^{180}. Since W^{180} was not known, he suggested that the beta-rays were conversion electrons from an internally converted gamma-ray from an excited level of Hf^{180} to ground.

Sizoo[30] has shown on simple theoretical considerations that if the middle one of three neighboring isobars is radioactive, and emits negative electrons, it is almost certain to transform some of the time by electron capture. (If it could not do this, i.e., if the isotope with smallest Z had greater mass, the latter would transform by negative electron emission to the middle one, and would therefore not be observed as stable.) On the chance that the negative electrons observed here were disintegration betas —that this case were an example of Sizoo branching—the author asked Professor A. J. Dempster to reexamine his mass spectra of W to see if there was any evidence for W^{180}. The old plates showed faint lines, and new plates confirmed the existence of this isotope.[31] Thus Oldenberg's activity is probably the first example of electron emission-electron capture branching.

K^{40}

Weizsäcker[32] has made the suggestion that the natural radioactivity of potassium may be a branching reaction of the Sizoo type. In addition to the well-known electron decay to Ca^{40}, he postulates electron capture to A^{40}. This would explain several anomalies in the abundance of argon. The position of argon in the periodic table is not given by its sequence in the list of

[30] Sizoo, Physica **4**, 467 (1937).
[31] Dempster, Phys. Rev. **52**, 1074 (1937).
[32] Weizsäcker, Physik. Zeits. **38**, 623 (1937).

atomic weights. This situation would be corrected if the percentage abundance of A^{40} were smaller. The abundance of argon in the atmosphere is about one percent, which is vastly greater than any of the other rare gases. This might be termed circumstantial evidence, but it does fit well with theoretical ideas about electron capture.

Zn^{65}

Barnes and Valley,[33] and Livingood and Seaborg,[34] have reported an 8-month zinc period which emits positrons, and which has an abnormally high ratio of gamma-rays to positrons. This is no doubt the longest of the four periods in zinc mentioned above, which emit x-rays. Absorption measurements on the x-rays in Ni and Cu were made at that time, which showed that the radiation was characteristic of copper.[35] This is what one would expect from Zn^{65} if it captured electrons. It seems relatively safe to assume that the two zinc activities are identical and to assume, as Barnes and Valley, and Livingood and Seaborg did, that this is the first case of positron emission-electron capture branching.

Hg^{197}

Heyn[36] has reported a period of about 43 minutes induced in mercury by very fast neutrons. No details of the radiations emitted were noted by him, but he made the reasonable suggestion that the reaction was of the $n-2n$ type. McMillan, Kamen and Ruben,[37] in their survey of the neutron induced activities in the heavy metals, investigated this activity in some detail. They showed that the period was due to mercury, and that the electrons were all negative. They noted that the beta-ray energy was too low for even the first Sargent curve, and in private conversations, they suggested that the electrons might be conversion electrons from a gamma-ray after electron capture. An absorption curve of the gamma-rays showed a "possible

complex structure with energies in the range 70–250 kev." On this basis, the assignment would have been to Hg^{197}. But these authors preferred the idea that the electrons were disintegration betas, and so assigned the activity to Hg^{203}.

An obvious difficulty with this assignment is the failure to observe the period with slow neutrons. Slow neutron capture is always energetically possible, and could not have escaped detection with such a short period. This would lead one to assign the period to Hg^{197}, since Hg^{196} is present only to 0.1 percent. The electrons then would not be primaries, which would eliminate the only point on a "negative Sargent curve." Since no positrons were observed, and since the gamma-ray spectrum does not extend to 500 kev, this is probably a case of electron capture.

Ag^{106}

Pool, Cork and Thornton[38] discovered an interesting case of isomerism in Ag^{106}. The 24-minute period emits positrons, while the one of 8-days half-life is electron active. Pool[39] has investigated the radiations from the isotope, and concluded that the 8-day period also captures electrons. If this explanation were correct, one would expect to find about three quanta of Pd K x-rays, and one Auger electron for every four gamma-rays. Pool's gamma-ray spectrum shows no electrons in this energy range, but they would have been a little difficult to measure. To check this point, a search for Pd $K\alpha$ was made in this laboratory. The source of radiosilver was covered with paraffin to absorb the beta-rays, and was thin enough to have negligible self-absorption for the x-rays. The gamma-ray ionization was measured in an electroscope filled with CH_3Br.[40] This type of instrument is much more sensitive to Pd x-rays than to gamma-rays, so almost all of the observed ionization should have been due to the former, if electron capture were taking place. Aluminum absorbers had very little effect on the ionization current, so it must be concluded that electron capture does not play an important part in the decay.

[33] Barnes and Valley, Phys. Rev. **53**, 946 (1938).
[34] Livingood and Seaborg, Phys. Rev. **54**, 239 (1938).
[35] The shortest of these four periods in unseparated, activated zinc was a new electron emitting isotope with a half-life of 15 minutes. The x-rays are definitely Zn K-radiation, and since this period is unknown, it might be due to Ga^{65} capturing electrons.
[36] Heyn, Nature **139**, 842 (1937).
[37] McMillan, Kamen and Ruben, Phys. Rev. **52**, 375 (1937).

[38] Pool, Cork and Thornton, Phys. Rev. **52**, 380 (1937).
[39] Pool, Phys. Rev. **53**, 116 (1938).
[40] I am indebted to Mr. Philip Abelson for the use of this instrument.

Cd[107] or Cd[109]

Ridenour, Delsasso, White, and Sherr[41] have found a 6.7-hour proton-induced activity in silver, which they ascribe either to Cd[107] or Cd[109]. The radiations emitted by this isotope are a soft electron group, a strong x-ray with the correct Al absorption coefficient, and a very weak gamma-ray. They conclude that the whole situation in this radioelement is similar to the case of Ga[67].

Be[7]

Roberts, Heydenburg, and Locher[42] have very recently observed an interesting example of electron capture in Be[7]. This new radioactive isotope may be formed in either of the following two reactions:

$$Li^6 + D \rightarrow Be^7 + n,$$

or

$$B^{10} + D \rightarrow Be^7 + He^4.$$

This isotope is the only one known which emits nothing but gamma-rays; their energy is about 425 kev. The neutron yield in the reaction leading to Be[7] from Li is ten times the radioactive gamma-ray yield, which is a large enough discrepancy to mean something. The interpretation given by these authors is that in 90 percent of the disintegrations, the neutrino carries away all the available energy, leaving the Li[7] nucleus in the ground state. In 10 percent of the cases, which are those observed, the Li[7] is left in the well-known level at 440 kev, from which it radiates to the ground state.

Mn[54]

Livingood and Seaborg[43] have found a long lived manganese isotope which may be prepared in three different reactions, and thus assigned to Mn[54]. The radiations from this isotope are a strong x-ray absorbed in Al as Cr $K\alpha$, a gamma-ray of about 1 Mev, and a very weak negative electron group which is probably of secondary origin. These data are best interpreted as evidence for electron capture. The gamma-rays

are internally converted to a very small extent, if at all.

The author wishes to acknowledge several valuable discussions of these problems with Professors E. O. Lawrence and J. R. Oppenheimer. The work has been materially aided by grants to the laboratory from the Research Corporation and the Chemical Foundation. The experiments were facilitated by assistance from the W.P.A.

APPENDIX

Additional data on Ga[67]

Since this paper is primarily concerned with electron capture, only those data about Ga[67] which bore directly on the problem were included in the section above. This appendix will complete the discussion of Ga[67], merely as an interesting case of artificial radioactivity. It was established earlier that the following radiations were emitted by this isotope: a line spectrum of electrons, the characteristic x-rays of zinc, a gamma-ray of about 250 kev and one at about 100 kev. In order to obtain a more complete picture of the processes involved, it is necessary to have some idea of the relative intensities of the various radiations.

An absorption curve in aluminum, of the total radiation from a thin, separated sample of Ga[67], showed that the ionization current due to the electrons was about 40 times that due to the x-rays. The chamber had a depth of 6 cm; from the absorption coefficient, one can calculate that 6 percent of the x-rays were absorbed in the active volume of the electroscope. The electrons have a range of about 10 cm in air, so about half of their energy was expended in the chamber. Each electron has ten times as much energy as an x-ray quantum. The relative intensity of x-rays and electrons is then

$$10/(40 \times 0.06 \times 2) \cong 2.$$

The fluorescent yield for Zn is 0.5, so the ratio of K electron excitations to conversion electrons is about 4.

The intensity of the gamma-rays may be estimated in the following manner: since the gamma-ray ionization was obtained with aluminum absorbers over the sample, it is necessary to calculate the equilibrium ratio of electrons to quanta in aluminum. The mass absorption coefficient of the gamma-rays is 0.11 cm²/g, and the range of the electrons is 0.045 g/cm², so 0.005 of the gamma-ray energy will be converted into electronic form in the effective upper layer of the aluminum. The range of the electrons in air is 0.045/0.001 = 45 cm, but only 6 cm of this range is effective in the chamber. The ratio of x-ray ionization to gamma-ray ionization was observed to be 3.5, so the relative intensity of these two types of radiation is then

$$\frac{3.5 \times 0.005 \times 6 \times 250}{0.06 \times 45 \times 9} \cong 1.$$

The relative intensities of K-excitations, gamma-rays, and electrons are therefore approximately 4 : 2 : 1. These

[41] Ridenour, Delsasso, White, and Sherr, Phys. Rev. **53**, 770 (1938).

[42] Roberts, Heydenburg, and Locher, Phys. Rev. **53**, 1016 (1938).

[43] Livingood and Seaborg, Phys. Rev. **54**, 391 (1938).

values suggest that there are two excited states in the Zn^{67} nucleus—one at 350 kev, and the other either at 100, or 250 kev above ground. Every case of electron capture leaves the Zn^{67} in the upper state, from which it cascades to ground in two steps. One of the transitions gives rise to the 250 kev gamma-ray, and the other is responsible for the internally converted radiation at 100 kev. The internal conversion coefficient is fairly high, since the transition curves show the 100 kev line to be weaker than the one at higher energy. The relative intensities of electrons and x-rays confirm this. If the line were totally internally converted, the K-excitation would be twice as great as the electron emission, since the K shell would be emptied by capture and conversion in each transition. The observed ratio is 4, which indicates that the conversion coefficient is about 40 percent. All the intensities are consistent with the scheme outlined above.

The question then arises:—how is it possible to have such a large internal conversion factor in an element as light as zinc? It has been quite generally thought by theoretical physicists that internal conversion would play a negligible part in artificial radioactivity, since the equations predict a Z^3 dependence. (Total conversion in a 0–0 transition would be possible, however.) When it first be-came apparent that Zn^{67} had an anomalously high internal conversion coefficient, Dancoff and Morrison[44] reexamined the theory, and found that in the energy range near 100 kev for zinc, the factor could be quite high. The 40 percent observed (internal conversion coeff. $\sim$0.7) would mean the gamma-rays were due to a quadripole transition. No detailed calculations of this nature had been made previously for light nuclei, as it had seemed certain from an inspection of the formulae that there were no terms entering which could make the ratio appreciable. All details of the picture, both experimental and theoretical, are now self-consistent.

It might be interesting to look for coincidences between the two x-rays ejected almost simultaneously in 1/8 of the disintegrations. (40 percent $\times$ (fluorescent yield)^2 = 1/8.) An apparatus for this experiment has been constructed, but has not as yet been used. The x-ray spectrum might yield interesting information about the relative lifetimes of x-ray and gamma-ray states. If the K shell were doubly excited, the wave-length of the characteristic radiation would be slightly changed, and this could be detected on a bent crystal spectrograph. No such case of a doubly excited K shell is known.

[44] Dancoff and Morrison, Phys. Rev. **54**, 149 (1938).

4

Isotopes of Mass 3

Hans A. Bethe

In his autobiography,[1] Luie W. Alvarez states that he was very much stimulated by my 1936–37 articles on nuclear physics, published at the time he was finishing his graduate studies.[2] Looking more closely, I find now that my articles mainly stimulated him to prove me wrong! Whenever I had stated that a certain experiment could not be done, he went ahead and did it. Of course, I was very happy about his action.

I had made no such statement, however, about the difference in mass of helium-3 and hydrogen-3. From the available data and some calculation, I concluded that helium-3 was the heavier nucleus by 0.08 ± 0.14 million electron volts (MeV), and should therefore be radioactive, a conclusion that was not justified because of the large experimental error.

Luie regarded this as a challenge and showed that helium-3 was the stable nucleus and hydrogen-3 (tritium) the radioactive one. He did this by an ingenious sequence of experiments. First, he used the then-existing 60 inch cyclotron at Berkeley as a highly sensitive mass spectrometer. He injected atmospheric helium into the cyclotron and produced doubly ionized helium ions. Then he adjusted the magnetic field and radio frequency to accelerate any helium-3 which might exist in the source gas. Indeed, he found helium-3 at the level of 1 part in 10 million of atmospheric helium. Using helium from natural gas wells, the fraction of helium-3 was only a tenth this amount. This was later interpreted as showing that helium in natural gas mostly originates from decay of radioactive substances close to the well, whereas helium-3 in the atmosphere probably arises from cosmic-ray protons interacting with nitrogen-14 nuclei.

Having thus proved helium-3 stable, Luie set out to show that hydrogen-3 is radioactive. He produced hydrogen-3 by collisions of hydrogen-2 (deuterons) accelerated in the cyclotron with a hydrogen-2 (deuterium) target and showed that it had a long-lived radioactivity. Circulating the product through active charcoal and diffusing it through hot palladium demonstrated that it was hydrogen.

Luie showed that the produced radioactivity had a very short range. Of course, its beta rays originated from the decay of hydrogen-3. From the range, he concluded that the energy of the beta-particle was about 10 thousand electron volts (keV). The modern value for the end-point energy of the beta-spectrum is 18.1 keV.

Both these discoveries, made with Robert Cornog and published in 1939, are reproduced in the accompanying articles. In 1940, they published two further abstracts, one giving the correct average energy of the beta rays but a wrong lifetime of 5 months.[3] The second stated that this lifetime was spurious, being due to the loss of hydrogen-3 from the vessel, and the true lifetime was probably about 10 years.[4] The modern value is 12.3 years.

The discovery that helium-3 is stable and hydrogen-3 radioactive is of great practical importance. Hydrogen-3 provides one of the strongest known neutron sources when it participates in the well known D-T (deuterium-tritium) interaction; hydrogen-2 collides with hydrogen-3 yielding helium-4, a neutron, and 17.6 MeV of energy. This process has an enormous resonance at low energy, due to the existence of an excited state of helium-5. The resulting large cross section, it is hoped, will soon make possible the design of a fusion reactor. It is, in fact, the main hope of using fusion energy for peaceful purposes.

Because hydrogen-3 is radioactive, it must be made in the laboratory. This can easily be done using neutrons from the D-T reaction, slowing them down, and then letting them be absorbed by lithium-6 to produce helium-4 and hydrogen-3. This process also has a large cross section. A practical design of a machine which uses this and the D-T reactions will still present quite a lot of problems. The hydrogen-3 because it is radioactive, will have to be used fairly soon after it is made. The D-T reaction is used in thermonuclear weapons, in which it has frequently been tested.

The relation between helium-3 and hydrogen-3 is also of great importance for nuclear theory, where it gives a very accurate measurement of the Coulomb energy of helium-3. It is equal to the neutron mass minus the atomic hydrogen mass (not the proton mass) plus the upper limit of the hydrogen-3 beta-decay spec-

1. L. W. Alvarez, *Alvarez - Adventures of a Physicist,* (Basic Books, New York, 1987).

2. H. Bethe and R. Bacher, Rev. Mod. Phys **8** (1936) Part A; H. Bethe, Rev. Mod. Phys. **9**, 69 (1937) Part B; and M. S. Livingston and H. Bethe, Rev. Mod. Phys. **9**, 245 (1937) Part C.

3. L. W. Alvarez and R. Cornog, Phys. Rev. **57**, 248 (1940).

4. L. W. Alvarez and R. Cornog, Phys. Rev. **58**, 197 (1940).

trum, the total being 0.7642 MeV. The accurate determination of this Coulomb energy gives a severe test of the wave function of the three-body nuclear system, which is the next simplest nucleus after the deuteron. Most theories have difficulty getting a value of the Coulomb energy as low as this experimental value.

In these ways Luie's discoveries of 1939 have been extremely important for both theoretical and practical nuclear physics. The experiments which Luie designed early in his career to prove his point were both simple and ingenious, which has been his style since.

Discovery of Hydrogen Helium Three

Robert Cornog

Working with Luie on the hydrogen-3, helium-3 problem was fast-paced and exciting. We discovered that helium-3 was stable and hydrogen-3 was radioactive within a week. Luie was quick. Once, while playing charades, he projected "The Voice of the Turtle" in three seconds flat. He was vigorous. Without warmup he could effortlessly press up into a handstand on a fence railing.

Luie was also articulate and direct. One Monday night at a meeting of the Radiation Laboratory group, Emilio Segrè described his plans to make element 85 by bombarding bismuth with alpha-particles accelerated in the 60-inch cyclotron. After the meeting, as Dale Corson and I walked together across campus, we talked of Segrè's proposed experiment. Unaware that Corson was already preparing to do some alpha-particle bombardments with the 60 inch cyclotron, I said, "You know, Dale, I have a lump of bismuth." "And I have a linear amplifier," Dale countered.

The next morning, bright and early, Corson and I bombarded bismuth with alpha-particles from the 60 inch cyclotron. We saw gobs of giant pulses when we placed the bismuth in front of our linear amplifier. I was elated but felt guilty as sin to have poached what I felt was Segrè's experiment, so I went directly to see Segrè. "Emilio, would you mind much if I had a try at that bismuth experiment you described last night?" I asked. After a short pause, Segrè replied, "No. There are other experiments that I can do." Now came the sticky part. "It's worse than that, Emilio. Dale Corson and I have already bombarded bismuth. We got giant pulses on his linear amplifier." Segrè paused somewhat longer. "I have only one request. Let me do the chemistry."

It was a day or two after these events that Luie was especially articulate and direct. He suggested that I work either on discovering element 85 or on discov-

ering the stability of hydrogen-3 and helium-3, but not both! So Corson, Mackenzie, and Segrè discovered element 85 and Alvarez and Cornog discovered hydrogen and helium of mass 3. All these events notwithstanding, Segrè continued to let me use his electrometer, an instrument which now resides in the Smithsonian collection.

Luie clearly describes how the first cyclotron beam of helium-3 ions was a short transient current that happened only when the magnetic field was being decreased.[5] Getting a constant helium-3 ion beam was relatively straightforward. Iron shims were inserted in the magnetic field so as to replace the magnetic effect produced by the transient eddy currents in the iron.

Getting a constant beam of hydrogen-3 ions was quite a different story. To accelerate hydrogen-3 in the 60 inch cyclotron without changing the radio frequency of the Dee voltage required that we increase the magnetic field by 50% over that used to accelerate hydrogen-2. Magnetic field power was supplied by an alternating current motor directly coupled to a direct current generator which had an exciter. This motor-generator-exciter set was housed in a concrete vault some distance from the cyclotron control room. I quickly determined that the generator, even at full rated output, would not quite produce enough power to give the required magnetic field increase. A possible solution occurred to me. Why not increase the exciter output, which would, in turn, increase the generator field coil current and the generator output?

I quickly jury-rigged a direct current source to substantially increase the exciter field current. Unfortunately, the results were not quite what I expected. Leaving the door to the motor-generator vault open, I went to the cyclotron control room to see how much

5. L. W. Alvarez, Phys. Today **35**, January, 25 (1982).

the current to the cyclotron magnet had increased by my changes. I noted with pleasure a substantial increase, but my elation was short-lived. Smelling smoke, I looked up to see clouds of smoke pouring out of the doorway from the vault. The motor-generator set had overheated. I was lucky that there was no permanent damage. I stopped trying to produce a beam of hydrogen-3 ions. As far as I know, more than thirty years elapsed before Muller produced the first such beam in a cyclotron.

We next tried to measure the beta-particle energies from hydrogen-3. I suggested one method which, although simple in concept, did not work in practice. The apparatus consisted of a thin-walled Geiger counter, a 0–50 kilovolt direct current power supply, and an electrode thinly coated with beta-emitting tritium material. I planned to measure the count rate decrease as I increased the retarding potential applied between the source and the counter. But the tritium beta rays were so soft that they would not go through the thinnest of aluminum windows. So from John Bachus I learned how to make plastic films a small fraction of a light-wavelength thick. I planned to use this ultrathin plastic film as the Geiger counter window. Alas, despite repeated attempts, these thin plastic windows always broke before I could make any measurements.

We had already obtained better results using a method suggested by Luie.[6] I admitted a cubic centimeter of tritium containing deuterium gas into a Geiger counter and measured the increase in count rate. To determine the energy of each disintegration, I placed a measured volume of this gas in a calibrated electrometer chamber. I then measured the resulting current on an old-fashioned, wall-mounted D'Arsonval galvanometer. Using the known value of electron volts per ion pair, I converted this current to ion pairs per second per cubic centimeter. With the Geiger counter disintegration rate measurements, I got a value of 9000 electron volts for the average decay electron energy. My calculations were overly simplistic. Some decay electrons lost part of their energy colliding with the chamber wall, and so the true average energy was somewhat higher.

We also investigated the half-life of tritium. One would think that this measurement should have been simple and straightforward. It wasn't. Our method, however, was simple. Put radioactive gas in a sealed electrometer chamber and observed the falloff of activity with time. So I filled a brass chamber with one atmosphere of the radioactive gas (deuterium and tritium) and sealed the small fill tube with a rubber tube and a pinch clamp. As a back-up I also filled two other chambers with the same radioactive gas. As a further check, I measured the decay in these two chambers

with a different electrometers, one belonging to Bob Livingston and the other to Gigi Wu.

Things did not turn out as planned. First, one of the backup chambers was made of Pyrex glass, with two electrodes supported on tungsten seals. In practice, the glass collected electrostatic surface charges and gave erratic readings. I followed a suggestion by Segrè and coated the offending surfaces with a grounded layer of conducting graphite.

Second, after a few weeks, first Livingston and then Wu decommissioned their electrometers. Also, for some reason, Segrè had to rebuild his electrometer and it now had a new sensitivity calibration. Furthermore, the background reading in Segrè's electrometer turned out to be significantly dependent on what level of radiation was being produced by the cyclotron in a different building, about 50 meters away. In spite of these irksome events, I managed to get a reasonably convincing decay curve using the brass chamber and Segrè's electrometer. Our results were published in the *Physical Review.*[7]

The dénouement came when, after many months of almost daily measurements, I decided to end the experiment and pump the gas out of the chamber. I was startled to discover that when I connected the chamber to a mercury manometer and opened the pinch clamp, I had a partial vacuum in the chamber! It very slowly dawned on me that instead of measuring the radioactive decay rate of tritium, I had in fact been measuring the rate of diffusion of deuterium gas through the walls of the short stub of rubber tubing used to seal off the chambers.

I was emotionally devastated. With heavy heart, I reported these results to Luie. "How can we bury this mistake when we've already reported the erroneous half-life?" Luie's response was immediate. "We don't bury the mistake. We report it." Within a couple of hours he composed a correction note, ran it past me for approval, and mailed it off to the *Physical Review.*[8] To put it mildly, I was relieved no end.

There have been other occasions when Luie has helped me out of difficulties. For example, early in 1947, while working as a project engineer for a defense contractor in Los Angeles, I had my security clearance withdrawn and my employment automatically terminated. In spite of my questions, no reason was given for the withdrawal. Three long years later, as a faculty member at the University of California at Berkeley, I was, with critical help from Robert Oppenheimer, able to arrange for a hearing in San Francisco before a representative of the newly formed Industrial Employment Review Board. Luie, learning of my impending hearing, offered to appear with me, and he

6. Ibid.

7. L. W. Alvarez and R. A. Cornog, Phys. Rev. **57**, 248 (1940).
8. L. W. Alvarez and R. A. Cornog, Phys. Rev. **58**, 197 (1940).

testified on my behalf. Six months later I was notified by the Board that my clearance had been improperly withdrawn. After two more years I again obtained a security clearance.

Now, back to helium-3. Having discovered the presence of the helium-3 isotope, we set about measuring its abundance relative to the helium-4. We used an atmospheric helium sample obtained from the Air Reduction Sales Company. Years later, Luie learned that in spite of the company's name, they normally obtain their helium from gas wells, as well-gas helium is much cheaper to extract. However, on this occasion he ordered atmospheric helium and he got atmospheric helium.

Our method was again simple and straightforward. We set the cyclotron magnetic field to accelerate helium-4 and measured the resulting beam current. We then reset the magnetic field to produce a helium-3 beam and again measured the beam current. The abundance ratio was the ratio of the two currents.

Our method had at least two important sources of error. First, we did not know that the cyclotron was equally effective in accelerating helium-3 and helium-4. Second, the helium-3 beam current was too small to measure with the microammeter used to measure the half-microampere helium-4 beam. At the same time, the helium-3 beam was much too intense to measure on our linear amplifier. Accordingly, we fitted a fist sized Victoreen R-meter chamber with a thin (5 centimeters air equivalent) iron window and placed it directly in the helium-3 beam. We measured the time it took the helium-3 beam to discharge the chamber as 5 seconds. We calculated the specific ionization of air by helium-3 ions. We applied a correction for ion-pair recombinations. We corrected for the portion of the helium-3 beam that did not enter the chamber, and for the variation of ionization density with path length. Putting it all together, we concluded that the helium-3/helium-4 abundance in atmospheric helium was 3 parts per ten million, a value about a fifth of that now accepted. We also measured well-gas helium, where we showed helium-3 to be 10 times less abundant than in the atmosphere, a ratio consistent with later measurements.[9]

With our discovery of helium-3 stability, I puzzled as to where the helium-3 came from. The nuclear production of helium-3 in the Earth's crust was unknown at that time; all the classical natural radioactive sources produce helium-4, not helium-3. I concluded that in the primordial synthesis of heavier elements from lighter ones, some helium-3 could have been left behind.[10] Recent measurements have indeed found that primordial helium-3 is intrinsic in Earth's mantle: it leaks through the crust to the surface. It is found in gases and lavas from volcanos, in hot springs, and in submarine hydrothermal vent discharge. Since the helium-3 in hot springs is from 7 to 30 times more abundant than in the atmosphere, Earth is still degassing primordial helium-3 inherited from its formation.

How, then, is the 10 times *smaller* helium-3 content of well-gas relative to the atmosphere explained? Easy. All well-gas helium-4 comes from the radioactive decay of uranium and thorium, while helium-3 is largely produced when neutrons are absorbed by lithium.[11] Measurements of this "excess" helium-4 are now being used as a tracer to help find uranium deposits.[12] Other sources of helium-3 have been discovered: cosmic-ray interactions in the atmosphere,[13] hydrogen-bomb tests,[14] and meteorites.[15]

In preparing these recollections I had good discussions with Luie, Harmon Craig, Martin Kamen, Alex Langsdorf, and Emilio Segrè. I also had occasion to refer to my Ph.D. thesis, I noted that of the people I there cited for helping me, no less than five subsequently won Nobel prizes: W. F. Libby, E. Segrè, M. Calvin, E. O. Lawrence, and Luie. From this I conclude that the easiest way to win a Nobel prize at Berkeley was to have helped me with my thesis work!

9. L. T. Aldrich and A. O. Nier, Phys. Rev. **70**, 983 (1946).

10. R. A. Cornog, Ph.D. thesis, University of California, 1940, p. 8.

11. L. T. Aldrich and A. O. Nier, Phys. Rev. **74**, 1590 (1948).

12. H. Craig, personal communication (1984).

13. W. F. Libby, Phys. Rev. **69**, 671 (1946).

14. W. J. Jenkins and W. B. Clarke, Deep-Sea Res. **23**, 481 (1976).

15. R. O. Pepin and P. Signer, Science **149**, 253 (1965).

Reprinted from THE PHYSICAL REVIEW, Vol. 56, No. 4, 379, August 15, 1939
Printed in U. S. A.

He³ in Helium

We have used the 60″ cyclotron[1] as a mass spectrograph to show that He³ is one of the stable isotopic constituents of ordinary helium. When the cyclotron was filled with helium, a linear amplifier chamber placed in the path of the ion beam was paralyzed at two values of the magnetic field, corresponding to the production of 8-Mev protons and 32-Mev alpha-particles. At a field midway between these two values, the amplifier showed the presence of a smaller, but quite definite, beam whose range was determined as 54 cm of air. He³⁺⁺ is the only ion which satisfies the three criteria of e/m, v, and R measured in this way. Further weight is given to this view by the observation that this beam did not appear when the tank was evacuated, or filled with deuterium.

It was not possible to produce a steady beam of 24-Mev "light alpha-particles," because the cyclotron was "shimmed" for normal alphas and was therefore out of adjustment at lower fields. But it was possible to produce pulses of ions by lowering the magnetic field rapidly and letting the induced currents in the pole pieces increase the central magnetic field. We plan to shim the 37″ cyclotron

for light alphas in the near future, to attempt to use the beam for disintegration experiments. The necessary intensity may be obtained by isotopic concentration or production of He³ in the D−D reaction.

Because of the shimming differences noted above, it is impossible to measure the relative abundance of the helium isotopes; but the orders of magnitudes of the two beams may be of interest. There were approximately 10^{12} alpha-particles per sec. in the He⁴ beam, and of the order of 10^3 per sec. in the He³ beam. We looked for a beam at the He⁵ magnetic field setting, but did not observe any effects.

We are indebted to Mr. William W. Farley for assistance in the experiment.

LUIS W. ALVAREZ
ROBERT CORNOG

Crocker Radiation Laboratory,
University of California,
Berkeley, California,
July 31, 1939.

[1] E. O. Lawrence, L. W. Alvarez, W. M. Brobeck, D. Cooksey, D. R. Corson, E. M. McMillan, W. W. Salisbury and R. L. Thornton, Phys. Rev. 56, 124 (1939).

Reprinted from THE PHYSICAL REVIEW, Vol. 56, No. 6, 613, September 15, 1939
Printed in U. S. A.

Helium and Hydrogen of Mass 3

We have now adjusted the shims of the 60-inch cyclotron so that it is possible to obtain a steady beam of 24-Mev He³⁺⁺ ions.[1] We have compared the isotopic ratio He³/He⁴ of tank (gas-well) helium to that of spectroscopically pure (atmospheric) helium, and find that it is about twelve times as great for atmospheric helium as for the gas-well variety. The absolute values have been approximately determined with the aid of a thin-walled Victoreen R-meter. These ratios are 10^{-8} and 10^{-7} for the two types of helium. When the cyclotron chamber is filled with atmospheric helium, the He³ beam has sufficient intensity to induce appreciable radioactivity in silicon. We have observed a 2.5-minute period with an initial intensity of 200 counts/ minute, on a background of 30 counts per minute. The activity could be followed for four half-lives; it is probably P³⁰ formed in the reaction.

$$_{14}Si^{28} + _2He^3 \rightarrow _{15}P^{30} + _1H^1$$
$$_{15}P^{30} \rightarrow _{14}Si^{30} + e^+.$$

When the silicon was bombarded under identical conditions except for the substitution of tank helium for spectroscopic helium, the activity was reduced to a small value consistent with the abundance ratios given above.

Since we have shown that He³ is stable, it seemed worth while to search for the radioactivity of H³. We have there-

fore bombarded deuterium gas with deuterons, and passed the gas into an ionization chamber connected to an FP-54 amplifier. The gas showed a definite activity of long half-life. We have now shown that this gas has the properties of hydrogen by circulating it through active charcoal cooled in liquid nitrogen and allowing it to diffuse through hot palladium. The radiation emitted by this hydrogen is of very short range as was shown by the almost linear form of the intensity $vs.$ pressure curve when the gas was pumped out of the chamber. When sufficient time has elapsed for us to make some statement regarding the half-life of this activity, we will submit the details of the work to this journal for publication.

We are indebted to Dr. S. Ruben for the use of his thin-walled counter and to Dr. A. Langsdorf for the loan of a d.c. amplifier. It is a pleasure to acknowledge the friendly interest of Professor E. O. Lawrence in these experiments, and to thank the Research Corporation for financial assistance.

LUIS W. ALVAREZ
ROBERT CORNOG

Radiation Laboratory,
Department of Physics,
University of California,
Berkeley, California,
August 29, 1939.

[1] L. W. Alvarez and R. Cornog, Phys. Rev. 56, 379 (1939).

5

The Neutron Magnetic Moment

Norman F. Ramsey

This contribution was to be written by Felix Bloch, the coauthor with Luis W. Alvarez of their fundamentally important paper on the neutron magnetic moment. At the time Alvarez and Bloch were working on this experiment in California, I was on the East Coast working with I. I. Rabi on molecular-beam magnetic-resonance experiments. Not only was Bloch's death a great loss to physics; but the account he could have given of the origin of the ideas for this research and of the difficulties that he and Luis had to overcome to reach a successful conclusion must remain untold.

The accompanying paper by Alvarez and Bloch is a masterpiece, not only for the importance of the experiment and the quality of the research, but also for the thoroughness and care with which the article is written. The quality of scientific research reports would markedly improve if authors would even now adopt this paper of Alvarez and Bloch as a model. Just as the reading of a book review is no substitute for reading the original book, a reading of the present review is no replacement for reading the original Bloch-Alvarez paper.

Alvarez and Bloch noted that Dunning, Powers, and others[1] had earlier found that neutrons scattered differently from a medium when magnetized and unmagnetized, and so neutrons must have a magnetic moment. Frisch, V. Halban, and Koch had attempted to measure the neutron magnetic moment by observing its precession in a fixed magnetic field between a polarizing magnetic scatterer and an analyzing scatterer. These experiments were difficult because of the small polarization achieved; and they were incapable of accuracy because the amount of precession was proportional to the length of time the neutrons were in the magnetic field, and this in turn depended on a velocity distribution that was both broad and unknown. One other attempt to measure the neutron magnetic moment was made during the summer of 1938 by Bloch, Bradbury, and Tatel. Although this experiment failed due to the low neutron intensity, it was the forerunner of the Alvarez-Bloch experiment.

To overcome the limitations of previous efforts, Alvarez and Bloch introduced many new innovations in their experiment, and most of these in turn have led

to subsequent new developments in neutron, atomic, and molecular physics.

The most important new innovation was the use of oscillator-driven magnetic fields to induce the resonance reorientations of the neutron at the Larmor precession frequency of the neutron. The same idea was independently invented by Rabi for measuring nuclear magnetic moments with molecular beams, and a paper announcing the success of the molecular beam magnetic resonance method was published[2] almost two years before the paper of Alvarez and Bloch. For this reason the invention of the method is generally attributed to I. I. Rabi, but, as noted in the paper of Alvarez and Bloch, the idea of the resonance method was independently conceived by Bloch for the neutron-beam experiments. It must have been a bitter disappointment to Bloch to learn that his independent idea had been anticipated by Rabi and later to learn that Rabi received the Nobel Prize for the research. However, the response of Alvarez and Bloch to this disappointment was the correct one. Instead of allowing the disappointment to inhibit their future work, each went on to other fundamental research and won separate Nobel Prizes in other fields.

A second major innovation in the Alvarez-Bloch experiment was the use of the proton cyclotron frequency to calibrate the magnetic fields used in the experiment. Alvarez and Bloch recognized that the accurate calibration of the magnetic field in which the neutrons precessed would be a major limitation of the experiment and that if the field were calibrated in terms of the cyclotron frequency of the proton, the results for the nuclear magnetic moment would be given directly in terms of the most widely used unit for nuclear magnetic moments, the nuclear magneton ($eh/4\pi Mc$). Subsequently, this method has been used as the fundamental calibration for all nuclear moments measured in nuclear magnetons,[3] and modifications of the method using the cyclotron frequency of electrons instead of protons[4] are used to determine nuclear, atomic, and

1. The extensive references given in the accompanying paper will not be repeated.

2. I. I. Rabi, J. R. Zacharias, S. Millman, and P. Kusch, Phys. Rev. **53**, 318 (1938).

3. J. A. Hipple, H. Sommer, and H. A. Thomas, Phys. Rev. **76**, 1877 (1949) and **82**, 697 (1951).

4. J. H. Gardner and E. M. Purcell, Phys. Rev. **76**, 1262 (1949); R. S. Van Dyck, P. B. Schwinberg, and H. G. Dehmelt, Phys. Rev. Lett. **38**, 310 (1977) and Bull. Am. Phys. Soc. **24**, 758 (1979).

electron magnetic moments in terms of the Bohr magneton ($eh/4\pi mc$).

Alvarez and Bloch for the first time used a cyclotron to provide a more intense beam and thereby more accurate measurements. This innovation had less far reaching consequences than their others, since there were no further attempts to measure the neutron magnetic moment until after World War II, when nuclear reactors provided more intense sources of slow neutrons. Alvarez and Bloch also studied with care the best thickness of magnetized iron for use as a neutron polarizer or analyzer and thereby obtained a further increase in sensitivity.

The result of the experiment was the determination that the magnitude of the magnetic moment μ_n of the neutron was 1.93 ± 0.02 nuclear magnetons. (The indicated experimental error 0.02 was the probable error, then customarily used. With present conventions it would be 0.03.) By present standards of nuclear magnetic moment measurements, the 1% accuracy achieved in the experiment is not impressive, but it was a tremendous step forward from the previous experiments, which had chiefly indicated that the neutron magnetic moment existed while doing little to determine its value.

An indication of the importance of neutron magnetic moment measurements is the rapidity with which new and more accurate measurements were undertaken immediately following World War II. These experiments combined the resonance techniques pioneered by Alvarez and Bloch with the powerful neutron sources available at nuclear reactors. Experiments markedly improving the value of the neutron moment were made in 1947 by Arnold and Roberts[5], in 1948 by Bloch, Nicodemus, and Staub,[6] and in 1956 by Corngold, Cohen, and Ramsey[7] and the sign of the magnetic moment was shown to be negative by Rogers and Staub.[8]

The value of the neutron magnetic moment determined by Corngold, Cohen, and Ramsey in 1956 was good to 30 parts per million and remained unchallenged for 23 years until the measurement of Greene and collaborators[9] diminished the uncertainty by an additional factor of 100. That experiment utilized most of the techniques pioneered by Alvarez and Bloch,[10] but it also incorporated the developments of the previous decades and a new calibration procedure that had never been used before. Two of the principal advances over the original Bloch-Alvarez experiment were the use of a much longer procession region for the neutrons (67

centimeters instead of 9) and the use of the separated oscillatory field method,[11] in which the oscillatory field is applied in two short coils with coherent oscillatory currents, one at the beginning and the other at the end of the neutron precession region. Much slower neutrons were used (100 meters per second instead of 2000), and the neutrons were confined by total reflection at glancing incidence to a glass tube. A major innovation in this experiment was the calibration of the magnetic field by measuring the proton precession frequency when water was transmitted through the same tube and the protons were coherently reoriented by the same two coils that were used for the neutron reorientation. With this procedure, the proton averaged the magnetic field in the same way that the neutron did. An additional, but similar, water tube for a proton resonance was operated with flowing water in it at all times to monitor the magnetic field for possible changes between the times neutrons and protons were in the tube. With these changes, the value of the neutron magnetic moment was found to be $-1.91304308 \pm 0.00000054$ nuclear magnetons. It is of interest to note that the original value obtained by Alvarez and Bloch agrees with the above result to within the experimental error of the original experiment.

The earliest consideration of the theoretical implications of the neutron magnetic moment were given by Alvarez and Bloch in the same paper that reported their measurement. They noted that to within the experimental accuracy, the deuteron magnetic moment as measured by Rabi and his associates was just equal to the sum of the proton and the neutron magnetic moments. They further observed that this result would be expected if the deuteron were in a pure triplet S-state and if the proton and neutron magnetic moments were purely additive, i.e., if their intrinsic values were not changed by the interaction between the neutron and proton, forming a deuteron. However, Alvarez and Bloch emphasized that the first assumption could not be exactly true, because of the D-state contribution required to give the deuteron quadrupole moment that had recently been reported;[12] and that the second assumption was implausible. This anticipation, that future experiments would fail to show additivity, was confirmed by subsequent experiments which showed that the sum of the proton and neutron magnetic moments exceeded the deuteron magnetic moment by 2.5%. Rarita and Schwinger used this difference to calculate the amount of D-state in the deuteron and showed that it was consistent with the observed deuteron quadrupole moment.[13] Subsequently, many theoretical papers have been written on this subject, making additional corrections and modifying the calculations

5. W. R. Arnold and A. Roberts, Phys. Rev. **71**, 878 (1947).

6. F. Bloch, D. Nicodemus, and H. H. Staub, Phys. Rev. **74**, 1025 (1948).

7. V. W. Cohen, N. R. Corngold, and N. F. Ramsey, Phys. Rev. **104**, 283 (1956).

8. E. H. Rogers and H. H. Staub, Phys. Rev. **76**, 980 (1949).

9. G. L. Greene, N. F. Ramsey, W. Mampe, J. M. Pendlebury, K. Smith, W. B. Dress, P. D. Miller, and Paul Perrin, Phys. Rev. **D20**, 2139 (1979) and Metrologia **18**, 93 (1982).

10. *See* Van Dyck et al., n. 4.

11. N. F. Ramsey, Phys. Today **33**, July, 25 (1980).

12. J. M. B. Kellogg, I. I. Rabi, N. F. Ramsey, and J. R. Zacharias, Phys. Rev. **55**, 318 and 595 (1939).

13. W. Rarita and J. Schwinger, Phys. Rev. **59**, 436 (1941).

to take into account the latest theory of nucleons and nuclear forces.

The anomalous moment of the neutron is of great interest in itself. If the neutron and proton were the simplest Dirac particles, the neutron would have no magnetic moment at all, since it has no net electric charge, and the proton would have the magnetic moment of a nuclear magneton. Neither of these simplistic predictions is true, and the departures from the predictions are called the anomalous moments. The earliest theoretical analyses on the basis of a meson exchange theory of nuclear forces predicted that the anomalous magnetic moments of the neutron and proton should be equal in magnitude but opposite in sign,[14] a prediction that is confirmed to ~1%. As new theories of nuclei and of nuclear forces have evolved, the values of the anomalous moments of the neutron and proton have continued as important tests of the theory.

The significance of the Alvarez-Bloch experiment is by no means limited to the measurement of the neutron magnetic moment and the theoretical studies based on the value of that moment. The experiment provided inspiration for a number of other important studies.

Felix Bloch's share[15] of the invention of nuclear induction and nuclear magnetic resonance (NMR) was to a considerable extent a result of Bloch's collaboration with Alvarez on neutron magnetic resonance. Not only did this early neutron magnetic resonance

work familiarize Bloch with the technique and the effectiveness of magnetic resonance, but also it stimulated his desire to measure the proton magnetic moment so the moments of the two nucleons could be compared. The subsequent applications of NMR to physics, chemistry, biology, and more recently medicine have been tremendous.

The pioneering neutron beam resonance work of Alvarez and Bloch provided the basic techniques later refined and applied to highly sensitive searches for a possible neutron electric dipole moment as a test of time reversal symmetry.[16]

Finally, the Alvarez-Bloch experiment is fundamentally the basis of Mezei's powerful new neutron spin-echo method for measuring very small neutron energy changes in neutron scattering.[17] In this method, polarized neutrons pass through a constant magnetic field region before the interaction region and through a similar region afterward but with the direction of the field reversed. If the velocity of the neutron is unchanged in the interaction region, the second field will exactly compensate for the effect of the first and the neutron spin will be returned to its original orientation, independent of its initial velocity. On the other hand, if the neutron loses even a few microelectron volts of energy, there will be a net reorientation of the neutron spin, and from the amount of the reorientation, the change in neutron velocity can be directly calculated. The method has been used, for example, to study proton excitations by neutrons in superfluid helium-4.[18]

14. H. Fröhlich, W. Heitler, and N. Kemmer, Proc. R. Soc. London, Ser. A **166**, 154 (1938).

15. F. Bloch, W. W. Hansen, and M. E. Packard, Phys. Rev. **69**, 127 (1946). See also E. M. Purcell, H. C. Torrey, and R. V. Pound, Phys. Rev. **69**, 37 (1946); and N. Blombergen, E. M. Purcell, and R. V. Pound, Phys. Rev. **73**, 679 (1948).

16. W. B. Dress, P. D. Miller, J. M. Pendlebury, P. Perin and N. F. Ramsey, Phys. Rev. **D15**, 9 (1977).

17. F. Mezei, Phys. Rev. Lett. **44**, 1601 (1980).

18. Ibid.

Reprinted from The Physical Review, Vol. 57, No. 2, 111–122, January 15, 1940
Printed in U. S. A.

A Quantitative Determination of the Neutron Moment in Absolute Nuclear Magnetons

Luis W. Alvarez, *Radiation Laboratory, Department of Physics, University of California, Berkeley, California*

AND

F. Bloch, *Department of Physics, Stanford University, Palo Alto, California*

(Received October 30, 1939)

The magnetic resonance method of determinng nuclear magnetic moments in molecular beams, recently described by Rabi and his collaborators, has been extended to allow the determination of the neutron moment. In place of deflection by inhomogeneous magnetic fields, magnetic scattering is used to produce and analyze the polarized beam of neutrons. Partial depolarization of the neutron beam is observed when the Larmor precessional frequency of the neutrons in a strong field is in resonance with a weak oscillating magnetic field normal to the strong field. A knowledge of the frequency and field when the resonance is observed, plus the assumption that the neutron spin is $\frac{1}{2}$, yields the moment directly. The theory of the experiment is developed in some detail, and a description of the apparatus is given. A new method of evaluating magnetic moments in all experiments using the resonance method is described. It is shown that the magnetic moment of any nucleus may be determined directly in absolute nuclear magnetons merely by a measurement of the *ratio* of two magnetic fields. These two fields are (a) that at which resonance occurs in a Rabi type experiment for a certain frequency, and (b) that at which protons are accelerated in a cyclotron operated on the nth harmonic of that frequency. The magnetic moment is then (for $J=\frac{1}{2}$), $\mu=H_b/nH_a$. n is an integer and H_b/H_a may be determined by null methods with arbitrary precision. The final result of a long series of experiments during which 200 million neutrons were counted is that the magnetic moment of the neutron, $\mu_n=1.93_5\pm0.02$ absolute nuclear magnetons. A brief discussion of the significance of this result is presented.

Introduction

THE study of hyperfine structure in atomic spectra has shown that a large number of atomic nuclei possess an angular momentum and a magnetic moment. Since, according to the theories of Heisenberg and Majorana, protons and neutrons are recognized as the elementary constituents of nuclear matter, their intrinsic properties and particularly their magnetic moments have become of considerable interest. The fundamental experiments of Stern and his collaborators[1] in which they determined the magnetic moments of the proton and the deuteron by deflections of molecular beams in inhomogeneous fields gave the first quantitative data of this sort. The approximate values which they gave for the two moments,[2]

$$\mu_p=2.5, \tag{1}$$

$$\mu_d=0.8, \tag{2}$$

suggested that in all likelihood, one would have to ascribe to the neutron a magnetic moment of the approximate value

$$\mu_n=-2. \tag{3}$$

The negative sign in formula (3) indicates that the relative orientation of their magnetic moments with respect to their angular momenta is opposite in the case of the neutron to that of the proton and the deuteron.

The technique of molecular beams has been greatly developed during the last few years by Rabi and his collaborators;[3] their ingenious methods have allowed them to determine the magnetic moments of many light nuclei with high precision, and to establish the existence of an electric quadrupole moment of the deuteron. Their values for the magnetic moments of the proton and deuteron are

$$\mu_p=2.785\pm0.02, \tag{4}$$

$$\mu_d=0.855\pm0.006. \tag{5}$$

They have also demonstrated that both moments

---

[1] R. Frisch and O. Stern, Zeits. f. Physik **85**, 4 (1933); I. Estermann and O. Stern, Zeits. f. Physik **85**, 17 (1933); O. Stern, Zeits. f. Physik **89**, 665 (1934); I. Estermann and O. Stern, Phys. Rev. **45**, 761 (1934).

[2] We shall throughout this paper give magnetic moments in units of the nuclear magneton $eh/4\pi Mc$ (e and M =charge and mass of the proton, c=velocity of light), and angular momenta in units $\hbar=h/2\pi$. The angular momentum of the deuteron is taken to be 1, that of the proton and the neutron to be $\frac{1}{2}$; the last is an assumption for which a direct experimental proof is still lacking, but which is generally accepted.

[3] I. I. Rabi, Phys. Rev. **51**, 652 (1937); I. I. Rabi, J. R. Zacharias, S. Millman and P. Kusch, Phys. Rev. **55**, 526 (1939).

are positive with respect to the direction of the angular momentum.

An experimental proof that a free neutron possesses a magnetic moment, and a measure of its strength, could also be achieved in principle by deflection of neutron beams in an inhomogeneous magnetic field. But while the great collimation required for this type of experiment may easily be obtained with molecular beams, it would be almost impossible with the neutron sources available at present. Better suited for the purpose is the method of magnetic scattering, which was suggested a few years ago by one of us.[4] It is based upon the principle that a noticeable part of the scattering of slow neutrons can be due to the interaction of their magnetic moments with that of the extranuclear electrons of the scattering atom. In the case of a magnetized scatterer this will cause a difference in the scattering cross section, dependent upon the orientation of the neutron moment with respect to the magnetization, and particularly in the case of ferromagnetics, it will cause a partial polarization of the transmitted neutron beam. The magnetic scattering of neutrons, and thereby the existence of the neutron moment, has been proved experimentally by several investigators, particularly by Dunning and his collaborators.[5] The magnetic scattering, however, can yield only a qualitative determination of the neutron moment since the interpretation of the effect is largely obscured by features involving the nature of the scattering substance.

Frisch, v. Halban and Koch[6] were the first to attempt to use the polarization of neutrons merely as a tool, and to determine the neutron moment by a change of the polarization, produced by a magnetic field between the polarizer and the analyzer. Such a change should indeed occur, because of the fact that the moment will precess in a magnetic field; by varying the field strength, one can reach a point where the time spent by the neutrons in the field is comparable

to the Larmor period. In this way, one could obtain at least the order of magnitude of the moment. Although these investigators have reported an effect of the expected type, yielding the order of magnitude 2 for the neutron moment, we have serious doubts that their results are significant. Their polarizer and analyzer consisted of rings of Swedish iron, carrying only their remanent magnetism ($B = 10,000$ gauss), while in agreement with Powers' results, we were never able to detect any noticeable polarization effects, independent of the kind of iron used, until it was magnetized between the poles of a strong electromagnet with an induction well above 20,000 gauss. Although we cannot deny the possibility that, due to unknown reasons, their iron was far more effective for polarization at low values of the induction than that used by other investigators, we think it more likely that in view of their rather large statistical errors the apparent effect was merely the result of fluctuations.

Although most valuable as a new method of approach, the experiment of Frisch, v. Halban and Koch could in any event give only qualitative results. The slow neutrons which one is forced to use emerge from paraffin with a complicated and none too well-known velocity distribution. The time during which they precess in the magnetic field will therefore be different for different neutrons and vary over a rather large range. Since it is that time which together with the field of precession determines the value of the moment, the latter will be known only approximately. A quantitative determination of the neutron moment therefore requires an arrangement which does not contain such features.

METHOD

Sometime ago, we conceived of an experimental method which could yield quantitative data of this sort. The method was independently proposed by Gorter and Rabi,[7] and most successfully used by the latter in his precision determinations

---

[4] F. Bloch, Phys. Rev. **50**, 259 (1936); **51**, 994 (1937); Ann. Inst. H. Poincaré **8**, 63 (1938); J. Schwinger, Phys. Rev. **51**, 544 (1937).

[5] P. N. Powers, H. Carroll and J. R. Dunning, Phys. Rev. **51**, 1112 (1937); P. N. Powers, H. Carroll, H. Beyer and J. R. Dunning, Phys. Rev. **52**, 38 (1937); P. N. Powers, Phys. Rev. **54**, 827 (1938).

[6] R. Frisch, H. v. Halban and J. Koch, Phys. Rev. **53**, 719 (1938).

[7] I. I. Rabi, J. R. Zacharias, S. Millman and P. Kusch, Phys. Rev. **53**, 318 (1938). During the summer of 1938 a first attempt to measure the neutron moment with such an arrangement was made at Stanford by Bloch, Bradbury and Tatel, using a 200-mg Ra-Be source of neutrons. However, it was unsuccessful due to instrumental effects which were hard to discover with the small intensity available.

of nuclear moments. Its principle consists in the variation of a magnetic field H_0 to the point where the Larmor precession of the neutrons is in resonance with the frequency of an oscillating magnetic field. The ratio of the resonance value of H_0 to the known frequency of the oscillating field gives immediately the value of the magnetic moment.[8]

The observation of the resonance point is based upon the fact that in its neighborhood there will be a finite probability P for a change of the orientation of the neutron moment with respect to the direction of the field H_0. Let this field be oriented in the z direction and let there be perpendicular to it, say in the x direction, an oscillating field with amplitude H_1 and circular frequency[9] ω, so that the total field in which a neutron is forced to move, is given by its components

$$H_x = H_1 \cos (\omega t + \delta); \quad H_y = 0; \quad H_z = H_0. \quad (6)$$

The solution of the Schroedinger equation for a neutron with angular momentum $\frac{1}{2}$ and magnetic moment μ gives the probability that a neutron, which at time $t = 0$ in such a field had a z component $m = \frac{1}{2}$ of its angular momentum, will be found at the time $t = T$ with a value $m = -\frac{1}{2}$, in the form

$$P = \sin^2 [(\mu H_1 T / 2\hbar) \\ \times (1 + \{2\Delta H / H_1\}^2)^{\frac{1}{2}}] / [1 + (2\Delta H / H_1)^2], \quad (7)$$

where

$$\Delta H = H_0 - H_0^* = H_0 - \hbar\omega / 2\mu \quad (8)$$

is the difference between the constant field H_0 and its value at resonance,

$$H_0^* = \hbar\omega / 2\mu, \quad (9)$$

for which the Larmor frequency $2H_0\mu/\hbar$ is equal to the frequency ω of the oscillating field. Since the time T which the neutrons spend in the oscillating field will, for different neutrons, vary over a wide range, it will be a good approximation

to substitute for the $\sin^2$ in the numerator of (7) its average value $\frac{1}{2}$. This means that, at resonance, complete depolarization of an originally polarized neutron beam will occur, and leads to the simplified formula

$$P = \frac{1}{2}[1 + (2\Delta H / H_1)^2]^{-1}. \quad (7a)$$

Formulae (7) and (7a) have been derived by neglecting in the differential equation under consideration certain rapidly oscillating terms with respect to slowly varying ones; this neglect is justified as long as the condition

$$H_1 / H_0 \ll 1 \quad (10)$$

holds. In our final experiments, H_1/H_0 was less than 2 percent; so we felt justified in applying formulae (7) and (7a).*

The dependence of P upon ΔH, and particularly its maximum value for which the field H_0 has its resonance value H_0^* given by Eq. (9), is detected by a polarizer-analyzer arrangement, similar to that used by Frisch, v. Halban and Koch. We shall show that a beam of neutrons passing through two successive plates of magnetized material will change its transmitted intensity by an amount proportional to P. Consider two plates F_1 and F_2 both magnetized in the $+z$ direction and call f_i^+ and f_i^- $(i = 1, 2)$ the fraction of neutrons with $m = +\frac{1}{2}$ and $m = -\frac{1}{2}$, respectively, transmitted by F_i. f_i^+ will be different from f_i^- since the magnetic scattering will be preferential for one of the two values of m, thus giving rise to a partial polarization of the neutrons. If there does not occur any change of polarization between F_1 and F_2, the fraction of the neutrons transmitted by both will be given by

$$f = f_1^+ f_2^+ + f_1^- f_2^-. \quad (11)$$

If, on the other hand, there is a probability P of a change of polarization between F_1 and F_2, the transmitted fraction will be

$$f' = (1 - P)(f_1^+ f_2^+ + f_1^- f_2^-) \\ + P(f_1^+ f_2^- + f_1^- f_2^+). \quad (12)$$

One will thus observe a relative change ΔI of the intensity I of the transmitted neutrons, given by

$$\Delta I / I = (f' - f) / f = -\alpha P, \quad (13)$$

---

[8] P. N. Powers, reference 5, has employed a modification of this method to determine the sign of the neutron moment. Instead of encountering an oscillating field between polarizer and analyzer, his neutrons were acted upon by a field which appeared to them, in consequence of their velocity, to be rotating with approximately their Larmor precession frequency. The sign of the moment was found to be negative, in agreement with theory.

[9] We shall always use circular frequencies although we sometimes refer to them merely as "frequencies."

---

* The corrections, introduced by this neglect are by no means of the order of 2 percent, but very much smaller, since they are essentially given by $(H_1/H_0)^2$. The mathematical theory of these corrections will soon be published.

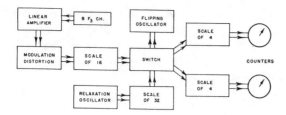

FIG. 1. Schematic drawing of electrical circuits.

with

$$\alpha = [(f_1{}^+ - f_1{}^-)(f_2{}^+ - f_2{}^-)]/(f_1{}^+ f_2{}^+ + f_1{}^- f_2{}^-), \quad (14)$$

where α is a constant, characteristic of the magnetic and geometric conditions of F_1 and F_2 only. We see that this constant α can also be determined independently of any possible changes of polarization. For two identical plates F_1 and F_2 with common thickness a and with the values $f_1{}^\pm = f_2{}^\pm = e^{-(\sigma \pm \Delta\sigma)a}(\Delta\sigma = 0$ for unmagnetized material) as they follow from the theory of magnetic scattering, one can see that α also has the significance of the relative increase of the total transmitted intensity which one observes by magnetizing F_1 and F_2 to the stage in which the fractions f_i have the values assumed in formula (14). Thus under ideal conditions, and if P assumes its maximum value $\frac{1}{2}$ at resonance, the relative change in intensity will just be $\frac{1}{2}$ of that observed when demagnetizing F_1 and F_2 in an experiment similar to that of Dunning and his collaborators.

Once the maximum value of (13) has been experimentally established, i.e., the value of H_0 has been found for which $\Delta H = 0$, or where according to (8) and (9)

$$H_0 = H_0{}^* = \hbar\omega/2\mu, \quad (15)$$

one will know the magnetic moment μ in the form

$$\mu = \hbar\omega/2H_0{}^*, \quad (15a)$$

provided that the values of $H_0{}^*$ and the frequency ω of the oscillating field are known.

One can of course measure the latter by means of an ordinary standardized wavemeter and the former by the usual method of flipping a coil in the magnetic field and comparing the ballistic throw of a galvanometer with that produced by breaking the current in the primary of a known mutual inductance. We have, however, thought of an alternative method, which not only can be

carried to almost unlimited accuracy, but also has the highly desirable feature of yielding the moment to be determined directly in units of the nuclear magneton. It consists in comparing the frequency $\omega = \omega_n$ and the field $H_0{}^* = H_n$ with the values ω_p and H_p, where ω_p is the circular frequency of an oscillator driving a cyclotron which accelerates protons in a magnetic field H_p. We have them from (15)

$$\omega_n = (2H_n\mu_n/\hbar)(e\hbar/2Mc) = (eH_n/Mc)\mu_n, \quad (16)$$

if instead of using absolute units as in (15), we express the neutron moment in units of the nuclear magneton, i.e., if we write

$$\mu = \mu_n(e\hbar/2Mc). \quad (17)$$

On the other hand, for the ideal resonance of protons in a cyclotron, we have the relation

$$\omega_p = eH_p/Mc. \quad (18)$$

Dividing (16) by (18), we now obtain

$$\mu_n = (\omega_n/\omega_p)(H_p/H_n). \quad (19)$$

This means that in order to determine the moment in absolute nuclear magnetons one has merely to determine the *ratio* of two frequencies and the *ratio* of two fields instead of determining ω_n and H_n absolutely. In addition to this great experimental advantage, we see that no knowledge of the universal constants of physics is required. The value of Planck's constant does not enter since the moment is expressed in nuclear magnetons and the value of e/Mc is eliminated by use of the cyclotron.

If the radiofrequency generator for the cyclotron were a power amplifier driven by a harmonic of the "neutron oscillator," the ratio ω_n/ω_p would be simply an integer. In this case one would merely have to measure the ratio H_p/H_n, in order to know μ_n. This measurement can be reduced to the determination of the ratio of the wound areas (area $\times$ turns) of two flip coils, which can in turn be reduced to that of two currents. If the two coils, located in the same position in a current system and both connected through the same galvanometer circuit give the same ballistic throw, then the ratio of the currents for which this condition is reached gives the ratio of their wound areas. Once that ratio is known, one also has that of any two fields which, using the two coils, give the same ballistic throw in the galvanometer. Gehrcke and v. Wogau[10] have developed this technique, by the use of null indicator methods, to a point where the ratio of two fields may be determined to five significant figures. Although we have determined the ratios of the frequencies and the fields by essentially these methods the errors, entering from elsewhere did not justify the application of all their refinements described above.

Description of Apparatus

Up to the present time, almost all of the precision measurements of neutron intensity have been made using Rn-Be, or similar sources not subject to the erratic fluctuations occurring with artificial sources. Although it is possible to keep the cyclotron ion beam constant to one or two percent for long periods of time, this is not sufficient for the present work, where an accuracy of the order of 0.1 percent is needed. Also, one cannot assume that the neutron intensity is proportional to the beam current, since the distribution of the latter over the target, and the proportion of stray ions striking the insides of the dees are both important in this respect. This rules out the possibility of using an integrating beam meter, or auxiliary BF_3 chamber, as a monitor. The counting-monitoring circuit which was finally adopted for this experiment is shown schematically in Fig. 1.

The neutrons were detected in a BF_3 chamber operated at 5000 volts. Its diameter and length were 9 cm and 15 cm, respectively, and the gas pressure was 50 cm of Hg. The arrangement of electrodes is quite similar to that of a chamber recently described by Powers.[3] A conventional linear amplifier of 4 stages fed the pulses into the modulation-distortion circuit, where all pulses of magnitude greater than a certain value were amplified to uniform size, and all below this height were suppressed. This circuit has been described in detail before, in connection with the production of monochromatic neutron beams,[11] but was not essential for most of this work, since modulated beams were not used in the final experiment. The uniform pulses from this stage

[10] E. Gehrcke and M. v. Wogau, Verh. d. D. Phys. Ges. **11**, 664 (1909).
[11] L. W. Alvarez, Phys. Rev. **54**, 609 (1938).

were fed to a scale-of-16 of the type described by Reich,[12] which uses hard tubes. After passing through the scale, the pulses entered a switching circuit which fed them to either of two scales of 4 plus a Cenco counter. The switching was carried out automatically at regular intervals of from 2–10 seconds, depending upon the adjustment. At the same time, the switching circuit turned a magnet on and off (from magnetic scattering experiments), or keyed the oscillator (for the resonance experiment).

The first circuits designed to carry out the switching used relays; these were found to be unsatisfactory due to differences in spring adjustments, so an all electronic switching circuit was devised. This is shown in Fig. 2. Negative pulses were fed to the grids of the two right-hand 6K7's, each of which had another 6K7 connected to it in the familiar Rossi coincidence circuit. If the grids of either of these left-hand 6K7's were negative, the plate current of its partner would be cut off, delivering a positive pulse to the corresponding scale-of-4. But if the grid was positive, the voltage pulse at the common plate connection would be too small to affect the scale. The circuit at the left is merely a device for making one grid positive and the other negative, and reversing this condition instantaneously at will. The two 885's are arranged in a conventional inverter, or scale-of-2 circuit. When current flows through the upper Thyratron and is blocked in the lower, pulses will pass through the switch into the upper scale-of-4. If a positive pulse is now applied to the grids of the 885's, the scale-of-2 will invert, and the lower scale-of-4 will receive the neutron pulses. The positive pulse which flips the scale-of-2 originates in a relaxation oscillator of the type used in cathode-ray oscillograph sweep circuits. It was found that the interval between

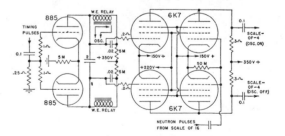

Fig. 2. Electronic switching circuit.

[12] H. Reich, Rev. Sci. Inst. **9**, 222 (1938).

pulses was not constant when R and C were large enough to make the time constant 5 seconds; so the frequency was increased, and the pulses from the oscillator were fed through a scale of 32 into the switching circuit. Fluctuations in the time interval were reduced to a negligible amount by this procedure. The length of the interval could be varied by means of a variable resistance in the oscillator circuit. Small relays in the plate circuits of the Thyratrons made it possible to control the magnets or R.F. oscillator in phase with the counter switching. A lag in these operations merely reduced slightly any real effect, while a lag in the counter switching would introduce a spurious effect of the same magnitude.

On first thought it would seem that the statistical errors would be considerably increased by the method of switching a scale-of-16 from one scale-of-4 to another. But calculation of the effect shows that it is of no importance, and our experimental fluctuations have always checked with those to be expected on the assumption that this arrangement is equivalent to two separate scales-of-64. We have also made runs to test the equality of the counting time under the two conditions, by feeding the output of a beat frequency oscillator into the distortion amplifier. It was always found that the two Cenco counters recorded equal numbers of counts per cycle of the switching.

We spent several months developing our polarizer-analyzer equipment. Our first attempts were along lines similar to those used by Frisch, v. Halban and Koch. Flat rings of iron with a diameter of 1 meter, a thickness of 1 cm, and a width of 10 cm were wound with copper wire in toroidal fashion to minimize the stray field. We were unable to observe any difference in the transmitted neutron intensity, which depended upon the magnetization of the iron. This is in agreement with the Columbia school which has shown that this iron (Armco) has a sharp magnetic effect threshold very near the saturation value, which most likely was not reached with our low magnetizing currents.

Since they also report that Swedish iron shows neutron magnetic effects at lower fields, we next wound bars of Swedish iron with flat copper strip through which 120 amperes could be passed. The bars were 1.5 meters long, 10 cm wide, and 1.5 cm thick. Although the H in the iron was 250 gauss ($B=20,000$), we were still unable to observe a difference in neutron transmission. We finally cut a piece from one of the Swedish iron bars, and magnetized it strongly between the pole pieces of a Weiss magnet and observed the magnetic effect. With an iron piece 4 cm thick, we obtained differences of about 6 percent in transmission. With half of the iron in the Weiss magnet, and the other half in a similar strong magnet, we confirmed Powers' results on double transmission, i.e., that the change in intensity was about the same for the pieces in one magnet, as when they were separated in two, at some distance. Many experiments were made to determine

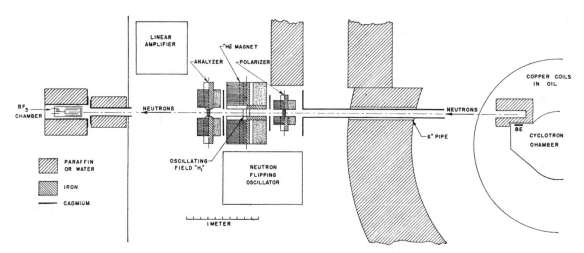

FIG. 3. Plan of the apparatus. Section taken 85 cm above floor of cyclotron room.

the optimum thickness of iron, and the magnetizing currents needed to obtain the maximum effects.

It was then possible to consider the construction of the precessing field, and the oscillator system. A plan of the apparatus is given in Fig. 3. Neutrons from the Be+D reaction were slowed to thermal velocities in the simple howitzer, and diffused down a cadmium-lined tube through the water tank to the polarizer. After passing through the precessing and flipping field, and the analyzer magnet, they were recorded in the BF_3 chamber. The large blocks of paraffin which defined the beam made the fast neutron background quite small. The iron plates reduced the thermal neutron intensity to about 5 percent, but the fast neutron background, as measured by covering one of the plates with Cd, was still only 25 percent of the total counting rate. This small background made it unnecessary to use the backgroundless modulated beam, as was originally planned. The somewhat increased magnetic effect at low neutron velocities was not enough to offset the decrease in intensity which attended the use of the modulation. In the earlier stages of the experiment, and actually in one of the resonance experiments, 120-cycle modulation was used successfully, but it was abandoned in favor of the greater intensity available without modulation.

The oscillator was of the conventional Hartley type, using an Eimac $150T$, self-excited. The frequency was found to remain constant to better than 0.05 percent for days. The solenoid which produced the oscillating magnetic field was wound with a 9-cm square cross section and 15 cm height. The conductor was flat copper strip $0.50'' \times 0.010''$ wound so that adjacent turns almost touched. Neutrons passed through the strip on entering and leaving the region of the oscillating field H_1. The solenoid was connected in parallel with a section of the tank coil chosen to give 11 amperes magnetizing current as measured on an R.F. ammeter in the solenoid circuit ($H_1 \approx 10$ gauss). The oscillator was keyed by means of a relay in the primaries of the plate supply transformers.

The steady precessing field H_0 was at first produced by a pair of Helmholtz coils plus a pair of cylindrical bars of iron to concentrate the flux. This field was quite inhomogeneous, and was used for the first two resonance dips only. For the later measurements, we used a pair of large coils which had energized the magnet of the original 1-mv cyclotron of Lawrence and Livingston. These magnet coils were loaned to us by Professor R. B. Brode, who had been using them in connection with a cosmic-ray cloud chamber. An arrangement of iron was found which produced a field of about 600 gauss ± 1 gauss throughout the volume occupied by the R.F. solenoid. The fluctuations in the laboratory d.c. supply were too great to allow its use in the coil circuit, so we utilized the constant current supply of the cyclotron magnet. This automatic stabilizer keeps the voltage fluctuations across the cyclotron magnet coils to less than 0.1 percent and gradually increases the value of the stabilized voltage to compensate for the heating of the coils. The magnet coils were then connected in parallel with the cyclotron coils through a variable resistance. The current was watched constantly, and the resistance was varied slightly when necessary, to allow for the differential heating of the two sets of coils.

A pair of large grounded copper plates were placed against the polepieces, so that the magnet coils were shielded from the R.F. fields of the solenoid. If this shield was not grounded, radio-frequency currents were induced in the d.c. lines and fed back into the stabilizer. A partial rectification of these currents would cause the stabilizer to change the magnet current, and therefore throw the cyclotron out of resonance. The resulting drop in the neutron intensity would be in phase with the switching cycle, and would therefore produce an apparent effect of the same sign and magnitude of that to be expected. Another cause of apparent effects was the interaction of the R.F. oscillator with the relaxation oscillator in the timing circuit. When the scale-of-32 was not properly grounded, it would count an extra pulse whenever the oscillator was turned on. This gave a 3 percent apparent effect of the correct sign and magnitude, due to the change in switching period. Several other spurious effects were encountered at various times during the experiment; it is due to the large intensities of neutrons available that we were able to trace down and eliminate them after comparatively short counting times.

The test for reliability of the apparatus in which we placed most faith was the following: A series of observations was taken at 20μa of deuterons, and it was desired to know whether the effect, e.g., difference in neutron counts with and without the oscillator on, was real. The current was then increased to 80 amperes, and the analyzer was covered with a cadmium sheet. The counting rate in all parts of the amplifier, from the BF_3 chamber to the final mechanical counters remained the same as before, and all relays, magnetic and electric fields were operating in the same manner. The only difference was that the counted neutrons did not for the most part pass through the polarizer and analyzer, and that part of high velocity which did could not exhibit any magnetic effects. Any difference between the relative counting rates under these two conditions was ascribed to magnetic effects of the thermal neutrons. It was also possible in this way to eliminate causes of systematic effects. The closing of relays would occasionally jar the BF_3 chamber, and the resulting noise would introduce extra counts in synchronism with the timing cycle. Several effects of this nature were discovered and eliminated after long blank runs. It was also important at these high counting rates to guard against saturation in the BF_3 chamber. The intensity was always adjusted so that the counting rate was still a linear function of the deuteron current.

These sources of trouble are mentioned to emphasize the great desirability of powerful neutron sources in experiments of this sort. In the final set of experiments, in which all these spurious effects were eliminated and on which our value of the neutron moment is based, only about 4 million neutrons were counted. This number of counts could be taken with Rn-Be sources of moderate strength in a reasonable length of time, but it would be almost impossible to track down obscure systematic effects in the manner outlined above. We counted more than 200 million neutrons in test runs before we felt sure enough of the apparatus to have confidence in the results outlined below. Perhaps the most convincing proof of the reality of the effect was that we varied H_0 and ω over a factor of almost 10, and

each time the frequency was increased, the magnetic field setting for the center of the new dip was found within the accuracy, to be expected at the value calculated from the previously determined value of H_0^*/ω.

MEASUREMENTS

In the first experiments with the small Helmholtz coils, the magnetic field was measured with the aid of a fluxmeter-search coil combination. Although somewhat crude, the measurements did show that H_0^*/ω was constant for two frequencies differing by about 20 percent, and that the effect could therefore be attributed to the magnetic moment of the neutron, as outlined above. The shallowness of the resonance dip observed in these first experiments ($\Delta I/I = 0.6$ percent) could be attributed to the inhomogeneity of the field H_0.

When the large magnet coils were installed, it was clear that a more accurate method of measuring the field would be necessary. A brass flip coil was therefore constructed in such a way that it could be alternately lowered into a standard position within the R.F. solenoid, or withdrawn to allow the neutrons to pass through. Stops were placed so that the coil turned through exactly 180°, as determined by optical means. The coil was turned by spring tension, after being cocked by hand. The leads from the coil were connected in series with a ballistic galvanometer of very long period, a Hibbert magnetic standard, and a search coil at the center of a standard solenoid. A knowledge of the constant of the standard solenoid plus the wound area of the flip coil and the search coil allowed the strength of the field to be ascertained. The current through the standard solenoid was measured by means of a standard resistance, standard cell, and potentiometer.

Before and after each counting period, the relation between the current through the coils, and the magnetic field was determined. This consisted in noting the ratio of the throws of the ballistic galvanometer d/s, where d was the deflection due to the turning of the flip coil, and s was the throw of the magnetic standard. Effects due to any nonlinearity of the galvanometer were minimized by adjusting this ratio close to one by rewinding the standard. d/s was determined as a

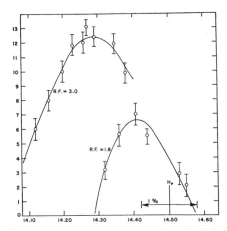

FIG. 4. Proton resonance curves. The potential drop in millivolts across the magnet current shunt is plotted as abscissa against the electromotive force in millivolts of the thermocouple used in the probe on which the proton current fell.

function of the magnetizing current, and the neutron intensity was also measured as a function of this current. The resonant field was then found from a comparison of these two curves.

The frequency of the oscillator was measured on a General Radio precision wavemeter. A knowledge of the field and frequency yielded the magnetic moment of the neutron by substitution in Eq. (15a).

The most trustworthy value of the moment is, however, given in absolute nuclear magnetons by the comparison method, outlined above. Although we ultimately trusted more this method it was gratifying to obtain well within the experimental error the same value for μ_n by the two entirely different methods of calibration. For this purpose it was necessary to construct a new flip coil which when turned through 180° in the proton resonance field of the 60-inch cyclotron would give a d/s almost equal to that obtained in the measurement of H_0^*. From a knowledge of the approximate values of the two magnetic fields involved, and the wound area of the original coil, it was a simple matter to construct such a coil. This coil was wound on a hard rubber form to ascertain the absence of magnetic impurities in it or in the brass used for the first coil.

It was then necessary to determine the ratio of the wound areas of the two flip coils. This measurement was carried out in the following manner: The two coils were mounted near the center of a pair of carefully constructed Helmholtz

coils, which were designed for accurate cloud-chamber beta-ray curvature measurements. The constant of the coils did not enter the calculation; it was merely assumed that the field was directly proportional to the current through the coils. This current was measured in terms of the potential drop across a standard resistance in the circuit, on a Wolff potentiometer standardized with a Weston cell. The small coil was flipped in the strongest field available, and the current was then reduced until the large coil gave the same ballistic throw as the small one in the higher field. (The effect of the earth's field was subtracted from all readings.) The ratio of the wound areas of the two coils was then given by the ratio of the currents as determined on the potentiometer. The two coils were connected in series with the galvanometer; so the total resistance of the circuit remained constant. The current ratio was found to be constant for many values of the fields. To test for magnetic impurities in the search coil, the coil was connected in series opposition, and the large coil was turned through such an angle that when a current was broken in the Helmholtz coils no kick was observed on the galvanometer. This condition prevailed for a wide range of magnetizing currents. A very sensitive galvanometer was used in these measurements; so this zero method gave us confidence that no appreciable ferromagnetic contamination existed in the flip coils, one of brass and the other of hard rubber.

The new coil, ballistic galvanometer, and magnetic standard were then carried to the Crocker Radiation Laboratory, where the magnetic calibration was made in terms of the 60-inch cyclotron field. The rod carrying the flip coil projected into the cyclotron chamber through a window so that the coil could rotate in the gap at a point about 25 cm from the center of the cyclotron. Tests with a large search coil connected to a fluxmeter showed the field to be constant throughout this central region to better than 0.1 percent. The magnetization curve of the cyclotron magnet was then measured in the region around the known proton field. The measurements consisted in a determination of d/s against the potential drop across the magnet current shunt, as measured on the Wolff potentiometer.

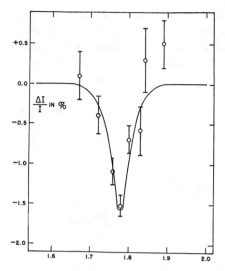

Fig. 5. Neutron resonance dip. The magnet current in arbitrary used is plotted against the fractional change ($\Delta I/I$) of the intensity of the neutron beam.

The cyclotron was then evacuated and hydrogen let into the chamber. The heating effect of the proton current to a probe placed at the point occupied by the flip coil was then measured as a function of the potential drop across the magnet current shunt. A comparison of these two curves gave the d/s value for the acceleration of protons at the frequency in use at the time. The effects of hysteresis on the resonance curve were found to be negligible, by observing it when the magnet current had been increased from zero to the proton value, and also when it had been carried through the cycle—zero, deuteron-field, proton-field.

The proton resonance curve, Fig. 4, was measured at two values of the dee voltage, and the center of gravity of the curve was found to move toward higher fields as the voltage was reduced. This was shown not to be caused by a change in the cyclotron frequency, by measurements taken during each run. The increase in width of the curves with voltage is due to the fact that at higher voltage, the protons have to circulate fewer times. The shift of the peaks is due mainly to the focusing properties of the cyclotron. Since the magnetic field was uniform out to the probe, the focusing was entirely electrostatic, and this is known to be dependent upon the phase of the protons with respect to that of the oscillator.[13]

[13] M. E. Rose, Phys. Rev. **53**, 392 (1938); R. R. Wilson, Phys. Rev. **53**, 408 (1938).

Professor E. McMillan and Mr. R. Wilson have made a study of these effects in this laboratory, and they agree that the proper method of choosing the magnetic field, H_p, would be to run a large series of resonance curves with decreasing dee voltage, and extrapolate the resonance field to zero dee voltage. Changes in the apparatus made this impossible, so they have suggested that we choose a point midway between the maximum of the curve taken at lower dee voltage and its high field intercept. Since we have seen both experimentally and theoretically that the extrapolated resonance curve will lie somewhere between these two points, separated in field by 1 percent, the maximum error introduced in this way could be only 0.5 percent.

The ratio of the cyclotron frequency to the neutron frequency was determined with the aid of a Super-Pro short wave radio receiver. The second harmonic of the cyclotron oscillator was tuned in on this receiver in the same frequency range as the sixth to tenth harmonics of the neutron oscillator. The agreement between the various values of the frequency of the neutron oscillator as determined from these measurements attests to the linearity of the frequency scale of the receiver. The receiver was thus used merely as an uncalibrated, linear wavemeter.

Results

Since the width of the resonance dip in gauss is constant for all values of the precessing field (Eq. (7a)), it is obvious that it is desirable to

Table I. *Data for neutron resonance curve (Fig. 5). Each reading is based on 120,000 counts. Magnetic field for neutron resonance, 622 gauss; frequency of oscillator, 1843 kilocycles; magnetic moment of the neutron, 1.94 nuclear magnetons; d/s for coil 1 in neutron field, 0.866; d/s for coil 2 in proton field, 0.844.*

Current	Percentage Effect	Mean Effect
1.67	-0.8, $+1.0$	$+0.1 \pm 0.3$
1.72	$+0.6$, -0.9, -0.8	-0.4 ± 0.23
1.76	-0.5, -0.9, -2.1, -1.1, -1.2, -0.8	-1.1 ± 0.16
1.78	-2.0, -2.2, -1.6, -1.5, -1.4, -0.5, -1.7, -1.2, -1.6, -1.5	-1.52 ± 0.13
1.79	-1.7, -1.7	-1.7 ± 0.3
1.80	-1.4, -0.5, -0.7, -1.5, $+0.1$, -0.3	-0.71 ± 0.16
1.81	-1.5	-1.5 ± 0.4
1.83	-0.8, -0.4	-0.6 ± 0.3
1.84	$+0.3$	$+0.3 \pm 0.4$
1.89	$+1.0$, $+0.0$	$+0.5 \pm 0.3$

perform exploratory experiments at low precessing fields, to find an approximate value of the magnetic moment. After the relatively wide dip has been found at low fields, it is possible to increase the precision of the determination by raising the field and the frequency in the same ratio. Our first measurements were made at 260 and 315 kc, and merely served to show that the value of the moment was close to 2 nuclear magnetons. When the large homogeneous field was installed, we repeated the measurement at 315 kc; after observing the dip again at 810 kc, we increased the frequency to 1843 kc, where the field calibrations were made. Since the resonance dips at all the lower frequencies were consistent with that at 1843 kc, we will not describe those measurements in detail. The curve at 1843 kc is shown in Fig. 5.

The experimental data from which this curve is plotted are given in Tables I, II and III. It will be noticed that two of the points were not plotted in the curve. These are the values at 1.79 and 1.81 amperes, which have very large probable errors, due to the small numbers of neutrons counted. These two points have no appreciable effect upon the position of the minimum, and although they lie far off the curve, their positions are quite consistent with their probable errors. The curve was obtained on two separate days a week apart, and since no systematic difference could be noted between the two sets of data, they have been plotted together. The value of the magnetic moment is then given from Eq. (19),

$$\mu_n = (1.856/9.685) \times (10.60 \times 0.844/0.886)$$
$$= 1.935 \text{ nuclear magnetons.}$$

The *probable errors* entering from various sources are: position of dip, 0.5 percent; galvanometer readings, 0.2 percent; coil calibration, 0.2 percent; frequency calibration, 0.2 percent; and cyclotron calibration, 0.6 percent. The final result may

TABLE II. *Current ratios in cloud-chamber Helmholtz coils for equal galvanometer throws.*

CURRENT	RATIO	CURRENT	RATIO
2.20	10.61	1.60	10.63
2.19	10.61	1.30	10.58
2.18	10.58	Ave.	10.60
2.10	10.60		

TABLE III. *Frequency calibration: reading, R, on radio receiver for nth harmonic.*

NEUTRON OSCILLATOR			CYCLOTRON OSCILLATOR		
n	R	R/n	n	R	R/n
6	11.14	1.856	2	19.37	9.685
7	12.99	1.856			
8	14.83	1.854	All values in megacycles.		
9	16.70	1.856			
10	18.56	1.856			

then be expressed as

$$\mu_n = 1.935 \pm 0.02$$

absolute nuclear magnetons.

DISCUSSION

The now rather accurately known values

$$\mu_p = 2.785 \pm 0.02 \qquad \mu_n = -1.935 \pm 0.02$$
$$\mu_d = 0.855 \pm 0.006$$

of the magnetic moments of proton, neutron and deuteron are of considerable interest for nuclear theory. The fact alone that μ_p differs from unity and μ_n differs from zero indicates that, unlike the electron, these particles are not sufficiently described by the relativistic wave equation of *Dirac* and that other causes underly their magnetic properties.

Whatever these causes may turn out to be one has to notice that there holds to well within the experimental error the simple empirical relation

$$\mu_d = \mu_p + \mu_n. \qquad (20)$$

This relation is far from being obvious and it would in fact seem rather surprising if it were rigorously satisfied. To explain it in simple terms one would have to make both the following assumptions:

(a) The fundamental state of the deuteron is a ^3S state so that there are no contributions to μ_d arising from orbital motion of the particles.

(b) The moments μ_p and μ_n are "additive," i.e., their intrinsic values are not changed by the interaction of the proton and the neutron, forming a deuteron.

The first assumption has been disproved by the recent discovery[14] that the deuteron possesses a finite electric quadrupole moment which is

[14] J. M. B. Kellogg, I. I. Rabi, N. F. Ramsey and J. R. Zacharias, Phys. Rev. **55**, 318 (1939).

incompatible with the symmetry character of a pure ^3S state. The second cannot be discarded on an experimental basis but it ceases to be plausible if one admits the possibility, that ultimately the same causes may underly both the magnetic properties and the mutual binding forces of the proton and the neutron.

It is conceivable that the departure from any one of the two assumptions (*a*) and (*b*) would separately cause a considerable deviation from (20) but that for unknown reasons both together cancel each other very closely. Until reliable estimates of these deviations can be obtained we consider it, however, more likely that neither of them amounts to more than a few percents.

ACKNOWLEDGMENT

We are indebted to Professor E. O. Lawrence for his interest in this experiment, and in particular, one of us is grateful to him for the opportunity to work as a guest in the Radiation Laboratory. Our sincere thanks are due Professor R. B. Brode for the loan of the magnet which made the precision determination of the magnetic moment possible. We also wish to express our appreciation to the members of the laboratory staff for their many hours at the cyclotron controls. The experiment was made possible by grants to the laboratory from the Rockefeller Foundation and the Research Corporation.

6

Neutron Scattering By Ortho- and Parahydrogen

Kenneth S. Pitzer

By the late 1930s the theory of neutron-proton scattering had advanced to include the spin dependence of nuclear forces. Therefore physicists were greatly interested in the difference in neutron scattering by molecular hydrogen where the spins were parallel (ortho) and antiparallel (para). Interactions of neutrons with protons of known relative spin allow this spin dependence to be studied. Bethe gives a simple and clear account of the relevant theory.[1] The first experiment, which used neutrons cooled by paraffin to liquid-air temperature and scattered by liquid hydrogen, showed a substantial qualitative ortho-para difference.[2]

Two improvements were needed to achieve quantitatively significant results. Single scatters on individual diatomic hydrogen molecules were required, so a gaseous hydrogen target was called for. The neutron velocity had to be lower and more uniform. Luis W. Alvarez discussed with me the feasibility of carrying out such an experiment, and we agreed that we could do it without great difficulty. Luie and I each had many other research activities at the time. Mine were primarily concerned with the statistical thermodynamics of flexible molecules, which I was not prepared to abandon or even greatly delay. I also was actively measuring heat capacities at liquid-hydrogen temperatures, so I was able to prepare the para hydrogen and design the refrigeration equipment rather easily. Luie produced the neutrons, which included velocity selection to eliminate those with energy greater than 50° K.

Our scattering chamber containing either para or normal (75% ortho) hydrogen gas was cooled to 20° K by boiling liquid hydrogen. This entire unit was protected by vacuum and a shield at liquid-air temperature. I obtained the para hydrogen by catalytic conversion on active charcoal for 36 hours at 20° K and verified the conversion by differential vapor pressure measurements. Our experiments were done carefully but quickly in the old wooden Radiation Laboratory cyclotron building. I was concerned about the fire hazard and took precautions to vent all the evaporate out of the building. There remained, however, the hazard of a sudden break in a container holding liquid hydrogen that might release hydrogen faster than it could be conducted outside. We took this chance, and no break occurred.

To some, the best remembered aspect of the experiment was my disposal of the liquid hydrogen that remained at the end of a run. I took the container well away from any building and simply poured the liquid hydrogen out on the pavement. It evaporated immediately and, being much lighter than air, the resulting hydrogen-air cloud rose rapidly. In this respect safe disposal of hydrogen is easier than that of heavy combustible vapors.

The results of our experiment were not easily interpreted. We found a large ortho-to-para scattering ratio, 19, and a small para cross section, 5 barns (10^{-24} cm^2). The most reasonable theoretical model predicted a still higher ratio and lower para cross section. Had World War II not intervened, we might have pursued this problem further.

After the War, a large team at Los Alamos carried out an extensive and thorough investigation of this problem.[3] While there are various aspects in which the Los Alamos study was superior to ours, one important matter was the conversion of ortho to para hydrogen in the scattering chamber. From the Los Alamos conversion data, it is clear that this effect explains our lower ortho cross section of 100 barns as compared to theirs of 125 barns. They also had more precise control of the neutron velocity, which may have been important in their lower para cross section of 4 barns.

Our "quickie" experiment advanced knowledge very substantially. It was exciting science, and the collaboration with Luie was lots of fun. It remained for others, however, to do the definitive experiment.

1. H. A. Bethe, *Elementary Nuclear Theory* (Wiley, New York, 1947), Chapter 10.

2. F. G. Brickwedde, J. R. Dunning, H. J. Hoge, and John H. Manley, Phys. Rev. **54**, 266 (1938).

3. R. B. Sutton, T. Hall, E. E. Anderson, H. S. Bridge, J. W. DeWire, L. S. Lavatelli, E. A. Long, T. Snyder, and R. W. Wiliams, Phys. Rev. **72**, 1147 (1947).

Reprinted from THE PHYSICAL REVIEW, Vol. 58, No. 11, 1003–1004, December 1, 1940
Printed in U. S. A.

Scattering of 20° Neutrons in Ortho- and Parahydrogen

Since our last report on this subject,[1] we have completely rebuilt our scattering chamber and have incorporated several new features in the neutron monochromator.[2] A rotating cadmium shutter was placed immediately in front of the BF_3 chamber to keep neutrons with energies above 50°K from entering the detector. The scattering hydrogen was in the gas phase to eliminate the complicating effects of liquid forces. A jacket of liquid hydrogen boiling at atmospheric pressure maintained the gas at 20.4°K. The scattering chamber was 10 cm in diameter and 46 cm long; it was placed midway between source and detector, which were seven meters apart. The intensity of the beam was monitored by passing it alternately through the scattering chamber and through a dummy chamber. The ratio of neutron transmission through these two chambers was measured for 20°K neutrons to be 1.00±0.01.

Fifty thousand counts were recorded during the experiment and the probable errors computed from the fluctuations agreed well with those expected from the total number of counts. The linearity of counting rate with cyclotron beam intensity was checked before and after each run, and all parts of the experimental set-up worked exceedingly well throughout the course of the experiment. This fact is worth noting in view of the complex nature of the apparatus. The neutron temperature calculated from boron absorption measurements, assuming a $1/v$ dependence, agreed exactly with that obtained from the velocity measurements, i.e., 20.2°K. (This is in sharp disagreement with the work of Fertel, Gibbs, Moon, Thomson and Wynn-Williams,[3] who were unable to observe a change in the boron absorption cross section with neutron velocity.)

The parahydrogen was prepared by allowing liquid hydrogen to remain in contact with active charcoal for 36 hours. The pressure difference between the charcoal chamber and a surrounding liquid hydrogen bath was observed at frequent intervals. This indicated that the equilibrium state was approached very closely. The parahydrogen was then vaporized and the gas passed immediately into the scattering chamber where it was maintained at 20.4°K for the duration of the experiment. We believe that the composition of this sample was certainly within 0.2 percent of the equilibrium value, 99.8 percent parahydrogen.

With one atmosphere of n-hydrogen in the scattering chamber $I/I_0 = 0.238 \pm 0.005$. With an equal pressure of parahydrogen, $I/I_0 = 0.868 \pm 0.01$. The corresponding molecular cross sections are $\sigma_0 = 103 \times 10^{-24}$ cm² and $\sigma_p = 7.40 \times 10^{-24}$ cm². To obtain the true scattering cross sections, the capture cross section of two protons per molecule must be subtracted. If one assumes a $1/v$ law, this is 2.24×10^{-24} cm² at 20°K. The scattering cross sections for 20°K neutrons in ortho- and parahydrogen are then

$$\sigma_0 = (100 \pm 3) \times 10^{-24} \text{ cm}^2$$
$$\sigma_p = (5.2 \pm 0.6) \times 10^{-24} \text{ cm}^2$$
$$\sigma_0/\sigma_p = 19.$$

The earlier work in this field[4-6] gave values of σ_0/σ_p from two to five. It may be concluded, as was suggested by those who performed these experiments that their ratios were reduced by excess scattering of faster neutrons in parahydrogen. The theoretical implications of these data will be discussed in a companion note by Dr. Schwinger. We had planned to repeat the work, to improve the statistical accuracy and to search for possible systematic errors, but pressure of other work now makes that impossible for some time.

Our thanks are due the Research Corporation for financial assistance in this work.

LUIS W. ALVAREZ
KENNETH S. PITZER

Radiation Laboratory,
  Departments of Physics and Chemistry,
    University of California,
      Berkeley, California,
        November 8, 1940.

[1] L. W. Alvarez and K. S. Pitzer, Phys. Rev. 55, 596 (1939).
[2] L. V. Alvarez, Phys. Rev. 54, 609 (1938).
[3] G. E. F. Fertel, D. F. Gibbs, P. B. Moon, G. P. Thomson, and C. E. Wynn-Williams, Proc. Roy. Soc. 175, 316 (1940).
[4] J. Halpern, I. Estermann, O. C. Simpson and O. Stern, Phys. Rev. 52, 142 (1937).
[5] F. G. Brickwedde, J. R. Dunning, H. J. Hoge, and J. H. Manley Phys. Rev. 54, 266 (1938).
[6] W. F. Libby and E. A. Long, Phys. Rev. 55, 339 (1939).

7

The Mercury-198 Lamp: A New Standard of Length

Jacob H. Wiens

My association with Luis W. Alvarez began almost by accident. Alvarez was active in the Radiation Laboratory when I was a second-year graduate student in the Physics Department. During my first year I worked in cosmic rays as an unpaid assistant for Robert Brode, and during my second summer as an understudy for a priest who was finishing his research in hyperfine spectroscopy using Fabry-Perot etalons. When the priest's work was completed, Harvey White told me that I could use the equipment, and I proposed to determine the nuclear spin of uranium-235. This project proved to be impracticable because uranium-235 constituted less than 1% of natural uranium. The spin was measured many years later by Oscar Anderson, with enriched uranium-235 from the nuclear explosive program and using my notes on the spectrograph settings.

My involvement with the mercury lamp likewise came about by accident. The Physics Department had regular Wednesday afternoon meetings where "new and interesting" subjects were presented. Actually, these meetings were often boring, the room too warm, and the topics interesting to only a few. At one lecture, W. E. Williams of King's College, London, talked about his experiments using high frequencies to excite very small quantities of gas. He said that with the addition of a second gas like helium, it was possible to excite minute quantities of the gas of interest. During the lecture he mentioned that if monoisotopic mercury could be made by absorbing neutrons from the cyclotron on gold, it might be possible to build a lamp that would replace the then standard cadmium lamp. Unknown to me, Williams had already discussed the need for a better length standard with Luie and Professors Birge, Jenkins, and White. Luie had proposed the reaction to make the mercury, but no one believed that this reaction could produce enough mercury to make a useful lamp.

I had just headed a preradar school where we had developed circuits that oscillated up to 1 gigahertz. Williams had talked about needing a 100 megahertz oscillator, while the best commercial one was 15 megahertz. I went directly from Williams's lecture to my laboratory, where I announced to Hans Welton, a fellow graduate student, that I had found my research topic. I had experience with glass blowing and etalons, while the Radiation Laboratory's neutrons were near at hand. Furthermore, I knew how to build oscillators that operated at much higher frequencies than Williams had mentioned.

Welton and I had just begun to discuss the details when Alvarez walked in. Although he was supervising Welton's work, he wanted to talk with me about the mercury lamp. He had been at Williams's lecture, and I assumed that this was why he brought the subject up. He gave no indication that the idea was his or that he had previously talked with Williams. I was taken aback when he suggested that he could get me the gold from the Chemistry Department as Physics had none. Since a graduate student does not say no to a professor, I accepted his offer to collaborate.

Alvarez placed the gold in the cyclotron area, and together we found its 2.7 day half-life. I then did not see Luie for two months, as he was involved in several war-related projects. I, however, was busy constructing a 1000 watt 150 megacycle oscillator, a dual-section oil diffusion pump, an electronic pressure gauge, and a manual-controlled heating oven.

I first tried a Pyrex glass tube in which to produce the mercury, but it proved too soft for proper outgassing and collapsed as I heated it. I changed to fused quartz. But when I got the unit too hot, the gold foil melted and I had to roll it out again. After several weeks of failures, I finally had the gold encased in a quartz tube on which I sealed a long, thin quartz capillary tube. I left the assembly only in the vicinity of the cyclotron, as our project had a very low priority. After three days I retrieved the box and heated the gold red hot while cooling the capillary tube with liquid air before sealing it off. From the gas in this tube I saw with a Hilgar spectroscope the green light of the first pure mercury-198 ever produced, but it only lasted a few seconds.

A 30-day exposure to neutrons enabled me to get the first Fabry-Perot etalon spectrographs of isotopically pure mercury-198. Jenkins, who was in contact with Luie, told me that he had already written the accompanying article. Luie agreed that I should publish my own papers without him after this first one, on which I would be the first author. I made the next spectrograph after I had solved a resolution problem and made the final spectrographs with mercury produced during an 8-month neutron exposure.

After completing the mercury lamp project for my Ph.D. degree,[1] I went to the Chairman of the Physics Department to ask if the University wanted to produce some longer lasting lamps. They decided not to continue my work at the University of California.

Years later, a *Life* story about the mercury lamp had a photograph captioned, "Mercury Lamps' Inventor, Dr. William Meggers of the Bureau of Standards, Makes it Glow by Holding it Near Two High Frequency Antennas."[2] Since Luie's and my article and my thesis predated Meggers' work, I took exception to this claim. I immediately wrote the editor of *Life* an intemperate but heartfelt letter.

After two lengthy telephone calls, *Life* published a diluted version of my letter,[3] adding "Dr. Meggers made the first practical mercury lamp and, with it, the first measurements.—Ed." To me, this was like crediting Lockheed with "inventing" the airplane because they made the first practical device that would carry 400 people from San Francisco to London. The Wright brothers' device, after all, was far less useful.

The fat was clearly in the fire. I began preparing a brief consisting of some twenty-five letters and com-ments when I received a long telephone call from Alvarez and Professors Jenkins and White. After much discussion we finally agreed that I would not publish my manuscript but Luie would use it to produce an article on the history of the mercury-198 lamp which Alvarez, Meggers, Williams, and I would sign. This paper was written in 1952, but Williams would not sign it and I am not sure what Meggers's response was. Although a fair appraisal of the matter, the paper was abandoned.

The lamp using mercury-198 was not adopted by an international committee appointed to make such a decision. I was not informed about this committee or where they met. After 1952 I was completely excluded from anything related to using mercury-198 as the primary standard of length. However, I did see an article 15 years later reporting that Bell & Howell scientists had succeeded in producing mercury-198 by bombarding gold with neutrons. I did not bother to reply.

A statement by Meggers, repeated in the long telephone conversation with Alvarez, Jenkins, and White did, however, come true. I did not recant and was never offered a position at any university.

I was in Alvarez's presence less than 10 hours during my life, but from our interaction came my most important piece of physics, as well as pain that I carry even now.

1. Also reported in Phys. Rev. **65**, 58 (1944) and **70**, 910 (1946).
2. Life, January 28, 1952, pp. 56–61.
3. Life, February 18, 1952, p. 56.

Reprinted from The Physical Review, Vol. 58, No. 11, 1005, December 1, 1940
Printed in U. S. A.

Spectroscopically Pure Mercury (198)

For some time, spectroscopists in the national laboratories have been searching without much success for a line more nearly monochromatic than the red line of cadmium, to use as a standard of length. Professor W. E. Williams pointed out to us that if it were ever possible by some means to separate the isotopes of mercury, the green line λ5461 produced by one of the even isotopes would be admirably suited for the purpose. There would be no hyperfine structure, no isotope shift, and little Doppler broadening because of the high mass.

We have bombarded gold with slow neutrons from the 60″ cyclotron, and have collected enough of the transmutation product, mercury, to observe its spectrum. Since gold has only one isotope, 197, slow neutron capture gives rise to a single radioactive isotope, Au^{198}. This artificially radioactive product emits negative beta-rays with a half-life of 2.7 days and therefore turns into Hg^{198}, one of the stable isotopes of Hg. The experimental procedure is as follows:

A cylinder of gold 15 cm long, 2.5 cm in diameter, and with a wall thickness of 0.2 mm is placed in a quartz tube of slightly larger diameter. To one end of this tube is fused a quartz capillary with an inside diameter of 2 mm. The whole system is evacuated and heated for 36 hours in a furnace almost to the melting point of gold. The gold is thus freed of any ordinary mercury contamination. Spectroscopically pure argon is then admitted to a pressure of 6 mm of Hg, and the quartz system is sealed off from the pumps. The gold cylinder in its quartz container is now placed in a paraffin-lined box near the target of the cyclotron, where it is bombarded with "stray" neutrons for about a month. At the end of this time the gold is again heated, while the end of the capillary tube is cooled in liquid air. After an hour of this treatment, a 3-cm length of the cooled capillary is sealed off. When the spectrum of the gas in this tube is excited by a 3-meter oscillator, the mercury lines are quite brilliant, but the argon spectrum is quenched. The mercury lines are visible after a neutron

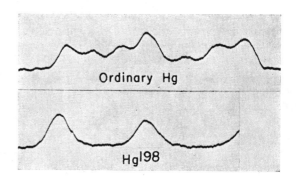

Fig. 1. Microphotometer traces for λ4047 from ordinary Hg and from Hg^{198}.

bombardment of a few hours, but they last for only a few seconds; under these conditions the Hg vapor is driven into the walls by the discharge. With a bombardment of a month, however, equilibrium between gas space and walls is apparently attained, so the spectrum is visible for some time. A microphotometer trace of a Fabry-Perot etalon spectrogram of the line λ4047 is shown in Fig. 1. The absence of the hyperfine components shows that the mercury is actually a transmutation product.

Since the Hg^{198} is a by-product of bombardments for biological purposes, no expenditure of "cyclotron time" is involved in its preparation. We will therefore be able to satisfy a reasonable demand for tubes filled with pure Hg^{198}, and we invite requests for such tubes. We gratefully acknowledge the support given to this work by the Research Corporation.

Jacob Wiens
Luis W. Alvarez

Radiation Laboratory,
Department of Physics,
University of California,
Berkeley, California,
November 9, 1940.

8

Heavy Ions

Cornelius A. Tobias

In the late 1930s, the Berkeley Radiation Laboratory (Rad Lab) and its 37-inch cyclotron was the world's center of nuclear physics research. I joined this enterprise as a twenty-one-year-old physics graduate student and have been there ever since. A number of exciting experiments were underway, many of them conceived by one of my thesis advisors, Luis W. Alvarez, who was only a few years older than I: the measurement of the magnetic moment of the neutron, the discovery of radioactive decay by K-electron capture, the identification of the internal conversion process, and the discovery of a number of new isotopes— a brilliant array. Among the new isotopes, the radioactivity of hydrogen-3 was discovered by Luie and Bob Cornog, and that of carbon-14 was characterized by Martin Kamen. Physicists and biologists alike immediately realized the enormous potential of isotopes for the study the complex molecular turnover in living bodies.

The possibilities for transuranium research also seemed limitless. However, the energies that could be excited in the heavy target nuclei with the low-mass cyclotron projectiles (protons, deuterons, or helium ions) were small compared to the self-energy of these nuclei. When Enrico Fermi announced that he had formed transuranium radioisotopes by neutron bombardment of uranium, attention quickly shifted to extending the periodic table of elements. Uranium was then the heaviest known element. Soon Hahn and his associates showed that Fermi's "transuranium" elements were actually isotopes of elements in the middle of the periodic table that were produced by the newly discovered nuclear fission process.

In a 1938 experiment, described in the accompanying paper, Luie introduced various carbonaceous gases into the ion source of the 37-inch cyclotron. A few energetic particles were brought out of the cyclotron to the air through a thin window, where they were identified by their measured range and ionization as 48 MeV carbon ions. Their ionization, seen by pulse height in a parallel-plate ionization chamber, was impressive, and Luie used it to verify that the energy transfer was proportional to the square of the ionic charge. The carbon-beam flux was, however, extremely low, only a few particles per second. Luie reasoned that the intensity was due to the low-voltage ion source, which

failed to strip the carbon ions of all their electrons. The accelerator resonance conditions required fully stripped carbon ions. The impression remained, however, that accelerated heavy ions would have a future in producing highly excited compound nuclei for nuclear structure studies and in the search for transuranium elements.

In 1941 I worked with Luie on one of his ingenious ideas, to demonstrate the neutron's radioactive decay by measuring in coincidence its proton and electron decay products. Robert Oppenheimer had calculated that the neutron half-life might be a few minutes. Our highly promising experiment was abruptly terminated when the 37-inch cyclotron became a mass spectrometer for separating uranium-235 from uranium-238. The neutron's lifetime was first measured a decade later by Art Snell at Oak Ridge National Laboratory.

Late in 1940, Luie took his leave of Berkeley to apply his talents to radar development at the newly established MIT Radiation Laboratory in Boston. I was left without a thesis topic and advisor until Emilio Segrè agreed to guide my work, with Luie and Ernest Lawrence, the Rad Lab Director and cyclotron inventor, as committee members. Following Luie's lead, I used the larger 60-inch cyclotron to accelerate carbon, oxygen, and neon ions to energies of 8 MeV per nucleon. Although these particles produced greater nuclear excitation energies than were previously possible, they fell short of what was believed necessary to liberate mesons. Since it was obvious that these heavy accelerated nuclei could be used to search for new fissionable isotopes, my thesis was promptly classified SECRET by Manhattan Project officials. The first artificial production of mesons had to wait until 1947, when the much higher energy helium-4 beams from the 184 inch synchrocyclotron were available.

Several Rad Lab staff members continued to be interested in accelerating heavy ions, among them Glenn Seaborg, who wanted to discover transuranium elements. There was also discussion about the possible existence of the magnetic monopoles postulated by Dirac. Luie was aware that monopoles might produce heavy-ion–like tracks as they pass through matter. I was becoming interested in bombarding living cells with these intensely ionizing heavy ions to study the responses of the cell nucleus. However, in the 1940s we did not have an economical method for accelerating intense heavy ion beams. Fermi's calculations in 1947 on nucleosynthesis after the Big Bang origin of the universe predicted that various nuclei might be floating about in space. It was the discovery of streams of fast heavy ions reaching us from the galaxy that rekindled interest in heavy-ion research. We only then realized that the heavy cosmic-ray component might present some hazards and limitations to extended space flight and manned exploration of the planets.

The presence of heavy ions in space and the damage they cause seriously threatened the validity of the "panspermia" idea. Arrhenius had suggested about a century earlier that dehydrated spores floating for long periods in interplanetary space might have permitted life on one celestial body to reach another. Recently, Luie and his son Walter have proposed that the extinction of the dinosaurs was caused by collisions of the Earth with other celestial bodies, raising another possible manner of exchanging biological matter.

In 1950, little was known about the biological effects of heavy cosmic rays except that they are very damaging. I predicted in 1951 that a single heavy particle crossing the retina could cause visual sensations, an effect observed 15 years later by the *Apollo 11* astronauts. During their journey to the Moon, these astronauts saw luminescent flashes and streaks with their eyes open and closed. Later in 1971, Tom Budinger and I verified that these sensations were caused by fast heavy ions when we allowed a few such particles produced by the Berkeley Bevatron to pass through our eyes. We now know that one hazard of extended space flights is heavy cosmic-ray primaries damaging brain tissue, where the nondividing neurons have difficulty regenerating.

The Rad Lab staff had an early and continuing interest in biological and medical problems. In the 1930s, Ernest Lawrence and his group built a high-voltage x-ray generator for Bob Stone, a radiology professor at the San Francisco Medical Center, to use in deep cancer therapy. The results were disappointing, since more dose was delivered to the body's surface than to the tumor beneath. So in 1934 John Lawrence, Ernest's brother, and Paul Aebersold demonstrated that the newly discovered neutrons had a much greater biological effect than x- or gamma rays. This resulted from the dense ionization of slow alpha-particles and protons produced by neutron reactions with the atoms in living tissue.

It has been demonstrated that single alpha-particles emitted from radium can kill cells when they penetrate the cell nuclei. Several hundred electrons are needed to accomplish the same task. As a result, various schemes were proposed to use highly ionized particles to study and control cancer. In 1939, Bob Stone and John Lawrence began the world's first fast-neutron therapy program. However, the absorption of fast neutrons reduces the number reaching the cancer and causes damage to tissue along their path, thus diminishing their therapeutic effectiveness. Since isotopes such as lithium-6 or boron-10 present very large capture cross sections for slow and thermal neutrons, they could be more therapeutically effective.

After World War II, some colleagues and I attempted to harness the energy released by highly ionizing fission fragments from uranium-235 bombarded by thermal neutrons for therapy. Neutron capture therapy is only effective if the tumor cells selective take up lithium, boron, or uranium. No one has yet achieved this; so the hope for neutron therapy remains unfulfilled. In the contemporary version of neutron therapy, monoclonal antibodies against tumor cells are prepared and tagged with large numbers of boron-10 atoms. These antibodies are then injected into the circulatory system, where they find the cancerous cells. The neutrons from an irradiation then interact with the boron-10, which emit alpha-particles, and may selectively kill the tumor.

After the War, many of the scientists who were engaged in national defense returned to peacetime research. One of these was Bob Wilson, a physicist who earned his Ph.D. at the Rad Lab. As a student, I enjoyed Bob's quick wit and original ideas. During a postwar visit, Bob suggested that high-energy protons could be effective in treating deep-seated tumors. The ionization of protons and of other charged nuclei had been first described by William Bragg in 1912. The use of the Bragg ionization peak would produce a greater dose inside the body as the protons come to rest than at the surface. Bob also realized that heavier ions, such as carbon, might be similarly useful if a sufficiently intense source could be developed.

The 184-inch synchrocyclotron was then under construction, and we expected a high-energy proton beam by 1947. When Bob returned to Princeton, Ernest Lawrence suggested that I measure localized depth-dose distributions of charged particles, which I did as soon as high-energy proton and deuteron beams became available. In 1951, John Lawrence (Director of Donner Laboratory), Hal Anger (the gamma-ray camera inventor), and I wrote an article showing our measured Bragg ionization curves and describing mice as "cured" after we exposed their tumors to deuteron beams that passed through their bodies. But many of the mice did not fully recover. We soon realized that low linear-energy-transfer radiations, such as electrons, could not eradicate certain forms of cancer. The linear-energy-transfer (LET) of fast heavy nuclei at a given velocity is proportional to the square of their charge. We soon realized that the yield of certain lesions in the genetic material, DNA, increases approximately as the square of the LET.

The first proton irradiation of patients took place at Berkeley in 1955. In the next 25 years, John Lawrence and our group statistically demonstrated that helium-particle beams have great advantages for treating acromegaly and Cushing's disease, highly localized pituitary gland tumors. More recently, the Bragg peaks of protons and of helium ions have been effectively used on highly localized malignant melanomas in the choroid of the retina. Further, particle treatment of sarcomas adjacent to the human spinal cord has also been successful. None of these tumors can be effectively treated with conventional radiation or surgery. Herman Suit of Harvard, Joseph Castro of Berkeley, and their teams, helped to prove the validity of Bob Wilson's initial suggestion.

When I first turned to biology, I benefited much from Luie's advice on the physical science aspects of my biological experiments. Luie, always a man of action, became increasingly impatient with the slow progress of my work. "We know the basic physical laws of atoms and molecules," he told me one day; "so what's holding up your understanding the physics of biological processes? After all, cells are made of atoms and molecules." It is true that physical scientists who have entered biology have helped to revolutionize our knowledge of biology, including molecular biology. However, physics has not been particularly helpful in explaining the basic tenets of biology, such as the evolution of species, physical mechanisms of self-reproduction, or the necessity for variability in biological systems. Biological events often proceed against physical potential gradients.

In the early 1950s, ideas emerged that would later allow heavy ions to be accelerated to high energies. Luie, who had just completed his proton linear accelerator, realized that partially stripped nuclei of various charges could be transported through high vacuums without charge exchange taking place. At Berkeley and at Yale, heavy-ion linear accelerators (HILACs) were built to generate intense ion beams for nuclear physics and chemistry research. The ion sources of these machines injected partially stripped heavy ions which were then accelerated to speeds similar to those of the electrons in their innermost shells. The remaining atomic electrons were then stripped when the beam passed through a gas curtain. Initially these beams had an angular divergence which was corrected by foil focusing, a technique Luie developed. The Berkeley HILAC, which accelerated light ions to energies of 7.5 MeV per nucleon, first operated in 1955. There many transuranium elements were discovered and Coulomb excitation studied, both activities which have added greatly to our understanding of nuclear structure.

Shortly after the HILAC was completed, my group started studying the systematic biological effects of heavy nucleus bombardment. Joseph Sayeg and Ann

Birge demonstrated that carbon ions were more effective than helium ions in killing yeast cells. We were engaged in friendly competition with Ernie Pollard's heavy-ion accelerator biophysics group at Yale, who were studying macromolecules and viruses while we concentrated on cells and genetics.

Heavy accelerated ions turned out to be excellent tools for studying basic genetic phenomena. These particles produced such dense ionization that the individual particles frequently broke both DNA strands simultaneously. Lightly ionizing radiations usually produced less drastic changes, for example, single-strand DNA breaks or base deletions. Single-strand breaks are easily and rapidly repaired by the cell's enzymatic apparatus, while double-strand breaks can be lethal. Mammalian cells repair some of the double-strand breaks by rejoining the broken ends. Misrepair, or misrejoining of the broken ends, also occurs. The coding sequence in misrepaired DNA is usually profoundly altered, and this can lead to cell mutation, chromosome aberrations, and transformation of normal cells into cancerous ones. Much later, in 1986, Ed Goodwin in our laboratory demonstrated that a single heavy ion, passing through the cell nucleus, is capable of producing a sequence of several chromatin breaks, wherever the chromatin spirals intercept the core of the track. This finding explains why heavy ions are much more efficient than X rays in killing cells, and why, at the same time, heavy ions are so very efficient carcinogens.

Between 1955 and 1968, evidence was accumulating at the HILAC for the value of using heavy ions in biomedical research. Paul Todd and the author voiced the need for heavy-ion beams with sufficient penetration to pass through the human body. In 1971, at Princeton and at Berkeley, high-energy nitrogen and neon beams were produced, thus proving the feasibility of the project. The HILAC was used as a preaccelerator for heavy ions that were passed through a long evacuated tube into the Bevatron. Beginning in 1975, this two-accelerator combination, the Bevalac, produced penetrating-ion beams as heavy as iron. In 1982 the HILAC was modified, and the Bevalac vacuum was improved so that every stable nucleus in the periodic table can now be accelerated.

With improving heavy particle beams came rapid progress in understanding cellular responses to ionizing radiations. In the 1950s, it was recognized that tissues deprived of oxygen are about three times more sensitive to conventional radiations than are well-oxygenated tissues. Since tumors that grow quickly often destroy the capillaries that provide oxygen and other nutrients to their cells, they also often contain islands of oxygen-deprived cells. Heavy ions, with their dense ionization, can abolish most of the radiobiological oxygen effect and also decrease various forms of repair. Argon beams appear optimal for treating shallow tumors, while silicon beams are most effective for deep-lying ones. Fast neutron and pion beams are also being studied in this regard, as well as for localized irradiation of deep-seated tumors.

The number and variety of biomedical applications of heavy ions has rapidly increased since the 1970s. In 1973, Eugene Benton and I demonstrated that heavy ions are excellent for soft-tissue radiography. Their range-energy relationship is so precise that images of the residual range distributions can be used to detect smaller tissue variations than other methods can.

Glancing high-energy collisions of heavy ions with resting nuclei can produce radioactive fragments that form relatively pure beams, Aloke Chatterjee and I demonstrated. These radioactive beams can be used for heavy-ion therapy and as instant tracers. Radioactivity deposited in patients can be imaged by gamma-ray cameras. Physicians can now verify that heavy-ion treatment radiation has been delivered to the intended regions. The ability to produce pure radioactive or inert fragment beams became the subject of interest for nuclear science; there are plans to use neutron-rich heavy-fragment beams for exploration of isotopes in the transuranium region.

We are also learning a great deal about the radiobiology of normal human cells and special cells of patients with certain genetic diseases that make them prone to develop cancer. For example, the cells of patients with ataxia telangiectasia are sensitive to heavy ions because the diseased cells have defects in their DNA repair mechanisms. There are about fifteen diseases of the nervous system in which an individual's cells are also defective in the ability to repair radiation injury. We do not understand why structure and function of the brain is related to resistance to radiation. It is known that normal neurons exist without cell division for our whole lifetime. In order to keep the genetic apparatus intact for many years, the biochemical repair mechanisms must apparently be in continually active and functional states.

The energy density in heavier ion tracks is high, which gives rise to novel biological effects. A single particle can cause irreversible lesions. Permanent damage to the cornea and retina exposed to these fast heavier ions has been seen. I believe these lesions develop as a result of the local high temperatures and free radicals produced by the ionizing tracks. In space radiation there are enough heavy particles to cause potentially harmful numbers of thermophysical lesions in the nervous system of space travelers on extended flights.

Much of the current interest in high-energy highly-charged particles was sparked by their success in biomedical science; however, physical and chemical research in nuclear science has been equally rewarding. Many new radioisotopes have been discovered in fragmentation products. Perhaps the most interesting research involves understanding close collisions which might produce compressed nuclear matter, which may be similar to the quark matter that exists in the interior of individual nucleons. The first evidence for the compression may have been seen. In 1984 niobium particles hit a niobium target, producing nuclear fragments with angular distributions indicative of "collective flow." With increased beam energy, these collective particle phenomena will become better studied.

I have been fortunate having had Luis as one of my Physics teachers and for the opportunity to exchange ideas with him this past forty-five years. The stream of his original ideas, his brilliant experimental innovations and his ability to critically and impartially examine scientific questions makes him a very important leader.

Studies with accelerated heavy ions have opened some of the frontiers of biology for me. In a futile manner, I have tried to explain biology with the basic laws of physics as we know these now. An opposite approach might be more fruitful; what can we learn from biology that will help to explain the physical world? A unique reason why we can observe and comprehend the physical universe is that we possess a highly developed nervous system and a brain. These represent complex machinery for information exchange, learning, logical operation and memory. The analysis of any physical system requires an appropriate informational machinery; yet, the laws of physics in their present form do not precisely claim roles for the observer and analyst as part of the system. Perhaps, when the informational and evolutionary aspects of biology are fully understood, then we can have a unified representation of the physical and biological sciences.

27. High Energy Carbon Nuclei. LUIS W. ALVAREZ, *University of California.*—The 37-inch cyclotron chamber was filled with CH_4 and a beam of 50 Mev $_6C^{12}{+}{+}{+}{+}{+}{+}$ ions was detected with a linear amplifier. To resolve these ions from alpha-particles, it was necessary to reduce the dee voltage and to adjust the magnetic field to the low side of the alpha-particle peak. Under these conditions, about 500 carbon nuclei entered the ionization chamber per minute. The pulses on the oscillograph were about nine times as high as those from Po alpha-particles, showing that their charge was about $6e$. Their range was between 6 and 11 cm of air. The theoretical value is about 8 cm. e/m and v are known from the cyclotron constants. If the helium contamination in the cyclotron were not present, it would probably be possible to use these ions in disintegration experiments, for one could then set on the peak of the carbon resonance curve. Attempts to observe 100-Mev carbon ions in the 60-inch cyclotron gave no conclusive results, as the carbon beam was masked by a low energy deuteron beam with the same range and resonance magnetic field. These spurious ions emerge from the gap between dees, and do not pass through the deflector channel.

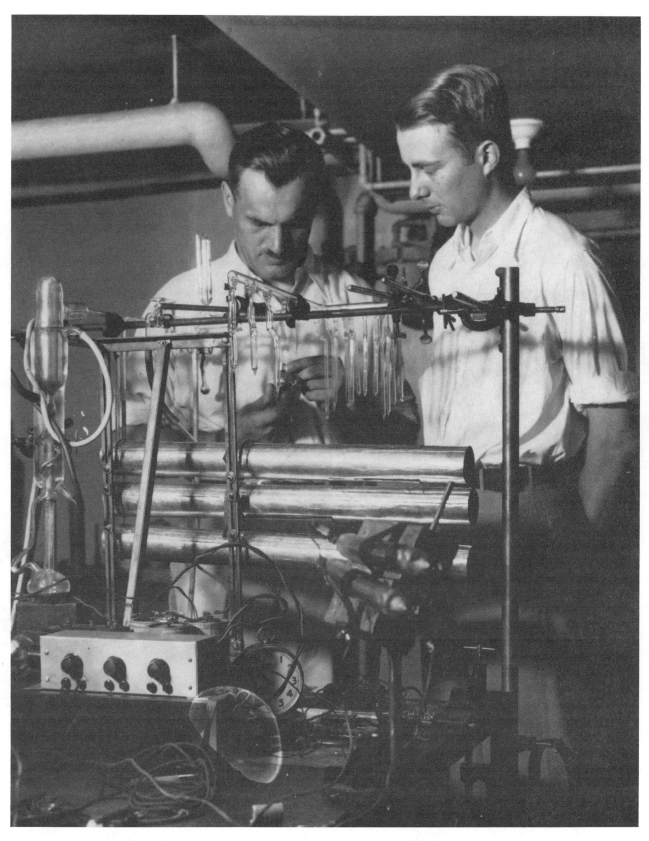

Alvarez showing his Geiger-Muller counters, among the first built in the
United States, to Arthur Compton (1932).

Alvarez as first-year graduate
student at the University of
Chicago serving as designated
stratospheric cosmic ray
observer for the Century of
Progress balloon flight (1933).

Four future Presidents of the
American Physical Society:
Alvarez, Robert Oppenheimer,
Willy Fowler, Bob Serber (1938).

Alvarez with personally built electronics and BF$_3$ ionization chamber (1938).

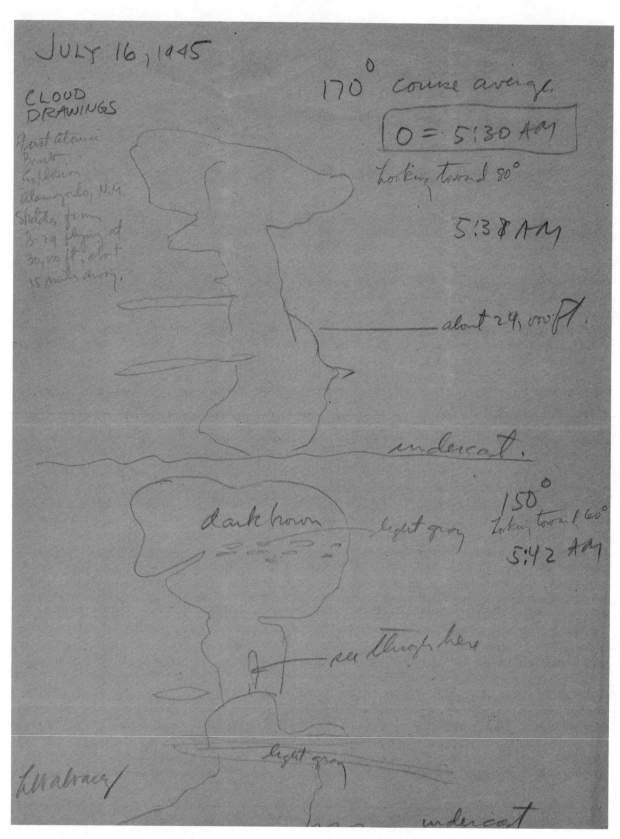

Alvarez sketches of first atomic bomb explosion made from B-29
over Almagordo, New Mexico (1945).

Emilio Segrè and Hans Bethe
with Luis Alvarez at Los
Alamos, after the war.

Recovered letter attached to
parachute-borne pressure
gauges dropped at Nagasaki
(1945).

Headquarters
Atomic Bomb Command
August 9, 1945.

To: Prof. R. Sagane

From: Three of your former scientific colleagues
during your stay in the United States.

We are sending this as a personal message to urge that
you use your influence as a reputable nuclear physicist,
to convince the Japanese General Staff of the terrible
consequences which will be suffered by your people
if you continue in this war

You have known for several years that an atomic
bomb could be built if a nation were willing to pay
the enormous cost of preparing the necessary
material. Now that you have seen that we have constructed the production plants, there can be no doubt in your mind,
that all the output of these factories, working 24 hours a
day, will be exploded on your homeland.

Within the space of three weeks, we have proof-fired one bomb
in the American desert, exploded one in Hiroshima, and
fired the third this morning.

We implore you to confirm these facts to your leaders,
and to do your utmost to stop the destruction and
waste of life which can only result in the total
annihilation of all your cities, if continued. As scientists,
we deplore the use to which a beautiful discovery has been
put, but we can assure you that unless Japan
surrenders at once, this rain of atomic bombs will
increase manyfold in fury.

Alvarez autographing Professor
Sagane's copy of the Nagasaki
letter (1949).

Alvarez with Pief Panofsky and 200 MHz linear accelerator test
cavity (1946).

Alvarez receiving Medal for Merit for his radar work during
World War II (1946).

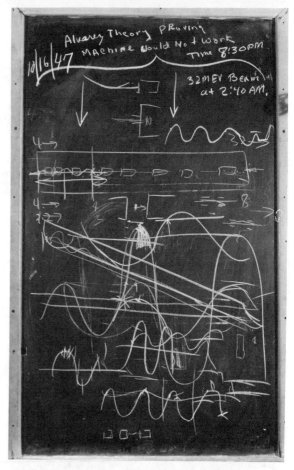

Search for the first beam from the proton linear accelerator (1947). Alvarez's 8:30 P.M. blackboard calculation "proving" that geometry must be changed and that the "machine would not work", with an added note that at 2:40 A.M., six hours later, they achieved the beam.

First proton linear accelerator, with lifting hoist (1947).

9

The War Years

Lawrence Johnston

In fall 1940, as a first-year graduate teaching assistant in the Berkeley Physics Department, my job was to help Luis W. Alvarez in his Modern Physics Laboratory course. Elsewhere, World War II was in progress. France had fallen, Britain was under aerial siege, and the United States, though not at war, was shipping Britain lend-lease equipment and supplies.

So I was not surprised when Luie told me in November that Ernest Lawrence wanted him to go back east for a month or so on urgent business and that I would carry the class alone. He said that he would return by Christmas, in time to assign the grades. Instead, in December he sent a telegram saying that I was needed at MIT as soon as possible. He simply assumed that I would come without knowing any details, and he was quite right; I went.

The MIT Radiation Laboratory[1] was an extraordinarily fruitful place for Luie to produce his radar ideas and get them quickly translated into working equipment. A major reason that the Laboratory could rapidly bring together the scientific manpower, government money, industrial support, and military cooperation was the Laboratory's top-level leadership. This group of men had worked together over the years encouraging large scientific endeavors and finding ways to fund them. There was so much preexisting understanding and trust among them that they were able to use their diverse talents and connections to tremendous effect.[2] Luie's close relationships with three of these men (Ernest Lawrence, Alfred Loomis, and Arthur Compton) greatly enhanced his effectiveness both in radar and later in atomic weapons development.

Sensing that scientists could be useful in the war effort, these men in June 1940 persuaded President Roosevelt to establish the National Defense Research Council (NDRC), with Vannevar Bush as the Head.

Bush then named Karl Compton to lead a Detection Division, and Compton promptly chose Loomis to run a Microwave Committee. Loomis had been experimenting with microwaves at his Tuxedo Park estate using the Varian brothers' new klystron tubes manufactured by the Sperry Gyroscope Company. Loomis appointed a committee of Ernest Lawrence, Edward Bowles, and electronics company and armed forces representatives.

No useful suggestions for military microwaves emerged until the visit of members of the British Tizard Mission, who described England's secret technology and brought with them a cavity magnetron transmitter tube for pulsed high-power microwave radar. This changed the situation completely. The Committee then sprang into action with the clear goal of mounting a crash program to make microwave radar sets for British night fighter airplanes.

In October 1940, MIT was chosen as the site for the new Radiation Laboratory and Ernest Lawrence began recruiting scientists, a task for which he was uniquely prepared. In the 1930s many nuclear physicists had come to Berkeley to learn how to build cyclotrons. There was close contact among the cyclotron laboratories they subsequently set up, and so there were excellent working relationships among people experienced in building big machines. Modifying these cyclotrons and doing experiments with them required a special brand of improvised, rapid-paced engineering and fabrication so that new ideas could be quickly tried out. Since Lawrence's Laboratory was the fountainhead of the new technology, he had a great deal of influence in this community. He recruited these people, and they often brought their colleagues with them. In at least one case an entire staff came with their expert technicians. From his own Laboratory, Lawrence brought Luie and Edwin McMillan, his two brightest young physicists. And Luie brought me, his graduate student. Much of the recruiting was done at a long-planned nuclear physics symposium held at MIT in late October.

Originally the Laboratory's staff physicists were to gather for a few weeks at MIT and plan the essentials of the radar system. They would then turn over the construction job to the resident engineers. Soon, however, they saw that this plan would produce inordinate

1. *Five Years at the Radiation Laboratory,* edited by C. Newton, T. Peterson, and N. Perkins (Andover, Cambridge, 1947).

2. A photograph of these men, seen in Phys. Today **36**, January, 30 (1983), shows Ernest Lawrence and Arthur Compton (physics professors); James Conant and Karl Compton (presidents of Harvard and MIT respectively); Vannevar Bush (president, Carnegie Foundation of Washington); and Alfred Loomis (investment banker/scientist) around a blackboard in the old Radiation Laboratory. All have boyish grins as Lawrence explains designs for the fantastic next cyclotron he wants to build, the 184 inch machine, while they make plans to obtain the necessary funding.

delays in making the necessary trials, failures, and corrections. The men who had the ideas had to stay on to build equipment and see the job through, as they did when they dreamed up a new nuclear physics experiment. They could have lots of help from machinists, technicians, and even engineers, but they had to assume the prime day-to-day responsibility of putting the system together and making it work.

I arrived at MIT in early January 1941, with no idea of what we would be working on, knowing only that it would be important and war-related. I guessed that it would be a nuclear project, since Lawrence and other Berkeley cyclotron people were involved. Needless to say, I was excited at the prospect of finding out. I drove Luie's car out from Berkeley, since he had come by train expecting that he would soon return. I somehow found a parking place near a massive MIT structure and inside I inquired for Room 4-133. Everybody knew where it was, and soon I was speaking to a friendly guard who called Luie. Luie appeared, took me inside, and began explaining what the Laboratory was about. I had no security clearance, but since a staff person vouched for my trustworthiness, I was told all that was necessary to get started. My formal clearance came later.

There were about fifty Laboratory employees then and no real organizational structure. You just did what you could do and what was needed. A Steering Committee met frequently to see that all the jobs got done. Luie was on the Steering Committee, and he kept me busy.

Room 4-133 was mostly occupied by people working on radar components: magnetrons, modulators, receivers, and the interconnecting plumbing, which at the time was all coaxial transmission lines. In a plywood building on the roof, people were assembling and testing radar systems. I arrived amid great excitement because the first experimental rooftop system had just shown radar echoes from downtown Boston buildings several miles away across the Charles River. This radar set used separate transmitting and receiving ''dish'' antennas to avoid burning out the receiver's crystal detector. A month later, the first echoes from an airplane were seen.

Another radar system was being assembled by Ed McMillan for the nose of a B-18 bomber to test the effectiveness of airborne night fighter microwave radar. However, by the time this system was installed, the Battle of Britain was ebbing and the British had lost interest in night fighters. But Alvarez and McMillan found that this set showed terrific signals from ships, thus motivating a group of airborne radars which were used with great effect against German submarines that preyed on our Atlantic shipping.

In most organizations the weekly staff meeting is an unwelcome intrusion on getting things done. But in the early MIT Radiation Laboratory days, people eagerly attended these Monday evening meetings. They usually started with a short talk by Lee DuBridge on general progress in the Laboratory, followed by Wheeler Loomis with housekeeping details. Then the systems people reported new performance records or new component failures, and the component groups told of their problems and progress. The first such meeting I attended was memorable because Taffy Bowen, our British liaison officer, showed aerial movies from night fighters as they shot down German bombers. These pictures, though blurry and without sound, suddenly brought us very close to the life-and-death struggle to which we were trying to contribute.

On another Monday evening some special, but unmemorable, business took up most of the meeting. As the hour was late, DuBridge departed from form and asked Luie to briefly summarize the state of things in the Laboratory. In five minutes he covered most of the news about systems and components, hardly pausing for breath. When Luie sat down, there was a pause and then a burst of spontaneous applause. I never found out if he knew that he would be called on, but regardless, his grasp of the business of the Laboratory was impressive! Only later did I learn that Luie's principal job then was coordinator for all the radar systems being developed. The Microwave Committee had ordered the critical components for five sets, and Luie saw to it that all the components were mutually compatible and that the systems people knew the delivery schedules.

The results which were attained attest to the fact that Luie did this job superbly. However, at the 1941 Christmas party there was a limerick contest to which Jackson Laslett, a good friend of Luie's, entered:

> What a wonderful guy is Luie.
> To him we say go away, Phooey!
> With all good intentions,
> He changed the dimensions
> And drove all the rest of us screwy!

Luie's first contributions to radar art resulted from his work on beacons. We had been introduced to radar beacons developed for the British night fighters. It was desirable for a pilot on radar approaching a bomber from the rear to verify that his target was not a friendly British plane before opening fire. So night fighters were equipped with small pulsed radio transmitters which responded to signals sent out by allied planes. Beacon pulses appeared to the probing radar receiver as strong return echoes. Thus, friendly airplanes showed up brightly on a fighter's radar screen, making them easy to distinguish from a foe. To discourage the Germans from defeating the system by carrying beacons of their own, the beacons were turned on and off in a coded program, like Morse code. On the interrogating radar screen a spot representing a friendly plane blinked brightly, spelling out the code, which changed nightly. This system was called Identification of Friend and Foe (IFF).

Luie realized that microwave pulses were so short that coding could be accomplished by having the beacon transmitter (now called the transponder) emit multiple microwave pulses for each interrogating pulse, so the plane's position became a series of bright spots displaying the code of the day stretched out in range. His technique worked well, but pilots were (and still are) reluctant to shoot down a plane solely on the basis of IFF radar information. A friendly plane could have an inoperative transponder that was silent or giving out a garbled pulse code. This type of beacon, however, has been a great success in civil aviation, where regional radar centers keep track of dozens of planes by their individually encoded signals displayed on the persistent screen of a radar tube.

Luie had an ingenious idea for using microwave radar in antisubmarine warfare. In 1942 German submarines were taking a heavy toll of allied ships, many of which were sunk close to our shores. The new airborne microwave radars were able to detect surfaced subs effectively, and a number were sunk by radar-equipped rocket-firing planes. The Germans soon provided their submarine crews with simple microwave receivers which warned them of approaching allied planes. In this game of cat and mouse, the submarine had a great natural advantage; signals from the airplane's transmitter were detected by the sub long before the aircraft's receiver could detect echoes bouncing off the sub. Direct signals received by the sub varied inversely in intensity as the square of the sub's range, while the radar echo from the sub at the plane varied as the inverse fourth power of the range.

Imagining himself as a U-boat commander, Luie realized that if he surfaced at night to recharge his batteries for the next day's underwater prowling and heard an airborne radar approaching, he could decide to dive at his leisure. If the signals did not increase in intensity, he would not submerge, since that would imply that the plane was not going to pass near his submarine.

Luie's strategy for the hunting airplane was that as soon as a submarine echo was detected, the radar operator should start decreasing the transmitter output power with the cube of the range. Thus, the plane approaching the sub created the impression that it was going away, so the sub would remain on the surface. His system was called Vixen because it "outfoxed" submarine commanders. Even if a sub commander figured out the tactic, prudence would force him underwater whenever he detected a radar signal.

Luie's radar systems were blind landing or Ground Controlled Approach (GCA), Microwave Early Warning (MEW), and Eagle precision bombing. All three made use of his revolutionary radar antenna, the linear dipole array. He did not invent his famous antenna and then go around looking for applications. With Luie, need was always the prime mover, and so the various antenna features evolved as needs were encountered.

Luie invented the linear dipole array for MEW. Luie realized that microwave radar should have a tremendous advantage over existing long-wave radar sets for long-range detection. The basic "Radar Equation:"

$$R^2 \backsim (A/\lambda^2)(\sigma_r P_0/P_r)^{1/2} ,$$

relates the maximum range R at which a target of area σ_r can be detected by a radar with transmitted power P_0 and received power P_r to the area of the antenna reflector A and the radiation wavelength λ. He saw that for a given-size antenna, the radar system's range is inversely proportional to the wavelength. The conventional wisdom then was that long wavelenghts were best for long-range detection, such as the Britons' Channel and our Pearl Harbor radar, which "saw" for a hundred miles.

Luie's first designs needed 600 square feet of reflector to get the required gain, in the form of a parabolic dish 25 feet in diameter. Scanning the resulting needle-like beam in a horizontal plane, only airplanes close to the ground would be seen. Luie then decided that he needed a fan-shaped beam several degrees wide vertically, but only a fraction of a degree wide horizontally. Such a beam could not be produced with a round parabolic reflector. Diffraction theory predicts that a wide aperture produces a beam narrow in angle and vice versa. So Luie decided that the reflector should be rectangular, 8 feet high by 75 feet wide! This would give the required area and the 3° vertical beam height needed.

Since such a reflector illuminated by a conventional single dipole radiator would be inefficient, Luie needed a radically new concept. So he decided on a long horizontal parabolic cylinder fed by cylindrical waves generated by a line source of coherent radiation. Luie didn't know exactly how to make this source but guessed that a 75 foot long "leaky pipe" placed on the focal axis of a cylindrical reflector could produce the required waves.

When Luie told people about his idea they didn't take him seriously, because it required so many new, extreme concepts. The mechanical problems of rotating a 75 foot reflector were difficult, but nobody had built such a linear source. His Steering Committee friends called it "Alvarez's Folly."

Further calculations convinced Luie that he could reduce the antenna's width to 24 feet and still get adequate performance. After charging up a series of blind alleys, he finally solved the radiator problem and produced his now famous linear dipole array. The basic ingredients of this solution came from our knowledge of field propagation in waveguides, which was provided in a beautiful set of lectures given by Stanford's Bill Hansen. Physical optics and optical interference devices, which are now a part of every physicist's basic education, were also essential to the solution.

Luie reasoned that waves traveling along a waveguide would produce a series of equally spaced points, each differing 360° in phase from its neighbor, at which he could attach small dipoles and excite them in synchrony. A few wavelengths from this line of dipoles the individually radiated waves would comingle into a single cylindrical wave concentric with the dipole line and would illuminate the reflector. Each dipole would be loosely coupled to the waveguide and extract only a small fraction of the available energy flux. When he built a model of this antenna, it produced the desired beam at 90° to the antenna plane, but it also had strong extra beams at 60° to each side.

Luie lost a lot of sleep figuring out a clean way to get rid of these side lobes. He realized that the three beams his antenna produced were analogous to spectral orders (zeroth and first) made by a diffraction grating, and the reason they appeared was that the radiators were spaced more than a wavelength apart. If his dipoles were closer together than a wavelength, the two first-order beams could be eliminated.[3] This posed a serious dilemma, since the equal-phase points in the waveguide had to be located somewhat more than a wavelength apart. Luie had a brilliant insight which saved the day. He doubled the number of dipoles, which halved the phase differences to full cancellation of 180°. But then he could twist alternate dipole prongs by 180° and be back in phase! The superfluous lobes were canceled, and all the radiated power was concentrated in the central beam. Luie said that once you know how to do this job, it seems obvious and hardly qualifies as an invention. However, he acknowledged that this was one of the most difficult inventions he had ever made. When Luie demonstrated his ideas in a working model, the "Alvarez's Folly" nickname became a badge of honor.

Radio engineers had produced dipole arrays for antennas before, for example in the "bedspring" antennas of the prewar 200 megahertz SCR-268 radar. These dipoles were fed by coupling networks from the twinax transmitter feed wire. The conventional wisdom was that the coupling network complexity increased rapidly with the number of dipoles in an antenna, limiting the number of dipoles. By using a waveguide and then letting the center conductor of each dipole couple slightly to the passing electric fields, Luie excited and phased large numbers of dipoles without any appreciable coupling problems.

MEW development faded with the threat of enemy planes attacking the continental United States, and it

was not in production when Alvarez left MIT in 1943. However, the inspired group of young men he had led finished the job. One of the first sets was put on the English Channel and directed the air armadas to the Normandy coast on D-Day. MEW provided air controllers with an unprecedentedly high resolution and long range coverage which undoubtedly contributed greatly to the operation's success.

The idea of making a radar blind landing system came to Luie as he watched an early gun-laying radar operate on the MIT roof in summer 1941. That radar had a 6 foot diameter parabolic antenna which automatically pointed at a moving airplane to continuously provide an antiaircraft battery with the plane's angular coordinates and range. Looking through a sight attached to the moving antenna, I was amazed to see the airplane bobbing around only slightly behind the telescope crosshairs. This radar was the progenitor of the SCR-584 gun-laying set which was widely and successfully used during the War. The SCR-584 teamed with Luie's MEW search radar was used to shoot down the German buzz bombs as they crossed the English Channel. MEW spotted the buzz bombs and prepared the 584 to lock onto one of them when it was close enough.

Luie reasoned that if radar knew a plane's location accurately enough to shoot it down, it should be capable of guiding a plane to a safe landing under conditions of poor visibility. He had heard enough from Taffy Bowen about British fighters flying in bad weather to know that many pilots and planes were lost because their landing field was socked in when they returned from a mission. Although radio-beam–type landing systems had been built, they depended on special receiving equipment in the plane and required pilots with considerable practiced skill in their use.

Luie envisioned someone at a landing field "seeing" the airplane with radar and giving the pilot precise and simple steering commands to bring him down safely on the runway. No special equipment was needed in the plane, and a weary pilot would have his navigation problems solved for him by a friendly, well-rested, and knowledgeable person on the ground. This he called the "talk-down" method.

Luie had me put together one of the first five Laboratory radars, which I adapted to go into a Douglas A-20 attack bomber converted into a night fighter. After I got the equipment working in a rooftop laboratory, I transported it to Wright Field in Dayton, Ohio. The A-20 delivery was repeatedly delayed because of the Air Force's diminishing concern with night bombing raids. So I set up my radar, showed the local people how to run it, and returned in midsummer 1941 to Cambridge. The set was finally installed, but we never heard of its being used.

When I got back to the Laboratory, Luie, who was recuperating from a gall bladder operation, was work-

3. In terms of actual radiation in the waveguide, the desired 90° beam was made by first-order interference, and the unwanted beams were zeroth- and second-order interference.

ing on ideas for MEW, beacons, blind landing, and assorted other things. By then the Laboratory was organized into businesslike divisions, each carrying responsibility for some radar component (transmitters, receivers, or indicators) or system (gun control, airborne, or naval). Fitting Luie into the Table of Organization posed a bureaucratic dilemma, since all the obvious turf was taken. Very wisely, a new Division was created whose sole responsibility was to work on Luie's radar ideas. Division 7, JBBL (Jamming, Beacons, and Blind Landing), later more broadly titled Special Systems Division, gave Luie the freedom and resources to realize the radar systems he invented.

Division 7 occupied a large open bay in a newly constructed frame building with Luie's office at one end. His better office in the Administrative Wing close to the Director went unused, as he preferred to be closer to the action. The big bay had a dozen desks and a drafting table in the middle, work benches around the perimeter, and a small conference room for meetings or quiet thinking. Everybody knew what everybody else was doing.

When I got back to the Laboratory, Luie told me about his talk-down blind landing system and asked me to be the Project Engineer. I was enthusiastic about saving lives, and I liked flying. He told me that Alfred Loomis also liked his idea, what a fine person Loomis was, and many interesting stories about him. I had no idea at that time that Loomis was Chairman of the Microwave Committee or even that a Microwave Committee existed.

We planned to use the successful SCR-584 gun-laying radar to provide the airplane coordinates which we would feed into a small analog computer called the "director." The director compared these coordinates to those of an ideal glide path and developed error signals on two meters monitored by the person talking to the pilot, the Controller. The Controller would give the pilot right-left steering instructions and adjustments to his rate of descent until he could see the runway and land.

Since it would be several months before we would have the gun-laying radars, Luie decided we should test the idea of guiding pilots by voice instructions. We drew a long straight chalk line on the floor of the old National Guard hangar at East Boston Airport. Then we put a set of earphones on a blindfolded man, and our Controller guided him first to the chalk line and then along it for some distance. We concluded that people were capable of following verbal instructions. So far so good.

Next we had to see if a pilot in an airplane could respond to voice commands from a ground Controller. We simulated many features of our ultimate radar landing system. We set up our equipment near the up-wind end and 100 feet to the side of the runway so we could get a good view of the plane as it was landing. With

theodolites we manually tracked the plane and measured azimuth and elevation. The range was determined by a simple radar system. Our director recorded two traces: azimuth angle versus range, and elevation angle versus range. We made records of ideal landings, from which our director calculated up-down and right-left error signals to use in actual blind landings. Our four-man crew handled all these functions. Luie was usually the Controller who radioed the error signals to the pilot, and a Navy Chief, Bruce Griffin, our favorite pilot, made landing approaches under a hood while his copilot kept watch as a safety check. Again, the system worked well, and Griffin eventually made several hooded landings to touchdown. The landing system success seemed assured, and all that we needed was to plug in the accurate coordinates available from the gun-laying radar truck, XT1.

Luie played a large part in an event involving my personal life. Before I left California in December 1940, I was engaged to Mildred Hillis. We planned to get married in June 1942, when she graduated from Berkeley. By February of that year our correspondence indicated serious problems in our relationship. My thoughts were far away, and when Luie asked what was bothering me, I told him. He said the obvious thing to do was to fly back to Berkeley and get the matter settled. Fly! In those days air travel was a luxury, and after Pearl Harbor nobody flew without special authorization. Luie arranged for me to be a special courier for an important letter to Ernest Lawrence. Before I left he said, "Bring her back with you." Two weeks later Mildred and I were married and on a plane to Boston.

We could not immediately try out the prototype XT1 radar, because it was urgently needed to develop and evaluate a full-scale production model to direct anti-aircraft fire. By May 1942, however, it was our turn, and the XT1 crew met us at a small navy airfield near Norfolk, Virginia.

What a disappointment was in store for us! With our two trucks parked side by side, the big radar dish tracked our plane when it was overhead. But as the plane came in on a landing path the dish bobbed wildly up and down. No adjustments the crew could make (and we tried many) could effect a cure. We finally faced the basic reason. The radar had a 3° wide beam which it rotated rapidly in a conical scan, and it achieved its 0.1° tracking accuracy by balancing signal intensities from each side of the cone's axis. When the beam partially illuminated the ground, a reflected airplane image was seen "under the ground" which competed with the direct signal from the plane. The radar alternately tracked the plane and its image, unable to decide which to follow. Discouraged, we drove our truck back to Boston, where we tried to figure out ways to see planes close to the ground.

Luie appeared one morning full of enthusiasm for a new approach to blind-landing radar. At dinner the night before with Alfred Loomis, he had come up with the solution. First, we would give up the automatic tracking system. To get greater angular accuracy we would make a high-precision radar whose coordinates we would read on a high-resolution radar screen. Since we needed two angle coordinates, we would use two high-precision radar sets, one to measure azimuth and the other elevation, the same as we had with our theodolites. A human operator, who could easily distinguish the direct signal from its reflection, would read the radar scopes.

To get the unheard-of angular resolution, Luie said we would use the newly available 3-centimeter wavelength (X-band) radar and a linear dipole array antenna we would build. We needed to know the elevation angle, which showed how close the plane was to the glide path, to half a degree, so a 12 foot high antenna was required. We needed azimuth resolution of three-quarters of a degree, and therefore an 8 foot long antenna. As in the MEW antenna, the reflectors would be parabolic cylinders, long in the array direction and narrow in the perpendicular direction, producing a beaver-tail–shaped beam. Luie pointed out that the blind-landing radar had one advantage over the gun-laying radar: In the former case the airplane and radar crews cooperated.

In addition to our other problems, we were now developing our own radar system! With Loomis's encouragement the Project Committee approved our enlarged enterprise, but since the Laboratory had many new commitments, they could provide little additional support. Luie recruited some of his former Chicago classmates and colleagues. He put me in the awkward position of being in charge of a dozen people, half being Ph.D. physicists who had known Luie much longer than I had. Luie must have made it clear that he wanted me in charge, because they were, with few exceptions, cooperative. Luie also had a small group designing the GCA and Eagle antennas. He took an active design role in his systems, but we always felt able to make contributions and let him know if he was on the wrong track. I designed the data-handling system from the moving antennas to the controller's error meters.

The prototype blind landing system, Mark I GCA, was to be housed in two specially built trucks. Both were to be parked beside the landing strip near the upwind end, where they had a direct view of the landing path to the touchdown point of the airplanes. Luie insisted that his system be able to do a complete job so it could be used at airports with no general search radar, control tower, or even electric power. So in addition to the two precision landing radars, a third general search radar was included which scanned 360° around the airport, as was a large diesel generator. The generator and three radar sets were in the larger truck with a radar engineer who kept everything running.

The smaller truck contained six operators, including three "human servos" who kept electronic crosshairs trained in the plane's radar signal in range, azimuth, and elevation. Three more people talked by radio to the airplane pilots so that planes could be landed in quick succession. Two operators watched radar scopes fed by the general search radar: the first put arriving planes into a holding pattern, making sure they did not collide, while the second "vectored" the next plane to be landed by voice onto a path starting toward the runway. About ten miles from the airport the plane's signals could be seen on the two high-resolution radars and the final operator, the Controller, took charge. He told the pilot how far he was above or below the ideal glide path, how much he was to the left or right of the runway centerline, and how close he was from the touchdown point. The Controller got his information from error meters for up-down and left-right and a range meter.

I have often used the lives of Luie and Ernest Lawrence to illustrate the meaning of faith, of which the last few paragraphs are a prime example. Luie had the faith and imagination to build a blind landing system complete in every detail when the most essential component, the antenna which used his newly invented linear dipole array, had only been tested in rudimentary form and those tests showed that it had serious problems. He clearly inspired faith in us who worked with him and in the administrators above him. He told me after the war that his most valuable possession was Ernest's trust.

Our Mark I GCA system was put together in an incredibly short 6 months, starting in early June 1942. We used standard X-band radar transmitter and receiver components, but almost everything else was custom-built. Tests at East Boston showed that the new antennas had the resolution to know the plane's height to 10 feet at touchdown, sufficient for landing. The Air Force held that blind landing was impossible, as the pilot must always see the runway before landing, so we had to change the name of our system. Luie cheerfully changed it to Ground Controlled Approach, as he figured that pilots needing to land would land, Air Force doctrine notwithstanding.

We worked the bugs out of the system at several east-coast airfields and found the best operational procedures. One exciting event happened at Quonset Point Naval Air Station in Rhode Island. We were set up beside their longest runway but unable to put our test planes in the air because a snowstorm had closed the field. We were in the trucks doing routine maintenance when the Base Commander called to say that a flight of PBYs desperately needed to land, as they were low on fuel and no nearby field was open. He asked if we could help in this emergency, and we agreed to try. Our radio was equipped with standard aviation frequencies, so we soon had our system going and were in radio contact with the distressed amphibians. Radar was super-secret, so when Luie told them that we could see them and if they followed instructions that we would land them, they did so entirely on faith. Using their radio navigation receivers they flew toward us, and at 25 miles out our search radar picked them up. As they were low on fuel we couldn't stack them in a holding pattern and bring one in at a time, so we had them come down the glide path one behind the next, maintaining visual contact with the plane ahead while Luie talked the leading plane down the glide path. There was a 100 foot ceiling so they touched down visually. These pilots, who were happy to be alive, provided additional evidence that GCA was effective with pilots who had no previous experience with the system.

The high point for us came in February 1943 when on short notice a series of evaluation tests was requested by Air Force General Harold McClelland. Pilots from all the services participated for several days in this evaluation at National Airport in Washington, D.C. We drove the Mark I GCA trucks from Boston to Washington, and during the trip a number of components were shaken loose. Most were quickly repaired and the system was soon functioning, but our large modulator tubes (Eimac 304TH), whose filament structures had apparently been weakened by vibration, were burning out at an alarming rate.

A large group of high-ranking military officers appeared at the field to witness the tests on 14 February, but we could not get everything going at the same time. For two days we kept postponing the demonstrations while working frantically on the equipment and wondering how long our audience would keep coming back. On the morning of the 17th we installed our last spare Eimac tubes and the whole system was working. Luie gathered the observers in a room in the airport building to brief them on how the system worked. Soon after he started I came in and whispered that our last set of tubes had blown out, but that help was on the way and he should keep talking to them.

Homer Tasker, our Gilfillan Radio Company representative, saved the day by locating tubes at the Anacostia Naval Air Station, just across the Potomac River from our airport. One of our pilots flew across the river and picked up a large package of the essential tubes and rushed them to our radar trucks. We first replaced the tubes in the search radar and to our joy they seemed to be lasting. So I went back to the briefing room and told Luie we could use the search radar and that we were working on the precision landing radars. He had held forth for a long time, and was relieved to announce that the demonstrations would now begin.

The first test pilot took off, and we began vectoring him around using the search radar data to convince him that we could see him. He was hooded, so only the check pilot could see out of the cockpit. Before long we also had the precision radar working, so we vectored the pilot onto the approach path and soon could see his signals on the high-resolution scopes. Luie as Controller started talking the pilot down the glide path. The assembled officers stood outside the truck listening on loudspeakers to this radio conversation while watching the plane respond to the voice information. At 50 feet elevation Luie told the pilot that he should be able to see well enough to touch the plane down.

This process was repeated many times with different pilots and various kinds and sizes of planes. Several pilots stayed under the hood to touchdown. After a few landings a pilot would join the assembled brass at the GCA truck and tell them how he liked the system. There developed a consensus that this was the way to land airplanes, and plans were made for Gilfillan to produce several hundred GCA sets.

GCA had heavy use in the Pacific and saved many pilots. It is still in use by the military services worldwide. One of its most spectacular successes occurred in 1948, when the Russians blockaded Berlin during the rainy fall and winter. This required the city to be supplied by air, but ILS type systems could not be used because the buildings that crowded near the airport bent their radio beams. So a military GCA set operated around the clock, bringing in a steady stream of planes, and was given major credit for making the airlift successful and breaking the blockade.

The third application of Luie's linear dipole array antenna was the Eagle blind bombing radar. Bombing required high resolution, which demanded a wide-aperature antenna which, if conventional, would be much too bulky to attach to an airplane. Luie reasoned that his linear dipole array didn't really need a reflector as it would put out cylindrical waves near the antenna which would spread to form a fan beam, narrow in horizontal angle and wide vertically. The dipole array was long and slim, and Luie decided it could be hidden behind the de-icing rubber on the leading edge of the wing.

The biggest remaining problem was how to scan the beam right and left to form a picture of the space ahead of the airplane. Luie merely shifted phase slightly between successive dipoles, and the cylindrical wave fronts became conical fan beams far from the airplane. To accomplish this he introduced a variable-width waveguide which changed the wavelength and relative phases of successive dipoles. But this was easier said than done, since the waveguide had large currents running perpendicular to its moving metal joints. His solution was to use quarter-wave choke slots, such as perform the same function when a microwave oven door is closed. The Eagle antenna scanning scheme was also used with the production of the Mark II GCA sets, so antennas no longer needed to be mechanically rocked, which saved space, machinery, and maintenance.

The Eagle antenna was never actually mounted inside the airplane wing, but under the fuselage, where it looked like a small extra wing. A number of B-29s were equipped with Eagle, and although it worked well in several raids on Japan, it came too late to be important for the war effort.

During his MIT years, Luie kept in touch with Ernest Lawrence and other physics friends and so knew the progress of all the important projects. In the fall of 1943 he accepted Robert Oppenheimer's invitation to join him at Los Alamos to work on the atom bomb project. Oppenheimer asked Luie on his way out to help Enrico Fermi at Chicago for a new months. Fermi was working out the neutron physics of the famous first chain-reacting pile. So Luie arrived at Los Alamos in Spring 1944.

By then the details of "Thin Man," the uranium-235 bomb, were pretty well worked out. The main effort was directed toward a plutonium-239 bomb, since the more plentiful plutonium was produced in reactors. Here a plutonium sphere is surrounded by a shell of high explosive, which when detonated implodes and compresses the plutonium, making it go critical. Essential to this device is a uniform spherical detonation wave moving inward toward the core, an unusual requirement in the world of explosives. George Kistiakowsky, a Harvard chemistry professor, headed up this effort, and Oppenheimer asked Luie to be Kistiakowsky's chief aide.

The system was to have explosive lenses placed against the outer explosive shell surface. The 32 lenses were then to be simultaneously exploded. The problem was how to initiate 32 explosions within a small fraction of a microsecond of each other.

When Luie appeared on the scene, the method being tested for starting the 32 points simultaneously used an explosive messenger cable called primacord. This looked like blasting fuse cord, but with the core made of the very fast high explosive PETN. Thirty-two 8 foot primacord lengths were formed into a harness like the ignition wires on an auto engine. To one end of each cord was attached an explosive lens and to the other ends a single blasting cap which would simultaneously start a detonation wave traveling down each cord. The test results were discouraging. Streak cameras showed that the detonation waves arrived at the ends of the cords many microseconds apart, due to variations in the propagation speeds.

Luie asked Kistiakowsky why they did not use a harness of electric cables, with an electric detonator at the end of each cable. Kistiakowsky replied that electric detonators had a long delay of several milliseconds before the explosion got started, and the best

electric detonators had a timing spread of a half millisecond. When Luie looked into this matter he found that such detonators then used primary explosives—those able to be ignited by a gunpowder fuse or a glowing electric wire. The temperature and pressure built up to produce a full detonation wave, which would then detonate the more difficult to trigger secondary explosive, such as TNT. A typical electric detonator chain started with a fine electric bridge wire which glowed when a few volts were placed across it. The bridge wire was immersed in powdered lead azide or mercury fulminate followed by a sensitive secondary explosive, like PETN, and a tetryl pellet, which would then send a strong wave into the main body of the charge to be detonated.

Luie reasoned that the slow deflagration of this sequence could be eliminated by exploding the bridge wire by discharging a large capacitor through it. This technique might also provide enough pressure to detonate high explosives directly. When I arrived at Los Alamos in late April 1944, Luie asked me to try out his detonator idea. I had never used explosives professionally but had enjoyed shooting firecrackers and in chemistry classes did experiments making forbidden explosive concoctions.

The explosive research areas at Los Alamos were located far from the main Tech Area and from each other. Since my explosions were to be small, they assigned me to South Mesa, to which I transported a portable hut, a 5000 volt power supply, a small gasoline-driven generator, and a few oil-filled capacitors. Within three days of my arrival, I started the first test of Luie's idea: to see if a secondary explosive could be directly detonated by an exploding wire. I opened a standard electric detonator and poured out its explosives. I then reloaded the shell with powdered PETN, which Kistiakowsky told us was the most sensitive of the secondary high explosives. Using the wall of the hut as a shield, I discharged a 1 microfarad 5000 volt capacitor through the bridge wire. The big bang signaled that the explosive had detonated, and the copper fragments driven into the hut wall were a record of the event.

I jumped into my jeep and drove to the Tech Area. Kistiakowsky smiled when I told him the news and said it was probably a low-order detonation which took a long time to get started. Luie was an immediate believer. We next tried to show how simultaneously two detonators could be shot, the ultimate criterion of success. Kistiakowsky had told Luie of the D'Autriche method of timing explosives, and so to each end of a 1 foot piece of Primacord I attached a modified detonator and to its middle I taped a 2 inch long lead plate. The Primacord detonation waves from each end would collide in the middle, producing extra high pressure (the Monroe effect) which would make a nick in the lead.

I taped the explosives and plate to a wooden two-by-four, facing it away from the hut. I connected the wires coming through the hole in the hut wall in parallel. The shot made a satisfying report, and when I surveyed the damage, I found the plank shattered and the lead plate 10 feet away. Every little string in the primacord had left its impression in the lead, and sure enough, there was a deep, dark notch cut about an eighth inch from the center of the plate. The explosive wave travels a quarter inch per microsecond, so the timing difference was 1 microsecond. When we then showed that the absolute time delay in the initiation of a single detonator was less than a microsecond, all the explosive experts became believers. Once convinced, these experts helpfully suggested how to uniformly load explosives around the bridge wires to improve their performance. Soon we were able to set off 32 detonators simultaneously to within a few tenths of a microsecond.

Our technicians were Army enlisted men, with some electronic, scientific, or mechanical skills. Although these Special Engineering Detachment (SED) men lived in army barracks, they were subject to much less rigorous discipline than most GIs. I, however, felt guilty living comfortably with my family at a good salary while my helpers were away from home and being paid soldier's wages. They were just thankful for not being shot at on foreign soil. Several SED men made the detonators required for our tests. But when several groups which test-fired explosive spheres adopted our detonation method, our staffing needs could only be met by Indian women from nearby pueblos. Our SED men trained and supervised these women, who were transported by bus to and from South Mesa, which began looking like a village rather than the bare sagebrush I had originally been given. Captain Parsons (USN), who was responsible for the configuration of the actual weapon, was pleased that our detonators did not contain primary explosives. They were safe from accidental detonation during the low-voltage testing of the bomb's wiring. This type of detonator also simplified arming, as the capacitors could be charged after the bomb was dropped from the airplane.

Most of the Los Alamos staff lived in green fourplex apartment buildings constructed along the main road through town. Each building was heated by a single wood-burning furnace which distributed hot air to the individual apartments. As there were no thermostats, windows were opened and closed to control the temperature. The furnaces were kept burning by Hispanic-Americans recruited from nearby villages, who also did most of the other menial labor. Having stoked a coal furnace in Boston, I found this arrangement liberating. We also had a fireplace. Some furniture was available from a warehouse, but we adapted packing boxes or anything else we could scrounge. Some of my two-year-old daughter's favorite toys were yellow cardboard tubes which had contained the electric detonators we used.

In each kitchen was a wood-burning "Black Beauty" stove, which required a lot of work and was hot in the summer. We could buy electric hot plates at the Commissary, and an electric roaster served as an oven. As the town grew, however, these appliances began to overload the diesel generator which supplied both the Laboratory and town power. About 11:00 every morning an announcement would come over the public address system in the Tech Area to turn off unnecessary electric equipment, as the system was overloaded. We turned off the lights, unplugged half the soldering irons, and maybe the cyclotron crew even turned off their machine, but I never heard anyone suggest that women cooking lunch for their husbands should use less electricity!

The staff housing policy would have made any Fabian Socialist happy. Rent was a certain fraction of your salary, and space was assigned in proportion to family size. "Expecting" families signed up at the housing office for a larger apartment. This list was on the wall, so everybody knew who was pregnant. When the Alvarezes moved to Los Alamos, their son Walter was about four years old. When Gerry Alvarez became pregnant, they signed up for a larger apartment. At the next Town Council meeting, the wife of a well-known European physicist took the floor to protest a Hispanic-American family moving into her apartment building. She must have been surprised when she found out who this family named Alvarez was.

The democratically elected Town Council heard complaints about civilian services and held court for traffic violations. I appeared there once for a speeding ticket, where the presiding Victor Weisskopf, a well-known nuclear theorist, fined me five dollars in a very kindly way.

Luie was at his best when working out his own ideas, but he also took on special assignments. Oppenheimer asked him to monitor the density of a metal sphere as it imploded. Luie figured out what equipment was needed, got various experts to build the parts, and assigned other people to put the pieces together and make them work.

This experiment was called Ra-La because a strong radio lanthanum-140 gamma-ray source was placed at the center of the metal sphere. Gamma-ray detectors located outside the detonating explosive shell saw signals that grew progressively weaker as the metal was compressed to greater densities. The detectors were blown up but not before their information was transmitted by cables to recording equipment located in a nearby Army tank. Luie was accompanied in that tank for the first shot by George Kistiakowsky, whose presence assured that the safety estimates were carefully done! The Ra-La experiment apparently provided useful data, as did several other methods used to diagnose implosions, but I remember a shared feeling that none of them gave clean-cut results. The uncertainty was only dispelled when the first implosion bomb was tested at Alamogordo.

By early 1945 it was apparent that we would soon have all the ingredients for an atomic bomb. It was equally apparent that the European War would soon end and that Japan would be our target. Luie told Oppenheimer that he would like to get involved in the Pacific operations. Oppenheimer needed someone to measure the explosive energy of the bombs which were to be dropped on Japan, and Luie took the job in April 1945.

The first atom bomb test, Trinity, was to take place at Alamogordo, New Mexico, in July. Luie decided to use this event as a dress rehearsal for his measurements over Japan. He had to tell from an airplane 10 miles from the explosion what energy was released. He decided to measure the air pressure wave by a parachute-borne microphone dropped from his airplane. But he needed a telemetering radio channel between the microphone and a recording system in the airplane.

Luie had recently visited the Caltech Rocket Project, which was to become the Jet Propulsion Laboratory. While waiting in an office, he had read a report by Jesse Dumond describing a system called Firing Error Indicator (FEI). Recalling the report, Luie returned to Caltech to talk to Jesse and his former graduate student who had built FEI, Pief Panofsky. Panofsky was glad to modify his microphones and FM transmitters and receivers and bring several of them to Los Alamos. Luie now had the critical components for his measurement system. When he got back to Los Alamos, he gathered Bernie Waldman, Harold Agnew, and me to complete the project. He also got a half-dozen SED men, since it was thought that only soldiers would be allowed to operate in a combat zone. As the time for the missions approached, however, Luie wrote General Groves and asked permission for the scientists to be on board. His ostensible reason was that plans might be changed while the planes were in flight, and scientists could best cope with the new situation. Another unstated reason was that we wanted to be a part of the action when all of our hard work was put to the test.

Luie asked for a B-29 bomber on which we would fly at 30,000 feet directly behind the plane carrying the bomb. When the bomb was dropped, Luie's plane would release three parachute-borne microphones which would telemeter their information to FM receivers in Luie's plane. The information would be recorded by three 16 mm movie cameras focused on cathode-ray tubes driven by the FM receivers. As might be expected of an Alvarez project, each data transmission stage was calibrated so that absolute pressure changes would be recorded on an accurate time scale. The equipment was designed so that it could be carried by any B-29 equipped to carry an atomic bomb. I was to design and build the recording equipment and operate it on the missions.

Our first measurement of a real explosion came on 7 May, a week after I transferred to Luie's new group. A 100 ton pile of TNT was to be exploded on a wooden tower at Alamogordo so that the instruments being set up on the ground for the forthcoming Trinity test could be calibrated. We dropped our parachute gauges over ground zero from a B-29 flying at 15,000 feet. Of our three parallel systems, only one succeeded in recording the N-shaped shock wave. We made many equipment and procedural improvements, but as it turned out, this was to be our only rehearsal before Japan. Luie made elaborate plans to position our B-29 at 30,000 feet directly over ground zero when the bomb was to explode on the July 16th Trinity test. He arranged to have an SCR-584 gun-laying radar track our plane from the ground. On the evening of July 15th at Kirtland Air Force Base near Albuquerque, where we were preparing our equipment, Luie got an urgent message to call Oppenheimer. Oppenheimer had decided that it would be too dangerous to have our plane close to the explosion and ordered him to observe 25 miles from ground zero. Luie was furiously angry because this meant that our parachute gauges could not be used, and we would not gain much-needed experience from the test. With the Alamogordo weather still uncertain at 4 A.M., we climbed into our B-29 and took off. Our crew consisted of Luie, Harold Agnew, Bernie Waldman, Pief Panofsky, Captain Parsons, and me.

Luie and Parsons stayed in the cockpit while the rest of us were in the compartment behind the bomb bays with our measuring and recording equipment. A long crawl tunnel joined the two compartments. We listened on the intercom earphones to the news from the cockpit and the timing tones radioed from the Alamogordo control room. As dawn broke we were at 30,000 feet looking down through our small side window at the clouds below as the countdown approached zero. Suddenly the sky turned bright, but since the plane was headed toward the bomb we could see little more. We then crawled several at a time forward to the cockpit and watched the mushroom cloud ascend as our plane circled it. I was tremendously relieved that our detonators had worked. I suppose everyone worried about the functioning of the particular bomb components on which he had worked.

After the Trinity test the huge steel tank, called Jumbo, remained intact on the desert floor, a monument to our success. This tank was capable of withstanding the pressure created by the detonation of the bomb's high explosive and still contain the precious plutonium should the bomb fail. By July 1945 the timing and symmetry of implosions had improved, thanks in no small way to Luie's detonators, so that Jumbo was not needed.

After the Trinity test, we had only a few days before we departed for the Pacific. Luie had everybody make a checklist of the operations we would perform on our flights, fill a tool box to service our equipment, and pack a personal wooden footlocker. We also hastily assembled spares of all our measurement-system components. I got little sleep that week.

The Adjutant-General's office issued each of us an official identification card so if we were downed over Japanese-held territory we would hopefully be treated as military personnel and not shot as spies. Luie was an Air Force colonel and Harold Agnew and I were captains. We were each handed a thick packet of mimeographed orders, told to make out a will, and given a series of immunizations.

On July 20th we left Los Alamos for Wendover, Utah, home of the 509th Bombardment Group Headquarters, whose elite airmen would carry us on missions over Japan. There we donned military uniforms and mailed our civilian clothes home. We boarded one of the Green Hornet Squadron's transports which was attached to the 509th to move our equipment and us to "Alberta," our Pacific base. We hopped across the Pacific: Hawaii, Johnston Island, Kwajalein, Saipan; and finally arrived on Tinian at 4:03 P.M. on July 26th.

Tinian, composed of extinct volcanic hills surrounded by large coral reefs, was ideal for airfields. Its long, crushed limestone runways were frequently rolled and sprinkled with sea water. Tinian, Saipan, and Guam were the principal bases from which the frequent massive B-29 bombing raids on Japan 2000 miles to the northwest originated. Each field had a number of parallel 10,000 foot long runways, so that many planes could land at the same time. The night sky resembled a fifty-lane freeway coming down a hill.

The 509th Group had its own real estate, personnel, and facilities and operated independently of the other island activities. We had our own sleeping tent areas, officer's mess, movie theater, officers club, latrines, and showers. We also had several large quonset huts for doing our bomb work and a service area for our B-29s that included a large hydraulic hoist for pushing the bomb up into the plane.

The first bomb was to be dropped August 2d, so we had time to prepare. Within a few days our parachute gauges and recording equipment were installed, recalibrated, and tested in our B-29, the *Great Artiste*. We went swimming at one of several sandless, sharp coral "beaches" where poisonous sea urchins kept our feet off the bottom. However, the marine life we viewed through a diving mask was really worth the effort. In the evenings we went to a movie or talked in the Of-

ficers Club, but since our enterprise was unknown to most, we could not publicly talk shop. We learned a great deal about the outlook of those who had worn a uniform much longer than we had and who could not soon look forward to getting home. Luie introduced me to William Laurence, the *New York Times* Science Editor, whom Luie and Dick Feynman had educated about the A-bomb so that when the President decided that the time was ripe, Laurence could have a big news release ready to go.

On August 3d we were told we would make the raid that night, weather permitting. Weather reports came from returning missions over Japan and from reconnaissance flights. Testing our equipment in the B-29 on the ground was difficult because the 400-cycle 100 volt power was only available when the engines were running. That day we got an auxiliary gasoline-driven generator going in the plane so we could run the equipment and a soldering iron. Testing the parachute gauge transmitters in the bomb bay also required power from the plane, but after they were dropped, they would run on their own internal batteries. After the evening mission briefing we were told that the mission was called off because of bad weather.

The same thing happened the next evening, but on August 5th it was the real thing. At the briefing we were told the target was Hiroshima, a city I had never heard of. On previous evenings different cities had been targeted. We were briefed on the weather, potential opposition, antiaircraft fire, our armaments (a tail gunner in each plane and one A-bomb), and so on. The last item was the chaplain's prayer for the crew members and for God to speed the end of hostilities. We then went to an equipment room, where we signed out a parachute, an inflatable life raft, a flack suit, a flack helmet, and a vest with all sorts of pockets containing emergency rations and shark repellent. I could barely carry my load to the plane from the truck that brought us out to the flight line.

On board the *Great Artiste* were Luie, Harold Agnew, and I and the pilot Charles Sweeney and his crew. We piled our equipment on our rear compartment floor and strapped ourselves into the bucket seats for a 2:45 A.M. takeoff. We were tired from being up late the last several nights, but as soon as we reached our 5000 foot cruising altitude, I began trying out our receivers and recording equipment. We hadn't decided how big an explosion we were expecting, and I needed to set the gain switches on the three cathode-ray tubes before starting the lengthy system calibration. I found Luie sound asleep in the padded tunnel over the bomb bays. He had been burning the midnight oil more than the rest of us, so I decided that the calibration could wait. I lay down on top of my flack suit and dozed intermittently. Occasionally I got bits of information from the tail gunner, who was listening to the pilot on the intercom. About 6 A.M. it began getting light. The tail gunner went to his post since we would soon enter enemy territory and start our climb. Since the gunner's turret could only be reached by passing through an unpressurized part of the tail, he made the trip before we pressurized.

I woke Luie from a sound sleep, and he let me know he was quite upset. Harold Agnew thrust the FM receiving antenna out into the airstream before we pressurized. We went through the calibration, tuned the receivers to their transmitters, and completed the rest of our check list.

Our three B-29s flew in a line as we neared Hiroshima. Colonel Tibbetts's *Enola Gay,* which carried the bomb and Captain Parsons, who armed the bomb, was in the lead. We flew 300 feet behind and were trailed by Bernie Waldman's plane, from which he was to take pictures of the expanding fireball with a high-speed Fastax movie camera.

Luie wore the intercom headphones and coordinated our activities. At 45 seconds before the drop, we received the bombardier's tone whose end signaled "bomb away," at which time our parachute gauges were dropped and Luie started his stopwatch. We were now very busy. As each parachute descended and its microphone diaphragm compressed, we tuned the corresponding FM receiver until its lead rate came to equilibrium with its fall rate. After 15 seconds, two calibration pulses produced by small spring-driven pistons on the microphones showed on the readout meters. All this was being recorded by movie cameras. Then our cabin was lit with a flash of white light coming through our one small window and striking the ceiling. We then had half a minute to make the final receiver adjustments before the pressure wave would reach our microphones. I saw my meter swing sharply positive and come back, and then my attention was diverted. Several seconds later the pressure wave hit the plane in a double jolt not strong enough for concern.

Our job was now over and we relaxed. I was relieved that the bomb worked and that our measuring system worked. I switched on the cameras so we could record calibration signals after the bomb records. When we looked out the window the bomb's hot gases had risen to several thousand feet and brown dust was spreading out on the ground, and we realized that thousands of people were dying at that moment. I had no strong emotions. We had long planned and worked for this blow against Japan and so had come to terms with the inevitable loss of life. We hoped for an early end to the War and its heavy drain of human life and potential. I was thankful that I could play a part in the process. Although we did not discuss these points on the flight, I knew they were also held by Harold Agnew and by Luie.

Harold Agnew had the foresight to bring an 8-millimeter home movie camera, which he gave to our tail gunner. Since we were flying away from Hiroshima when the bomb went off, the gunner had a perfect vantage point and recorded the rising cloud starting a few seconds after the detonation. For some reason Bernie Waldman's Fastax did not give any images, so ours was the only movie of the explosion. Flying back to Tinian, Luie wrote the letter reproduced below (p. 69) to his young son Walter and then slept. I wrote a letter to my daughter Virginia and also got some sleep. We arrived at Tinian about 4 P.M. after a 13 hour flight.

The next day, when President Truman announced that we had used an atomic bomb on Hiroshima, reporters from Guam interviewed those of us who had been on the mission. A reporter from the *Los Angeles Times* took my story, which they and the *Hollywood Citizen News* published. Our stories also appeared in our hometown papers. There was great excitement on the island when everybody heard what our B-29s had done. The big question remained, would the Japanese surrender or would we drop more bombs?

The next bomb was already on the island, and we kept its plutonium core in a special magnesium box, 10 inches on edge, in our quonset hut. Only a thermometer to measure alpha-decay protruded. On learning of Hiroshima, the special guards who protected this box treated it with a great deal more respect. The evening before the next raid Luie and two friends were talking about ways to end the war more quickly. Luie suggested attaching a letter to the Japanese on the gauge canisters which would be found near the bomb. They addressed the letter to Luie's friend Professor Sagane at Tokyo University and asked him to confirm to the Japanese authorities the nature of the bomb and advise them to surrender before more cities were wiped out.

When we were told on August 8th that the mission was on, we had to decide who would go. Luie thought that the scientists should not be exposed to further dangers and that the enlisted men could handle the job. We soon realized we had not trained our SED men in the use of the onboard recording equipment, so I volunteered to go with Sergeants Walter Goodman and John Kupferman. Our primary target was Kokura, a naval base northwest of Nagasaki. We met up with the plane carrying the bomb off the Kyushu coast, but the observation plane was nowhere to be found. After circling for an hour, we proceeded to Kokura. In the meantime the weather had started to deteriorate and we approached Kokura on radar. We hoped for a hole in the clouds, as we were ordered to only drop the bomb on a visual contact with the target.

The bombardier could not see the ground at the critical moment, so the bomb was not released. Our bombing run ended directly over the Yawata steel mills, reputed to have the heaviest antiaircraft defenses in Japan. Our bombardier reported over the intercom "two-tenths flack" underneath us as we left Yawata.

We made two more passes at Kokura but could not see the target. Each time we flew over Yawata, the antiaircraft fire got more intense and more accurate until on the third run the plane bounced around as flack bursts came much too close for comfort. We gave up on Kokura and headed for our secondary target, Nagasaki.

The plane carrying the bomb was now getting critically short of fuel, even assuming that we could refuel at Okinawa. So the planned over-water approach to Nagasaki was abandoned, knowing that we would still only have one pass. Nagasaki was also beginning to get cloud cover, so again we approached on radar. We had been poised on each Kokura pass to drop our parachute gauges, with Sergeant Goodman calibrating and operating the recording equipment so he would be experienced if there were missions after Nagasaki. After the last Kokura run, we got apprehensive and I ran the equipment when we dropped the Nagasaki bomb. Goodman took movies of the explosion from our small window with our 8-millimeter home movie camera. These were the only movies of the Nagasaki explosion, as the plane loaded with celebrities and cameras never made rendezvous with the mission.

Once the bomb was dropped on Nagasaki, we headed straight for Okinawa, our air base closest to Japan. Ordinarily B-29s couldn't use this field since its runways were too short, but our planes were equipped with the newest, secret, reversible-pitch propellers. Plane number one, now almost completely out of fuel, radioed for an emergency landing and fired its rocket pistols to emphasize the emergency. The ground crew, their firetrucks and ambulances ready, knew a B-29 could not stop. They were amazed as both our planes landed with runway to spare.

While our planes were being refueled, we had something to eat in the Officers' Mess. Nobody asked about our mission and we didn't say anything. We easily took off from the short runway, since both our planes were loaded with only enough fuel for the 1500 miles to Tinian. With both A-bombs delivered we relaxed, since it would be many weeks before another bomb could be completed and delivered to us. Luie had the pressure data films developed, and our local experts made some preliminary estimates of the bomb's energy. We kept listening for news of surrender negotiations with Japan.

By the time the Japanese surrender was signed in Tokyo Bay, we were well rested and anxious to go back to peacetime pursuits, about which each of us was actively thinking. Luie sent several telegrams to General Groves requesting permission to return. General Groves finally replied tersely that we were not to leave. Presumably he wanted us there until it was certain that the American occupation of Japan would go smoothly. Luie chafed at his confinement and said it made him appreciate the meaning of freedom.

Tinian offered lots of diversions, so for the next few weeks we slept late, swam, and explored with jeep and camera. Luie often spent afternoons in our air-conditioned quonset hut playing hearts with some buddies. He also took part in late-night poker games. After dinner there was a choice of conversation in the Officer's Club or several outdoor movies, to which we would wear raincoats.

I was interested in exploring the tide pools and beaches. Of special interest was a species of inch-long amphibious fish, which sat on the damp rocks on the edge of tide pools and could only be persuaded to jump into the water if threatened. If a wave came close, they hopped across the surface of the water and landed on another rock higher above the water line. There were moray eels, starfish, abalones, and many kinds of brightly colored tropical fish.

When I noticed that the post exchange was selling necklaces made of exquisite local cowry shells, I decided that I must find where they came from. During one early-morning search at low tide, I found several hiding in cavities in the rocks. The people in our group were excited by them, but none relished the idea of a prolonged search. We heard that the Micronesian natives, who used the cowries for currency, dove for them. Luie immediately remembered that Harold Agnew had brought a large supply of hotel-sized soap bars specifically for trading with the natives. So Luie suggested that we should form a "Cowry Cartel" and send a representative to bargain, reducing the bidding against each other. Luie was chosen as cowry negotiator. The first day he returned with a bag of a hundred cowries purchased for twenty soap bars. As the days went by the price went up, until at one cowry per two bars, we felt that we had cornered the market. Each of us got several dozen beautiful shells. Luie had one very large cowry fashioned, with five smaller ones, into a turtle ornament for his wife.

During one of my explorations I found a large number of cowries in 10 feet of water at the main garbage dump! These well-fed cowries were choice specimens. They were hard to spot, however, because they kept their mantle wrapped around their shell and so looked just like a small sea slug. When you touched them they retreated into their shell, exposing the gorgeous spotted chocolate-brown colors.

I also discovered a good way to separate the animal from its shell when I left a bunch of cowries under a bush near the beach. When I came back the next day they were scattered over a 10 foot circle, their shells clean and empty. The cleaners were hermit crabs some 10-centimeters long who, having outgrown all the available snail shells in the tide pools, had moved into the sugar cane fields.

On the trip home from Tinian we were again faced with freedom, and we could choose what to do next. Luie was full of plans to put our new knowledge of high-power microwaves to work accelerating particles to unheard-of energies in a linear accelerator. I wanted to continue my graduate work at Berkeley, so I asked Luie to be my major professor. That was the start of another five years of exciting accomplishment.

The Radiation Laboratory we found in 1945 was a far cry from the one we had left in 1940. It had moved up on the "Hill" and had buildings of its own, professional machinists, shops, and engineers. It also had at least three major accelerator developments under way and lots of government money. For the next two decades Berkeley was the acknowledged capital of accelerator development and high-energy physics. Luie was a stellar member of the team which made that happen.

August 6th 1945
10 miles off the Japanese
Coast at 28,000 feet

Dear Walter:

This is the first grown-up letter I have ever written to you, and it is really for you to read when you are older. During the last few hours I have been thinking of you and your mother and our little sister Jean. It was tough to take off on this flight, not knowing whether I would ever see any of you again. But lots of other fathers have been in the same spot many times before in this war, and I had a job to do, so I can't claim to be any sort of hero.

I wonder if you will remember the time in Albuquerque, when we climbed all through a B-29 Superfortress. Probably you will remember climbing thru the tunnel over the bombbay, as that really impressed you at the time. Well, I have been in this B-29 for eight hours so far, and we won't be back for another five or six.

The story of our mission will probably be well known to everyone by the time you read this, but at the moment only the crews of our three B-29s, and the unfortunate residents of the Hiroshima district in Japan are aware of what has happened to aerial warfare. Last week the 20th Air Force, stationed in the Marianas Islands, put over the biggest bombing raid in history, with 6000 tons of bombs (about 3000 tons of high explosive). Today, the lead plane of our little formation dropped a single bomb which probably exploded with the force of 15,000 tons of high explosive. That means that the days of large bombing raids, with several hundred planes, are finished. A single plane disguised as a friendly transport can now wipe out a city. That means to me that nations will have to get along together in a friendly fashion, or suffer the consequences of sudden sneak attacks which can cripple them overnight.

What regrets I have about being a party to killing and maiming thousands of Japanese civilians this morning are tempered with the hope that this terrible weapon we have created may bring the countries of the world together and prevent further wars. Alfred Nobel thought that his invention of high explosives would have that effect, by making wars too terrible, but unfortunately it had just the opposite reaction. Our new destructive force is so many thousands of times worse that it may realize Nobel's dream.

After that little sermon, I'll try to describe what it is like to go into combat for the first time. I had not made up my mind to go on the mission before I left the states, but I was pretty well convinced that I would end up by going. I thought the thing through on at least a dozen nights, while I was trying to go to sleep. I think these mental trips were the worst part of the deal.

When I arrived in the Marianas, I told the commanding officer that I thought I should go. I got cleared after a lot of radio messages to and from Washington. The mission was held up for several days by weather, and this was tough. We would get keyed up and ready to go, and then the weather experts would call it off. Finally we got the go-ahead sign and then worked most of the day checking instruments. We had several briefings which were quite exciting. I had attended bombing briefings in England for the RAF, but it is quite different when you are to go on the mission yourself. Data on anti-aircraft batteries and enemy fighters becomes of great personal concern.

One of the planes of our squadron had come home with large flack holes in its wing two days before, so we felt some concern on that score. We were told a lot about parachuting out at various altitudes over land and sea, and about landing the plane in the ocean. The big worry, of course, was landing on the Empire and being captured by the Japs. They have been particularly savage with ordinary pilots, and I am sure they would have a special reason for disliking us intensely.

We were to take off at 2:45 A.M., and this last waiting was the worst part. We saw a movie until 9:30, and then packed up last minute supplies for the plane. Then we got equipped with our combat flying suits, which weigh about seventy or eighty pounds. First comes a survival vest, with fish hooks, drinking water kits, first aid packages, food, and a host of other things useful to a man forced down on the ocean. Over that was our parachute harness, to which could be clipped a chest chute pack, and a one-man liferaft. With this equipment, it is possible to go into the water from a plane, some distance from anyone else, and survive. Over this already bulging mess, we wore our flack suits, to protect our bodies from flying shell fragments. This is a very heavy and clumsy thing, like a suit of armor, but we were glad to put up with the discomfort during our 65 minutes over the Empire. Finally, we wore a cloth helmet with an oxygen mask attached, and over that a flack helmet to protect our heads.

We arrived at the plane an hour and a half early, as there were lots of historic pictures to be taken with the aid of a big battery of lights. It looked just like the opening of a gas station in Hollywood. We had our pictures taken in front of the plane which held the big bomb in its bomb bay, and then went to our own plane. By this time all my tension had gone away and I haven't felt any since, with the exception of a little tingling sensation when the Japanese shores appeared on the horizon. All of the civilians had thought we would be scared over the empire, but I can say truthfully that I was completely at ease, and so were my two companions. We weren't excited, as we were too busy with our work. After the bomb was dropped we made an exceedingly sharp turn to get away from the blast. We got 2 g's, which made our 80 pounds weigh 160.

A few seconds after we completed the turn, the plane was hit with the blast wave from the explosion. It gave the ship a couple of good jolts, but only about what we expected. We went to the portholes to see the result of the explosion. It was awe-inspiring. Already the smoke cloud was up to 35 or 40,000 feet. The ground was covered with a layer of smoke so that the city was blotted out from view. I forgot to mention the most spectacular effect of all—the light flash. It was many times brighter than the sun when we were seven miles away. I had looked at it directly, through dark glasses, on the trial shot in New Mexico last month.

Well, here we are over Iwo Jima, and on the home stretch, so I'll stop writing and go up and talk to the pilots. I wanted to tell you about this while it was still fresh in my mind.

With much love from
your Father

P.S. When I saw the pilots, they said they saw flack bursting a mile below us. The Japs apparently didn't have their good anti-aircraft in this region.

The Collier Trophy, awarded to Alvarez in 1945.

10

Building the Proton Linear Accelerator

W. K. H. Panofsky

Although the idea of a linear accelerator probably originated with the Swedish physicist Ising in 1924, Luis W. Alvarez can justly be called the father of the modern practical proton linear accelerator. Before World War II, radio-frequency linear accelerators had been built in rudimentary form by Sloan and Lawrence, but these accelerators produced heavy ions with very modest energies. Moreover, at that time there was no evidence whatever that linear accelerators offered any advantages over the conventional cyclotron, then in its prime of productivity.

When I joined Luie during the last year of the War at Los Alamos, we discussed his preliminary plans for a proton linear accelerator, which he believed would eventually beat out the cyclotron. His rationale was simple. The conventional cyclotron operated under a scaling law in which its linear dimensions grew approximately linearly with energy; therefore its cost increased roughly with the third power of the particle energy produced. In contrast, a linear accelerator presumably led to a linear scaling relationship. Thus, eventually, as one went to higher energies, the linear accelerator would replace the cyclotron on economic grounds. The only question, therefore, was, "When is eventually?"

In hindsight, these arguments were entirely correct at the time. However, due to changing events they were overshadowed by other factors, so that the conclusions from these arguments were not as far-reaching as they then appeared. The first factor, which emerged immediately after the War, stemmed from new inventions in the circular accelerator field. These converted the scaling law from the cubic relationship of the conventional cyclotron to the linear relationship pertaining to the proton synchrotron. As a result, at least for protons, the linear accelerator has not proved to be competitive with circular machines as producers of the highest energies now available. Rather, the proton linear accelerator has been largely relegated to the role of the universal injector into circular proton machines. The most notable exception is the Los Alamos "Meson Factory," LAMPF, which is operating highly successfully as the leading tool in medium-energy physics. Moreover, Luie's prediction may yet prove true for the highest energy regions in the very far distant future. There are recent developments in linear accelerators

that may eventually put proton linear accelerators into a competitive position at the very highest energies.

Luie's second motivation to push forward with proton linear accelerators was the use of surplus radar equipment. Again, as it turned out, this was an inspired idea that proved to be a strong motivation for getting the program going and ultimately led to the successful construction of proton linear accelerators. Yet the history of that program saw the use of surplus gear only at first. As soon as the operating inefficiencies and unreliabilities engendered by the radar equipment became evident, these surplus components were gradually replaced by equipment designed specifically for the purpose.

Luie noted that the SCR-268 early warning radar system operating at a frequency near 200 megahertz had become obsolete. Therefore, large quantities of components of that system were in government surplus. Using the well-known formulas for the shunt impedance of radio-frequency cavities at 200 megahertz, Luie calculated the power requirements for a proton linear accelerator up to perhaps 100 MeV. He concluded that by hooking up a large number of the modulators and transmitters from the SCR-268 radar system, proton energies well in excess of those available with conventional cyclotrons could be obtained at affordable cost.

The accelerator program started by Luie in Berkeley immediately after the War produced the first proton linear accelerator, which, in contrast to many other accelerators, is well documented in the accompanying article. Organized by subsystems and other technical elements of the device, this article gives a coherent description of the facility. I will only supplement this article by giving my subjective impression of the historical evolution of the activity, led by Luie, which brought this facility into successful operation. Immediately upon his return from Los Alamos to Berkeley after the War, Luie collected a large research group. Some of his collaborators were colleagues from the wartime years; others were specialists whom Luie recognized as being able to make important contributions. I will recall some of the collaborators and their special roles. The team included C. M. Turner, who was put in independent charge of constructing a 4 MeV proton

Van de Graaff injector for the linear accelerator. John R. Woodyard was a gifted radio-frequency engineer who was well acquainted with the microwave work going on at Stanford University and in industry. Frank Oppenheimer participated in the early phase of the work, dedicating his time primarily to microwave measurements. Several young people joined the team directly from the Armed Services. In particular, J. Donald Gow and Walter Selstedt helped in several different capacities. Hugh Bradner, Lauriston Marshall, Hayden Gordon, and C. A. (Slim) Harris contributed in many specialized ways, as did many others.

In 1945 it was decided to build a 40 foot machine as a pilot project. Characteristic of those times, it was impossible to estimate the costs and therefore produce a budget for this venture. Further, there was no estimate of the total number of man-years that might be required, only a schedule and a design goal. There were numerous arguments about whether the design energy of 32 MeV would be sufficient for physics, or whether the only mission of this machine would be to serve as a technical pilot project. Extensive discussions with theorists provided encouragement and assurances that 30–40 MeV proton energy would constitute a great leap forward for proton-proton scattering. Clearly, at that energy *P*-wave effects would become very much in evidence. Therefore, the 40 foot proton linear accelerator was to be both a physics tool and a pioneering step in technology. Both these predicted missions of the instrument have been realized. Not only were proton-proton scattering experiments carried out with this machine using two different experimental arrangements, but many other experiments were made possible using the 32 MeV proton beam. Among these are various experiments in nuclear chemistry and in the study of nuclear reactions. In addition, the basic device has remained the prototype for all subsequently designed and constructed proton linear accelerators.

Again, characteristic of those early days, intensive development of components for construction proceeded while a great deal of the theory of the proposed machine still needed elaboration. In particular, the orbit dynamics of protons in the linear accelerator were just beginning to be understood.

At the same time work on the linear accelerator commenced, McMillan at Berkeley and Veksler in the Soviet Union discovered the principle of phase stability. It was recognized immediately, with particular contributions by Robert Serber, that the phase stability principle was applicable to linear accelerators. However, a change in sign of the phase angle of acceleration which corresponded to stability was necessary. In other words, phase-stable acceleration takes place in the linear accelerator at a phase angle where the accelerating field is increasing. The opposite is true for circular

machines, with the exception of proton machines at highest energies. In the linear accelerator we found, from the orbit mathematics, that phase stability and radial proton orbit stability are incompatible conditions. McMillan later published a general theorem which indicated that this incompatibility was fundamental for structures of cylindrical symmetry.

The only way to beat the McMillan theorem, we determined, was to introduce charge into the beam in such a way that the accelerating fields simultaneously generated converging lines of force, thus giving rise to radial focusing. The McMillan theorem can be circumvented in two additional ways: first, by accelerating configurations which do not have azimuthal symmetry (this is a principle reduced to practice using the radio-frequency quadrupole); or second, by alternating the regime of acceleration between radial-focusing, phase-unstable and radial-defocusing, phase-stable conditions. The latter method is based on the strong focusing principle discovered later, and I calculated that a very small dually stable regime for acceleration in a proton linear accelerator was indeed possible.

Designing the 40-foot proton accelerator, we decided that charge must be introduced into the beam, and Luie calculated that a thin foil of beryllium across each accelerating gap would do the trick. Beryllium was chosen because it extended the best hope of minimizing the multiple Coulomb scattering which counteracted the focusing action. However, for this system to work the foils had to be only a few microns thick. Hugh Bradner successfully undertook the difficult development process of fabricating these extremely thin foils—so thin that once they got away, they would float around the room. Elaborate holders were designed to stretch and clamp these foils across each accelerating gap. It is interesting to note that in those affluent times the machine shop would fabricate without question almost anything that was requested. A drawing went into the shop for a "beryllium foil holder," but the draftsman forgot to specify that it was to be made of copper. The shop, therefore, fabricated the entire assembly involving threads, pins, and knurled handles of beryllium!

The final episode of this story is that all these beryllium foils were introduced into the machine and disappeared almost instantly as the accelerating field was turned on. The foils simply did not survive the conditions of high electric fields and electron bombardment to which they were exposed. The foils were replaced by grids calculated carefully to produce minimum beam intercept for the focusing action required, and to avoid excessive field concentration.

Because phase stability is important for both linear and circular accelerators, the following event is worth mentioning. As the proton linear accelerator at Berkeley was being designed, the late William Webster Hansen was proceeding with the design of an electron linear accelerator at Stanford University. The Stanford accelerator was to bring electrons up to fully relativistic speeds very rapidly. Therefore, electron linear accelerators cannot be operated in a phase-stable manner because the velocity becomes immutable. Rather, relativistic electron linear accelerators operate, for all practical purposes, in a phase-neutral mode. I remember an occasion when Luie and Hansen were sitting together outdoors on a bench arguing vigorously about whether adequate mechanical tolerances on an electron linear accelerator could be maintained so that phase-neutral operation would be feasible in practice. Luie said no, while Hansen said yes. History has shown that in this particular instance Hansen's prediction proved correct.

It is also interesting to note that operation of the 40-foot proton linear accelerator proved feasible even when the focusing grids were removed. This meant, presumably, that the machine could also operate in a phase-neutral or even slightly phase-unstable mode. Although operation without grids was indeed possible, adjustments were very sensitive and critical. Therefore, as a practical matter grids were much preferable when it came to actual operation for particle physics.

The basic problem was to design the resonant cavity structure of the accelerator itself. This required that the entire structure be operated at the same frequency while preserving the phase relationship between the accelerating particles and the accelerating gaps. Stemming from the reentrant cavity data used in Klystron design, the basic solution was generated by constructive interaction among Luie, Woodyard, and Oppenheimer. Oppenheimer and others modeled these structures at 3 gigahertz and developed the necessary design tables governing the proportions. Short sections of such accelerator structures operating at 200 megahertz were then built and were successfully excited by the surplus radar transmitters, as originally conceived by Luie. Encouraged by these tests, we decided to go forward with final design of a cavity structure.

Luie proposed that it would be best to separate the vacuum envelope and the radio-frequency cavity boundary. To maintain the high precision required for the radio-frequency structure, he suggested that it be fabricated using the so-called hydroforming technique then widely used in the aircraft industry. He arranged for cooperation with Douglas Aircraft Company, and the first 40-foot radio-frequency envelope—a rather delicate structure—was delivered. This hydroformed 40-foot long cavity was inserted in the vacuum envelope, a simple steel tank. The radio-frequency structure was fitted with drift tubes which shielded the protons from the radio frequency fields during the time the phase of these fields was in the decelerating direction. The surplus transmitters were refitted with coupling devices to feed radio-frequency power into the structure at multiple feed points, and initial test operation began.

Putting this entire structure into operation, we immediately ran into a variety of problems. First, the lifetime of the tubes powering the surplus radar transmitters was excruciatingly short, and a store of literally thousands of these tubes had to be maintained to keep the sockets filled. Although these tubes did not cost the Laboratory anything, the total amount of labor and down-time involved in this continuing stream of tubes became prohibitive. The second problem was that the windows that isolated the accelerator vacuum from the coaxial transmission lines had a disturbingly high failure rate. Third, the electric field distribution along the tank, as measured, did not at all correspond to the uniform acceleration gradient for which the structure was designed. Finally, when the accelerator structure was excited *in vacuo* the field refused to build up beyond a rather low level.

Each of these problems would have kept the machine from operating properly, and each had to be attacked in turn. The tube failure problem was solved by throwing away all the transmitters and building new ones using more modern tubes which also could be obtained surplus. The lifetime of these replacement tubes was still poor, but at least the total tube traffic proved manageable. In addition, we threw away the individual radar modulators and built a single large power supply and modulator with a rotating spark gap switch. All that remained from the original surplus radars were the filament transformers. The window problem was gradually solved by careful choice of window insulators and tailoring of the edge field around the windows. Again, this continued to be a source of trouble, but at a tolerable level.

Adjusting the field profile in the machine required the development of a mathematical solution. The longer a single resonant tank becomes, the closer the spacing is between the fundamental and higher modes. This next higher mode corresponds to excitation in which a longitudinal standing wave is generated with a node at the midpoint of the tank. Thus, if the mechanical variances are such that the resonant frequencies of each of the subcavities are dispersed, then the lowest mode does not generate a uniform field but instead generates a field distribution that is a linear superposition of all modes. A simple algorithm was developed from which the resonant frequency errors of each section could be deduced using the measurements of field distribution. In turn, corrections could be made to these radio frequencies by the insertion of appropriate tuners. This process led to a satisfactory flattening of the longitudinal field distribution, starting from the initial measurements at the edge of the long cavity.

The fourth difficulty was easy to diagnose but not to cure. The problem was caused by what is now called multipactoring. Electrons are emitted from the metal surfaces and deflected in the electromagnetic field of the cavity so that they strike other parts of the walls. This in turn produces secondary electrons which again execute complex trajectories. Under certain circumstances, this phenomenon can regenerate and thus load down the available power. Generally, this phenomenon occurs at a relatively low excitation of the cavity. In those days, without the availability of computers, it was clearly impossible to calculate the vast number of possible trajectories. Rather, the matter had to be attacked empirically. Part of the cure proved to be radio-frequency processing and cleaning of the metal surfaces to reduce the secondary electron emission coefficient. This was only partially successful. Occasionally it appeared possible, somewhat accidentally, to break through the multipactoring threshold. Since this approach seemed to be unproductive, a second remedy was attempted. A direct current voltage was put on the drift tubes, thereby biasing the field distribution in an asymmetrical manner and, hopefully, making multipactoring less likely. Technically, putting a direct current voltage on the drift tubes is difficult; it requires insulation of the drift tubes from the cavity walls while at the same time a continuous radio-frequency path from the drift tube stems to the walls must be maintained. In addition, the cooling water circuits in the drift tubes have to be interrupted by insulators.

All this became a bit of a technical and reliability nightmare. The drift tube stems were fitted with titanium oxide bypass condensers, while glass inserts were introduced into the cooling circuits. The titanium oxide bypass condensers were beset by high failure rates. The glass insulators would break so frequently that it was suggested that a float valve be inserted into the vacuum tank to shut off the water supply in case the water in the vacuum tank rose too high! Yet, after struggling with all these problems, we succeeded in applying a reliable direct current bias to the drift tubes; indeed, after some processing the multipactoring disappeared and the tank could be excited to its full rated value. The only trouble was that after scoring this enormous success we disconnected the bias voltage and the system worked just as well! We have never fully understood to this day why multipactoring initially proved to be such an insuperable obstacle while it eventually just disappeared.

The actual turn-on of the accelerator was interrupted by an unfortunate accident. Handling the radio-frequency cavity separately from the vacuum tank proved to be a delicate operation. It was lifted by two small electric hoists, and one of the engineers installed mechanical fine-adjustment links into these hoists. Unfortunately, under certain conditions this gadget would lock the electric switch that also served as a limit switch in the "on" position. Consequently, one of the hoists suddenly decided on its own to ascend to the roof. Since the limit switch was not operative, the power of

the hoist broke the lifting chain and one end of the cavity assembly smashed to the floor. Fortunately, no one was hurt. There was no choice but to order a new cavity from the Douglas Aircraft Company. At the same time, a 10-foot section from the damaged 40-foot cavity could be salvaged and used for various tests while the new 40-foot liner was being fabricated.

After all these trials and tribulations, actual beam tests with the machine could commence. Unfortunately, the limiting factor for quite some time turned out to be the energy available from the Van de Graaff injector. Since the 32 MeV proton linear accelerator operates at essentially nonrelativistic energies, exact maintenance of the design injection energy is critical. The injector was beset for a substantial time by many electrical breakdowns and other problems which made a 4 MeV proton beam available only at infrequent intervals.

Once these difficulties were overcome, acceleration was demonstrated immediately, but it took a considerable amount of time for reasonably high beam intensities to be produced. It is amusing to note that for this reason the proton linear accelerator immediately acquired a reputation as a very low intensity device. I recall that when the scintillation counter was discovered and first introduced at Berkeley, Ed McMillan remarked, "Now we have a device to detect the beam from the linear accelerator." This situation is particularly amusing when the next step in accelerator development at Berkeley is considered. Prompted by fears of possible cutoff from overseas sources of uranium, Ernest Lawrence was searching for an accelerator to breed fissionable isotopes, in particular plutonium, from depleted uranium stockpiles. After some study Lawrence concluded that a large linear accelerator would be the optimal source of the very high intensity deuterium beams required for the breeding of fissionable material, and the so-called MTA project at Berkeley and Livermore was the result. Based on the MTA pilot plant at Livermore, the plans for a production breeder of fissionable material in the Midwest called for a 1 GeV, 1 ampere average current linear accelerator. These plans never materialized, since the perceived need for accelerator breeding disappeared. The anticipated shortage of uranium in fact turned into a liberal supply. However, this story indicates that confidence in the operational reliability and high current potential of a linear accelerator was fully established by the operation of the Alvarez machine. In fact, average currents between $1/4$ and 1 microampere of 32 MeV protons could be reliably delivered, with peak currents above 100 microamperes. Once construction of the machine was completed, operation was turned over to a team headed by Robert Watt, who did a superb job of converting this pilot machine into a well-functioning instrument for particle physics.

Operation of the Alvarez machine demonstrated its usefulness for physics, and plans were developed to build a follow-up machine of similar design at the University of Minnesota under the leadership of the late Professor John H. Williams. This machine was brought into successful operation and contributed substantially to our knowledge about proton scattering and other phenomena in the region above 60 MeV. In the meantime, it was decided to discontinue use of the 32 MeV Alvarez machine, since the focus at Berkeley had shifted to energies well above the threshold for producing pi mesons using the 184 inch synchrocyclotron and the 300 MeV synchrotron. A rather controversial decision was made to transplant the Alvarez machine to the University of Southern California. This move was controversial because, clearly, transferring the machine would engender moving and reinstrumentation costs quite comparable to the original effort involved in building the machine. Nevertheless, the transfer was made, and some further operation of the pioneering machine conceived by Luie and constructed under his direction resulted.

Beyond the direct contributions of the 32 MeV Alvarez machine and its "daughter" machine in Minnesota, the proton linear accelerator has seen widespread use as an injector into every high-energy proton synchrotron in the world. For some time there remained uncertainty whether the leading candidate for an accelerator to serve as a "meson factory" should be a proton linear accelerator or various forms of the sector-focused cyclotron. Although high-intensity machines of the latter type have been built very successfully in Canada and Switzerland, the leading machine in this country is now LAMPF. It operates at an energy of 800 MeV with average currents approaching 1 milliampere. Thus, the expectation that the linear accelerator would be a high average current machine— conceived during the fissionable materials breeding days—became a reality. A similar machine is currently under construction in the Soviet Union.

The question is still open as to whether the proton linear accelerator might make a comeback as a source for an extremely high energy proton collider. Recently, studies have been conducted at the Stanford Linear Accelerator Center and at the Nuclear Physics Institute at Novosibirsk exploring the possibilities of very high energy (multi–TeV) electron linear accelerators. The results look encouraging, although much research and development—particularly on components—remains to be done. At these energies even protons can be considered to be fully relativistic, and these design exercises can just as well apply to protons as to electrons. Moreover, it appears that in the multi–TeV range synchrotron radiation becomes a serious obstacle even for proton circular accelerators operating at high magnetic fields. Therefore, the original expectation expressed by Luie in 1945—that at the very highest energies the proton linear accelerator might become the leading proton machine—may yet prove to be correct.

Reprinted from THE REVIEW OF SCIENTIFIC INSTRUMENTS, Vol. 26, No. 2, 111–133, February, 1955
Printed in U. S A.

Berkeley Proton Linear Accelerator*

LUIS W. ALVAREZ, HUGH BRADNER, JACK V. FRANCK, HAYDEN GORDON, J. DONALD GOW, LAURISTON C. MARSHALL,†
FRANK OPPENHEIMER,‡ WOLFGANG K. H. PANOFSKY,§ CHAIM RICHMAN, AND JOHN R. WOODYARD
Radiation Laboratory, Department of Physics, University of California, Berkeley, California
(Received July 8, 1954)

A linear accelerator is now in use which increases the energy of protons from a 4-Mev Van de Graaff injector to a final energy of 31.5 Mev. The accelerator consists of a cavity 40 feet long and 39 inches in diameter, excited at resonance in a longitudinal electric mode with a radio-frequency power of about 2.1×10^6 watts peak at 202.5 Mc. Acceleration is made possible by the introduction of 46 axial "drift tubes" into the cavity, which is designed so that the particles traverse the distance between the centers of successive tubes in one cycle of the rf power. The proton bunches are longitudinally stable as in the synchrotron, and are stabilized transversely by the action of converging fields produced by focusing grids. The electrical cavity is constructed like an inverted airplane fuselage and is supported in a vacuum tank. Power is supplied by 9 high-powered oscillators fed from a pulse generator of the artificial transmission line type. Output currents are 3×10^{-7} ampere, average, and 60 μa, peak. The beam has a diameter of 1 cm and an angular divergence of 10^{-3} radian.

I. INTRODUCTION

1. Historical Summary

THERE was general agreement among physicists before the war that radio-frequency linear accelerators of the Sloan-Lawrence[1] type were of historical interest only. This feeling arose largely because the cyclotron was such a reliable device, with beam intensities far beyond the most optimistic hopes of its originators. At the same time, it was recognized that to make a competitive linear accelerator for protons or deuterons would require far higher power than was then available at very short wavelengths. Although pre-war rf linear accelerators were used to gain our present knowledge of the production of x-rays by high-speed heavy ions, they played no part in increasing our knowledge of the nucleus. (Kinsey[2] reported in 1936, however, that high-speed Li ions impinging on hydrogenous material give the well-known alpha particles first observed by Cockcroft and Walton.)

Interest in linear accelerators was revived toward the end of the war, when vacuum tubes had been developed that could produce megawatts of pulsed rf power down to the microwave range. It had been apparent for some time that there was an upper energy limit for particles accelerated by a cyclotron. The 184-inch cyclotron was originally designed to extend that limit as far as possible, but the goal was only about 100-Mev deuterons, even though a dee voltage in the neighborhood of 1.5×10^6 was to be used.

Because no significant theoretical limit is apparent for linear accelerators, it was felt that they should be reinvestigated as a means of reaching energies in excess of 100 Mev. Similarly, for electrons, the betatron was known to have a practical energy limit only a few times greater than that of the cyclotron. No electron linear accelerators had been built before the war, but a description of the type now operating in a number of laboratories was given in 1941 by D. H. Sloan.[3] (Such accelerators have been treated in detail by several authors,[4] and are not discussed in this article.)

The entire accelerator situation was drastically altered in 1945 by the introduction of the synchrotron concept.[5] In theory this removed the upper limit for

* This work was done under the auspices of the U. S. Atomic Energy Commission.
† Now with the Link-Belt Company, Indianapolis, Indiana.
‡ Now at Pagosa Springs, Colorado.
§ Now with the Department of Physics, Stanford University, Stanford, California.
[1] D. H. Sloan and E. O. Lawrence, Phys. Rev. 38, 2021 (1931).
[2] B. B. Kinsey, Phys. Rev. 50, 386(A) (1936), and private communication.

[3] D. H. Sloan, Patent No. 2,398,162.
[4] W. W. Hansen, Rev. Sci. Instr. 19, 89 (1948).
[5] V. I. Veksler, J. Phys. (U.S.S.R.) 9, 153 (1945); E. M. McMillan, Phys. Rev. 68, 1943 (1945).

cyclotrons, and raised it considerably for betatrons (to about 2000 Mev). But the linear accelerator had other apparent advantages. Although recent studies have shown that the original arguments for undertaking the construction of a proton linear accelerator were not basic, the state of the art at that time did not permit convincing contradiction.

The argument was essentially as follows: The cost of a relativistic magnetic accelerator varies roughly as the cube of the energy, so long as the basic design is merely scaled in its linear dimension, which is proportional to the energy. On the other hand, the cost of a linear accelerator varies directly as the first power of the energy. If these cost-vs-energy curves are plotted on double logarithmic paper, they are straight lines with slopes three and one. There will always be an intersection of the two lines, and for energies greater than the "crossover" energy, the cost of a linear machine will be less than that of the circular machine. Because the cost of either machine is high, it was felt that even though a linear accelerator might be more complex than a synchrocyclotron, the design decision might have to be made on economic grounds.

The new consideration that has altered our thinking in this manner is that beyond a certain energy the magnetic machines can be changed in basic design. Instead of accelerating with constant field and changing radius, we can reverse the two conditions. The importance of this change is that a ring magnet can then be used. This ring-shaped magnetic machine, which is really a proton synchrotron, was proposed by Oliphant in 1944 (before the work of Veksler and McMillan), but for some time the feeling in this country was that, although Oliphant's plan was most attractive in many ways, so many unsolved and serious problems were involved in its practical realization that alternative methods of attaining high-energy protons should be explored. Recent critical examination of the whole problem by two groups in this country has shown that all the problems are solvable. Therefore, both the Brookhaven National Laboratory and the University of California Radiation Laboratory have built such machines for protons.

This factor alters the economic conclusions originally reached because not long after the "magnetic cost line" has crossed the "linear cost line," the former breaks off (when the ring is adopted) and starts over at a much lower cost value for the same energy, and then rises again with slope equal to three. The arguments of dimensional analysis are still theoretically sound; there will still be a crossover at higher energies. But long before this point is reached, the cost of either machine is so high that both are excluded on economic grounds.

In the fall of 1945, we started the design of a "pilot model" proton linear accelerator to explore the possibilities of the method. No plans were made at that time for extending the length of the machine beyond the

original 40 feet, but it was assumed that this question would be explored after successful operation of the first section. If it subsequently appeared wise to continue on to higher energies, such a decision could then be made. In view of the present status of the proton synchrotron at Berkeley, no extension of this machine is planned here. It should be pointed out that we do not at present know of any technical reason why this extension should be impossible or even difficult. In fact, J. H. Williams at Minnesota is now building a 65-Mev proton linear accelerator of the same type which is 100 feet long, and the British Atomic Energy Establishment is actively engaged in the design of a 600-Mev proton linear accelerator.

For the moment, then, we are concentrating on using the 40-foot accelerator as a physics research tool. It has a number of advantages over the synchrocyclotron for that purpose. In particular, the characteristics of the external beam of the linear accelerator are most attractive. Eighty-five percent of the beam is concentrated within a 3-mm circle; the angular divergence of the beam is approximately 10^{-3} radian, and the energy homogeneity is about 3×10^{-3}. The average external-beam current is about 10^3 times that of the 184-inch cyclotron, and the average external-beam current density is about 10^6 times as great.

2. General Design Characteristics

These general design characteristics pertaining to linear accelerators have been discussed in detail by Slater.[6] In particular, Slater points out that the attainable voltage V of a linear accelerator of length l being fed at a peak power P is given by

$$V = K P^{\frac{1}{2}} l^{\frac{1}{2}} \lambda^{-\frac{1}{2}}, \qquad (1)$$

where λ is the free-space wavelength. The constant K depends on the detailed geometrical arrangement, and numerical values for the Berkeley accelerator will be discussed later. A linear accelerator contains essentially different types of equipment whose cost is proportional to either length or peak power or average power or energy per pulse. All these factors affect the choice of wavelength, length of the machine, and duty cycle of operation. The pulse length of a resonant accelerator, τ, is given by

$$\tau \propto Q \lambda \propto \lambda^{\frac{3}{2}}, \qquad (2)$$

in order to permit full build-up of field. In terms of energy per pulse, U, Eq. (1) therefore becomes

$$V = K U^{\frac{1}{2}} l^{\frac{1}{2}} \lambda^{-1}. \qquad (3)$$

The wavelength dependence of power requirement for a given voltage and length is therefore small ($\lambda^{\frac{1}{2}}$), but the wavelength dependence of energy per pulse, on which the cost of the pulse equipment depends, is large (λ^2). From the latter point of view, it is particularly

---

[6] J. C. Slater, Revs. Modern Phys. **20**, 473 (1948).

desirable to choose a small wavelength. However, there are several reasons for using long wavelengths, so one must make a compromise. As is shown in the next section on cavity modes, the difficulty of maintaining the proper field pattern in a standing-wave accelerator increases as the square of the ratio between cavity length and wavelength. (This ratio is called the electrical length of the cavity.) A second reason for using long wavelength comes from the necessity for a sufficiently large drift-tube aperture. This reason dominated our thinking when foil focusing was to be used, but now that grid focusing is used, aperture is of lesser importance than electrical length.

The present accelerator is designed with $\lambda = 150$ cm (200 megacycles). An important reason for choice of this wavelength was the availability of surplus radar equipment in this range. Although this radar equipment is no longer used to power the accelerator, it appears that the original choice of wavelength was very fortunate.

It might appear that the balance between length and power would be made by matching the cost of the length-proportional items and the power- (or energy-) proportional items. Actually, in this and all other machines under design, as much power per unit length is applied as is feasible in consideration of electrical breakdown or available power sources. For a machine of the type discussed here, this limitation originally precluded energy gains greater than two to three million electron volts per meter. At higher gradients, x-ray emission from the machine requires extensive shielding, and sparking difficulties are encountered. (The x-ray yield increases about as the sixth power of the gradient.) Recent studies at this laboratory have shown that at energy gains of 3 Mev per meter, surface layers of pump oil are responsible for most of the electron emission.

The choice of duty cycle is dictated by considerations of power consumption, cavity cooling, and tube power dissipation. Increase in repetition rate does not appreciably increase the cost of the pulse equipment; the pulse length is therefore chosen to give a duration of the pulse equal to several "build-up times," and the repetition rate is then determined by available power.

The injection energy chosen for the Berkeley accelerator was 4 Mev. The reason for this choice was twofold: (a) 4 Mev is a reasonable voltage to attain with an electrostatic generator, and the construction of such a machine as a general research tool was desirable

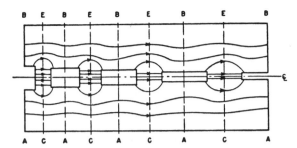

FIG. 2. Linear accelerator produced by introducing drift tubes into cavity excited as in Fig. 1. Division into unit cells.

at this laboratory; (b) at the time of design, it was intended to accomplish focusing by means of thin beryllium foils,[7] and multiple scattering in these foils[8] called for a high injection energy.[9]

II. CAVITY DESIGN

1. Basic Geometry

A linear accelerator for protons starting from low velocities ($\beta = v/c = 0.092$, 4 Mev) cannot be practically constructed on the basis of loaded wave-guide geometry, as is done for electrons.[4] Loading that would reduce the phase velocity to this low figure would lead to excessive rf power loss. For this reason, loading as such is not used. Instead, the phase velocity is made infinite, that is, the electric field is everywhere in the same phase, and small hollow conducting "drift tubes" are provided to shield the protons while the field is in the wrong direction. (The drift tubes cause a small amount of field perturbation, which must be taken into account in designing the resonator.)

The resonant cavity is essentially a long cylindrical cavity operated in the TM_{010} mode, that is, the mode in which an axial electric field without azimuthal or transverse nodes is produced, as shown in Fig. 1. Efficient acceleration in a simple cavity is possible only if its aixal length is shorter than $\beta\lambda/2$, where λ is the free-space wavelength corresponding to the driving frequency, which, except for very low-voltage machines, is not feasible. For this reason, the drift tubes previously mentioned are introduced coaxially in the cylinder, as illustrated in Fig. 2, with the distance between center lines AB of successive drift tubes equal to $\beta\lambda$. If the gap g between drift tubes is short compared to $\beta\lambda/2$, the voltage gain of the particle can be nearly equal to the peak rf voltage developed across the gap. Note that the particle takes one full rf period to travel between midpoints of successive drift tubes. Each drift tube is charged oppositely at each end (no net charge on any tube), and all drift tubes are excited in phase. In the

[7] H. Bradner, Rev. Sci. Instr. **19**, 662 (1948).
[8] R. Serber, Phys. Rev. **73**, 535(A) (1948).
[9] Use of grid focusing makes it possible to use much lower injection energy, permitting a more simple injector. The 10-Mev linear accelerator used as an injector for the Berkeley bevatron is designed to accept protons from a 500-kev Cockcroft-Walton-type machine.

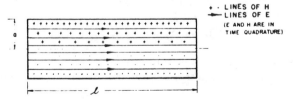

FIG. 1. Long cylindrical cavity excited in TM_{010} mode.

long-wavelength Sloan-Lawrence accelerator,[1] the drift tubes have alternately plus and minus net charges, and the distance between drift-tube midpoints is $\beta\lambda/2$, rather than $\beta\lambda$ as in the cavity accelerator.

The general field geometry in the "unit cell" $ABBA$ is very nearly the geometry of a doubly re-entrant symmetrical cavity excited in the lowest mode, such as is used in klystron resonators, TR boxes, etc. The entire accelerator can therefore be considered as a series of such cells in juxtaposition, repeatedly excited in such a phase that the currents flowing on opposite sides of the joining faces AB (see Figs. 2 and 3) cancel. Such a picture would be exact if each successive cell were identical to the previous one; however, owing to the progressive change in β corresponding to the gain in energy, the field distribution of the actual accelerator does not exactly correspond to what the distribution would be if the walls AB were actually present. For purposes of design, however, experimental and theoretical data based on single-cell-structure models are entirely adequate.

The joining of the unit cells is possible if each cell is accurately tuned to the same frequency. (The tolerance of the tuning of the individual cells is discussed later.) Published design figures[10] on the resonant frequencies of this type of geometry are not of sufficient range that they could be used here; furthermore, the details of the mechanical support structure of the drift tubes and similar deviations from ideal geometry made it necessary to derive the data pertaining to the resonant frequency of the unit cell from models. The resulting data are shown in Fig. 4, using the notation as indicated in Figs. 2 and 3. The data are plotted in terms of dimensionless ratios, to permit easy scaling. These data can be fitted by the empirical equation,

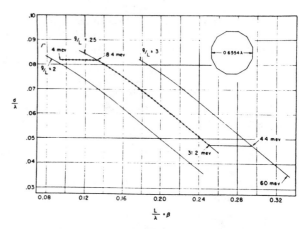

FIG. 4. Results of model tests on resonant frequencies of re-entrant cavities. Dotted curve indicates design points for 32-Mev accelerator, each dot corresponding to a unit cell. [Oppenheimer, Johnston, and Richman, Phys. Rev. 70, 447(A) (1946).]

$$\frac{g}{L} = (-1.271) + (1.63)\frac{D}{\lambda} + (1.096)\frac{L}{\lambda} + (3.58)\frac{d}{\lambda}, \quad (4)$$

in the range of application used here.

It is clear that as the energy increases, and hence also β and L/λ, the diameter d of the drift tube decreases; this decrease will eventually lead to a point where the design becomes impractical because of insufficient beam aperture and excessive curvature at the ends of the drift tubes, with consequent high surface fields. For this reason, the diameter D of the outer cavity must be chosen small enough to conform to this limitation for the highest values of β; if it is chosen too small, the drift tubes on the low-β end will become too large in diameter, with consequent increase in losses. In the 40-foot linear accelerator described here, it was possible to compromise between these two extremes by the choice $D/\lambda = 0.66$. (Note that an unloaded cylindrical cavity is resonant at $D/\lambda = 0.766$.) If low-energy injection had been used it would probably have been impossible to find a satisfactory compromise on the diameter of the cavity that would have permitted the use of "reasonable" drift-tube geometry at the two ends of the cavity. The cure for this would have been the use of a tapered cavity, which would have increased the mechanical problems in fabrication. To build a multicavity accelerator for high energies, one would follow the Minnesota scheme using cavities with smaller diameters in the higher-energy sections.

2. Voltage Gain and Input Power

The voltage gain per unit cell is determined by three factors, (a) geometry, (b) power input, and (c) phase of particle transit.

Let B_0 be the crest value of the magnetic flux

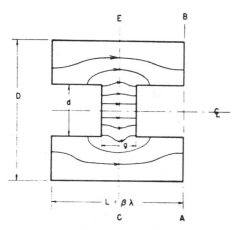

FIG. 3. Fields in "unit cell" of accelerator. Note that a conductor across EC would not change distribution.

[10] Theodore Moreno, *Microwave Transmission Design Data* (McGraw-Hill Book Company, Inc., New York, 1948).

circulating in the unit cell.[11] It can be shown easily that the voltage gain in the unit cell is given by

$$V = \omega B_0 \cos\phi \frac{\sin(\pi g/L)}{\pi g/L} \qquad (5)^{12}$$

if the electric field lines are assumed to be parallel across the gap. Here ϕ is the transit phase angle relative to the phase of a particle crossing the center of the gap at the time of maximum voltage. On the other hand, ωB_0 is given in terms of the power losses in the walls by a relation of the form

$$\omega B_0 = (2PZ_1)^{\frac{1}{2}}, \qquad (6)$$

where Z_1, the shunt impedance/unit cell, is determined by the geometry and by the conductivity of the cavity wall material. For a continuing structure, Z_1 is proportional to the length; e.g., for a cylindrical unloaded cavity of length l excited in the TM_{010} mode, the shunt impedance is given (neglecting end losses) by

$$Z_0 = 0.367 l(\sigma f)^{\frac{1}{2}} \text{ ohms}, \qquad (7)$$

where σ is the conductivity of the walls. To judge the merits of comparative drift-tube structures and to estimate the voltage gain as a function of power input, it is therefore necessary to evaluate the shunt impedance per unit length (Z_1/l) of the loaded cavity relative to the similar quantity for the unloaded cylindrical structure as given above. This evaluation was done semiempirically by mapping the magnetic field B across the azimuthal plane of the cavity, by means of an exploring loop, as shown in Fig. 5. The shunt impedance Z_1 and the Q are then obtained by numerical evaluation of the integrals

$$Z_1 = \left(\frac{\mu_0}{\epsilon_0}\right)^{\frac{1}{2}} \frac{4\pi}{\lambda\delta} \frac{\left(\int\limits_{\text{Azimuthal plane}} B dA\right)^2}{\int\limits_{\text{Conductor surfaces}} B^2 dA'} \qquad (8)$$

and

$$Q = \frac{4\pi}{\delta} \frac{\int\limits_{\text{Azimuthal plane}} B^2 r dA}{\int\limits_{\text{Conductor surfaces}} B^2 dA'}, \qquad (9)$$

TABLE I. Q and shunt impedance for 40-foot accelerator.

	Q	Z (ohms)	Power for $V = 28$ Mev[a] (watts)
From flux plot	106 000	457×10^6	1.40×10^6
From Q measurement	72 000	311×10^6	2.06×10^6

[a] Taking $\Phi_s = -30°$, $g/l = 0.25$ [see Eq. (5)].

[11] Since the magnetic field is everywhere in phase in the mode used here, B_0 can be defined uniquely as the surface integral of the crest value of the magnetic induction.

[12] mks units are used throughout.

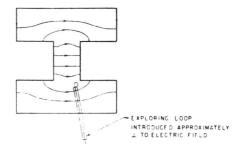

FIG. 5. Schematic diagram showing magnetic field mapping of the unit cell. The field maps permit evaluation of the integrals in Eqs. (8) and (9).

where $\delta = (\pi f \mu \sigma)^{-\frac{1}{2}}$ is the skin depth. The resulting values of Z_1, Z_0, and Q require an additional correction factor of approximately $1/(1+a/l)$ to correct for end losses in the cavity.

The loss values arrived at by the flux-plotting method are lower limits, because losses in the drift-tube support structure, pumping slots, joints, etc., are not taken into account. A simple measurement of the experimental Q of the cavity, however, gives a measurement of the true shunt impedance also, since Z and Q are reduced in the same ratio. The final values, arrived at by experimental measurement of Q, are given in Table I, computed for the entire accelerator.

3. Modes of a Long Cavity

In designing the accelerator cavity from data obtained on unit cells, one has to make a definite assumption pertaining to the voltage gain per cell. In order to join the cells without partition, the wall currents—and therefore the wall magnetic fields—must be continuous. Magnetic flux plots showed that the ratio of total flux per unit length (and therefore voltage gain per unit length) to the magnetic field at the edge varies by only 20 percent from the injection end to the output end of the accelerator (the ratio being higher at the high-voltage end). It was therefore decided to design the accelerator for constant voltage gain per unit length, which results in a 20-percent "taper" of magnetic field along the cavity walls. Constancy of the voltage gain per unit length is of course desirable also, in order to equalize the surface gradient along the accelerator and to reduce the tendency to spark.

In order to assure equality of the mean electric field in a coupled structure involving 47 individual resonators, the individual resonators must have very accurately the same resonant frequency. This can only partially be assured by the model "unit cell" measurements referred to above, because (a) the extension from unit cells to the long accelerator of varying units is not exact, (b) slight mechanical changes from the model geometry interfere with exact transfer of the data, and (c) the accuracy (approximately 0.05 percent) of the model frequency measure-

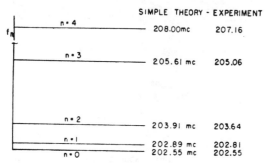

FIG. 6. Mode spectrum of linear accelerator cavity. n=number of longitudinal modes in the field pattern.

ments is not quite sufficient. For this reason it is necessary to adjust the final field distribution to the value required for acceleration by corrections applied to the cavity as a whole.

The behavior of the cavity can best be described by means of its mode spectrum. The operating mode of the machine is on the lower edge of the pass band of the cavity used as a wave guide, and therefore the mode separation varies quadratically as the limit is approached. Specifically, for an unloaded cavity, the spectrum is given by

$$f_n = f_0 \left[1 + \left(\frac{nc}{2f_0 L} \right)^2 \right]^{\frac{1}{2}}, \qquad (10)$$

where f_0 is the frequency of the lowest mode. The low end of the spectrum of the 40-foot accelerator cavity is plotted in Fig. 6. Owing to the small mode separation (0.17 percent) near the limit, two problems have to be considered: (a) the possibility of "mode jumping" to an adjacent mode, and (b) the tolerances in geometry required to assure that the lowest mode has the desired field pattern. These appear at first sight to be independent, but it is shown shortly that they are different aspects of the same problem. The first is solvable by proper disposition of the exciting oscillators or by careful tuning, and is discussed later. The second can be analyzed by a simple perturbation calculation in which we assume that the actual field distribution excited in a single mode can be expanded in terms of the modes of a cavity whose mode distribution is the desired one,[13] hereafter called the "ideal" cavity.

A detailed derivation of the perturbation theory equations[14] would be out of place in this paper; instead, an outline of the method of attack is given together with the final working equation. The long cavity is treated as a hollow structure of circular cross section; the radius varies as a function of z, and is given by the Fourier series

$$a_z = a_0 \left\{ 1 + \sum_{i=1}^{\infty} P_i \cos \frac{i\pi z}{L} \right\}. \qquad (11)$$

[13] Such an expansion is always possible since the normal modes in a cavity form a complete orthogonal set of functions.
[14] Panofsky, Richman, and Oppenheimer, Phys. Rev. 73, 535(A) (1948).

In an actual linear accelerator cavity, the natural frequencies of the various unit cells vary as a function of z. The connection between the theories of the two types of cavities comes from the fact that in the hollow cavity the natural frequency of a given section is related to the radius by the equation

$$f_s = 0.383c/a_s. \qquad (12)$$

We express the axial electric field in the hollow cavity as

$$E_z = J_0(k_z r) e^{j\omega t} E_0 \left\{ 1 + \sum_{n=1}^{\infty} \epsilon_n \cos \frac{n\pi z}{L} \right\}. \qquad (13)$$

This field must satisfy the differential equation

$$\frac{1}{r} \frac{\partial}{\partial r} \left\{ r \frac{\partial E}{\partial r} \right\} + \frac{\partial^2 E}{\partial z^2} + \frac{\omega^2 E}{c^2} \left(1 + \frac{j}{Q} \right) = 0. \qquad (14)$$

The boundary conditions are introduced through the requirement that $J_0(k_z a_z) = 0$ (tangential electric field=0 at a conductor). This requirement introduces Eq. (11) into the problem.

The solution of the problem shows that the coefficient ϵ_n depends only on the coefficient $P_i = P_n$; a cavity with only an nth-harmonic radius variation will have only an nth-harmonic perturbation in the field pattern of its operating field. The magnitude of ϵ_n is related to that of P_n by

$$\epsilon_n = (8N^2/n^2) P_n, \qquad (15)$$

where N is the "electrical length" of the cavity, L/λ. To show the relation between field distortion and mode separation in an ideal cavity, Eq. (15) may be combined with Eq. (10) to yield

$$\epsilon_n = \frac{f_0}{f_n - f_0} P_n. \qquad (16)$$

(All these equations are approximations that are good for low modes in long cavities.)

N is 8.3 for the present accelerator, so

$$\epsilon_n = 550 P_n/n^2. \qquad (17)$$

It is apparent from Eq. (17) that a first-harmonic tuning error of a given amplitude, $-P_1$,[15] would give rise to a first-harmonic field error 550 times as great. The "amplification factor," $8N^2/n^2$, is of such a form that tuning errors in the high harmonics do not distort the field patterns appreciably. As an example of the high degree of accuracy with which the various cells in the accelerator must be tuned, we calculate the permissible first-harmonic tuning error, if the field is to be "flat" to ±5 percent ($\epsilon_1 = 0.05$); $P_n = 0.009$ percent. This means that the average frequencies of the two halves of the accelerator must be the same to

[15] It should be noted that P_n was defined as the Fourier coefficient of the radius variation; by Eq. (12), it has opposite sign when used as the tuning-error coefficient.

within about one part in 10^4. (The corresponding tolerance on drift-tube lengths is about 0.001 inch!) This is too strict a tolerance to be met mechanically, even if the design data were accurately enough known. In practice, "end tuners," or variable-length half drift tubes at the two ends of the cavity, are used to alter P_1 and P_2. Such alterations can be made by motor controls while the cavity is excited at its operating field strength. These adjustments are stable for weeks during operation, and need be made again only after the accelerator has been opened for adjustments.

Keeping the accelerator in adjustment is by no means so serious a problem as first putting it into such condition that Eq. (15) may be used as a guide for the final adjustments. If the two halves of the cavity were detuned from each other by one percent, the value of ϵ_1 would be greater than unity. The equations were developed from a perturbation theory ($\epsilon_n \ll 1$), and are no longer applicable when $P_1 \approx 1$ percent. But such errors were present in the newly constructed accelerator cavity, and they had first to be eliminated by another procedure. A movable "end wall" was placed between two centrally located drift tubes. The resonant frequencies of the two halves of the cavity were then measured independently. The positions of the drift tubes were altered by small amounts so that the two halves of the cavity had the same natural frequency. The same procedure was repeated for quarter-length segments of the accelerator.

After these preliminary adjustments, the cavity could be excited to give a perturbed "zeroth mode" field pattern [Fig. 7(a)]. This field pattern was Fourier analyzed, and the coefficients ϵ_n were determined. From Eq. (16) the tuning-error coefficients P_n were determined. When these coefficients were used a Fourier synthesis yielded the tuning error as a function of z [Fig. 7(c)]. Model work yielded df/dl, the variation in natural frequency of a given unit cell with drift-tube length. The tuning-error curve together with df/dl gave a "drift-tube-length error curve." Mechanical shims were constructed according to this last curve, and inserted under the removable ends of the individual drift tubes. The new field pattern, which was acceptably flat, is shown in Fig. 7(b).

The above discussion has been so simplified that the axial electric field is treated as proportional to the surface magnetic field. This is very nearly so, but the constant of proportionality changes slowly along the length of the accelerator. It is a simple matter to "tilt the field pattern" under operating conditions, by the use of the end tuners, so one can ignore the small lack of constancy of $[E_z/H_\phi](z)$. In practice, one tilts the field to yield a proton beam at the lowest-threshold electric field in the cavity.

If the accelerator cavity were to be made 3.15 times as long as it is, N^2 would be 10 times as great. This would cut the frequency separation of the lowest two modes to one-tenth of its present value (340 kc/10 = 34

kc). The modes would still be distinct, as the width of the tuning curve is of order $f/Q = 2 \times 10^5$ kc/$8 \times 10^4 \approx 3$ kc. Mode jumping would therefore not be a problem, even though the modes were 10 times as close. However, a cavity 10 times as long as the present one would have problems connected with mode jumping. (One could not be assured that the field pattern would be constant

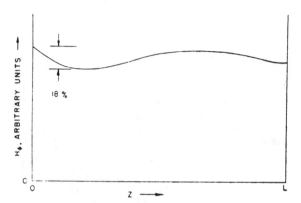

FIG. 7(a). Application of Fourier analysis to correction of cavity field [original $H_\phi(z)$].

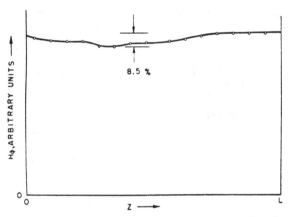

FIG. 7(b). Application of Fourier analysis to correction of cavity field [$H_\phi(z)$ after introduction of shims].

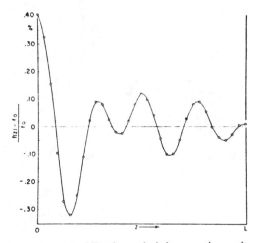

FIG. 7(c). Effect of Fourier analysis in correcting cavity.

in time, because the tuning curves of the lowest modes would overlap.)

The effect of increased length on the field pattern in the operating mode also goes as N^2, or inversely as the mode separation. If the cavity were $10^{\frac{1}{2}}$ times its present length, the tolerances could probably be met by attention to detail, but if the cavity were 10 times as long, the problem of maintaining a flat mode would probably be too severe. Mode-jumping difficulties would come in at about the same length. It might be thought that these problems could be circumvented by making several independent cavities, tightly coupled. The same problems would exist in that case, however. If accelerators of this type are to be built with large values of N, they will probably have several independent cavities fed by a master oscillator and power amplifiers, rather than by the self-excited oscillator used with our accelerator. The independent cavities could easily be kept tuned to the master oscillator by a servo system, and then individual short lengths would make the field-flatness problem easy of solution. The Minnesota accelerator is designed according to these ideas.

4. Experimental Field Plots

A three-dimensional picture of the complete field plot as obtained by the magnetic loop method is shown in Fig. 8. Note that the fields are a mixture of fields of the coaxial type ($1/r$ dependence near the drift tubes) and the TM_{01} type ($[J_1(kr)]$ dependence in the gaps).

For use in transit-time and focusing calculations, the electric field $E(z)$ along the axis is needed. As is explained later, the entrance of each drift tube is closed with a focusing grid and the exit end is open. For this reason, theoretical analysis of the field is

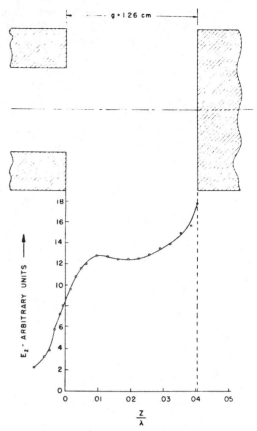

Fig. 9. Typical axial electric field plot along drift-tube axis.

difficult and an experimental procedure is used. This procedure consists in measuring the frequency shift produced by placing a small metallic object of volume δV at various points in the field. The disturbed frequency is given[16-18] by

$$f^2 = f_0^2 \left\{ 1 + A \int_{\delta V} (H^2 - E^2) dv \right\}, \qquad (18)$$

where A is a constant depending on the geometry of the metallic object.

H^2 and E^2 are normalized to unity over the total volume of the cavity. If the measurement is made along the axis, $H = 0$ and

$$\frac{f^2 - f_0^2}{f_0^2} = - \frac{2K \int_{\delta V} E^2 dV}{\int_{\text{volume of cavity}} E^2 dV}, \qquad (19)$$

and, if the frequency deviation is small,

$$\Delta f/f = K E^2 \delta_V. \qquad (20)$$

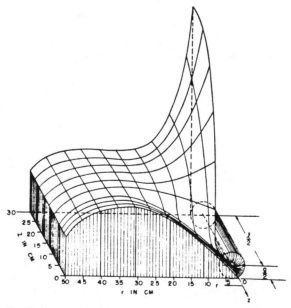

Fig. 8. Contour plot of magnetic field in typical cavity section.

[16] J. C. Slater, Revs. Modern Phys. 18, 441 (1946).
[17] W. W. Hansen and R. F. Post, J. Appl. Phys. 19, 1059 (1948).
[18] L. B. Mullett, "Perturbation of a resonator," AERE-G/R-853 (Harwell, England, February 1952).

An axial field plot for a typical drift-tube geometry is shown in Fig. 9. These data were taken on a 1000-Mc scale model using a heterodyne frequency-measuring method. The metallic sphere was supported on a stretched thread.

III. BEAM DYNAMICS[19]

1. General Equations of Motion

A particle of charge e and rest mass M_0, moving in a linear accelerator, is acted on by both radial and longitudinal forces. The longitudinal forces are caused by the axial component of the radiofrequency electric field; the transverse forces are caused by (a) the transverse radio-frequency electric field and (b) by the radio-frequency magnetic field.

In order that the motion may be known precisely, the electric field components $E_z(r,z,t)$, $E_r(r,z,t)$, and the rf magnetic field $B_\phi(r,z,t)$ have to be known. For a sinusoidal time variation the equations of motion are

$$\frac{d}{dt}\left[\frac{M_0}{(1-\beta^2)^{\frac{1}{2}}}\frac{dz}{dt}\right]=e\left[E_z^0(r,z,t)\cos(\omega t+\Phi)\right.$$
$$\left.+B_\phi^0(r,z)\frac{dr}{dt}\sin(\omega t+\Phi)\right], \quad (21)$$

$$\frac{d}{dt}\left[\frac{M_0}{(1-\beta^2)^{\frac{1}{2}}}\frac{dr}{dt}\right]=e\left[E_r^0(r,z)\cos(\omega t+\Phi)\right.$$
$$\left.-B_\phi^0(r,z)\frac{dz}{dt}\sin(\omega t+\Phi)\right]. \quad (22)$$

The superscript ^{0} denotes the amplitude of the respective fields. If exact results are needed one has to integrate these equations numerically using empirical fields. Many general facts about the nature of the motion can be learned, however, without exact integration.

In general, as Eqs. (21) and (22) show, the radial and longitudinal motion of a particle on the linear accelerator are coupled. This coupling is important if the energy gain per gap depends markedly on the radial position, which would be the case if one had relatively large drift-tube apertures; also, in the early stages of the accelerator where the phase motion (z motion) has not yet damped out, there is considerable coupling between the radial and axial motions. If the drift-tube apertures are small, which is the case of interest, one can find some reasonable approximations to the phase motion by integrating the equation at constant r. Because the resulting phase motion damps rapidly with N, it is also possible to arrive at a useful solution for radial motion.

[19] For details of the calculations presented in this section see Wolfgang K. H. Panofsky, University of California Radiation Laboratory Report No. UCRL-1216 (February, 1951).

2. Synchronous Conditions

The basic geometry of the linear accelerator is shown in Fig. 10. At a given time the electric fields are everywhere in phase and the magnetic field is in time quadrature with the electric field. Let L_n be the "repeat length" of the nth gap and g_n be the gap length. In the crossing of the nth gap the velocity of the particle changes from $c\beta_{n-1}$ to $c\beta_n$ (Fig. 10). The coordinate z is measured from the "electrical center" of each gap; this center is defined by the condition

$$\int_L E_z(z)\sin\left(\frac{2\pi z}{L}\right)dz=0. \quad (23)$$

For a symmetrical gap, this definition corresponds to the geometrical center. The phase of a particle Φ_n is the number of radians in time by which the particle crosses the electrical center of the nth gap relative to the time at which the electric field is at its maximum value. Let

$$\Phi_n>0 \quad (24)$$

correspond to a particle crossing after the maximum field has been reached and

$$\Phi_n<0 \quad (25)$$

to a particle crossing before. These definitions agree with Eqs. (21) and (22).

In order to simplify the discussion we assume that the machine is constructed so that for certain injection conditions the phase Φ_n of the particle at each gap is independent of n. This particle is called a synchronous particle and all quantities associated with this particle—energy, momentum, etc.—are denoted by a subscript s. The phase angle Φ_s is called the synchronous phase. In principle, the machine cannot be designed to have a synchronous particle without knowing the motion; the motion cannot be determined without knowing the fields in the machine. In particular, if the fractional velocity change per gap is large, the design can only be done by successive approximation. However, if the fractional velocity change is small, then the synchronous condition is

$$\frac{L_n}{\lambda}=(\beta_{n-1,s}+\beta_{n,s})/2. \quad (26)$$

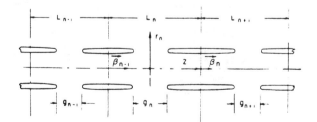

FIG. 10. Basic geometry of the linear accelerator.

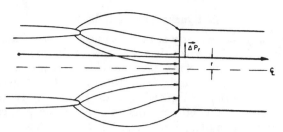

FIG. 11. Electric field distribution between drift tubes with grid or foil focusing.

The synchronous particle increases its total relativistic energy by

$$W_{n,s} - W_{n-1,s} = \int eE_z{}^0 \cos\left(\frac{\omega z}{V_s} + \Phi_s\right) dz. \quad (27)$$

V_s is the synchronous velocity. In general, using the definition of the electrical center, we can write Eq. (27) as

$$W_{n,s} - W_{n-1,s} = e\lambda T E_0\left(\frac{\beta_{n,s} + \beta_{n-1,s}}{2}\right)\cos\Phi_s \quad (28)$$

$$= eT E_0 L_n \cos\Phi_s,$$

where

$$E_0 = \int_{gap} E_z{}^0(z)\,dz \Big/ \int_{gap} dz \quad (29)$$

is the mean effective field, and

$$T = \int E_z{}^0(z) \cos\left(\frac{2\pi z}{L_n}\right) dz \Big/ \int E_z{}^0(z)\,dz \quad (30)$$

is the "transit-time factor." For a "square-wave" field that is uniform in the gap and zero in the drift tubes

$$T = \sin\left(\frac{\pi g_n}{L_n}\right) \Big/ \frac{\pi g_n}{L_n}. \quad (31)$$

If one also considers the radial variation of T, one finds that for the square-wave field and a drift-tube bore of $2A_n$ one has

$$T = \frac{I_0\left(\dfrac{2\pi r}{L_n}\right)}{I_0\left(\dfrac{2\pi A_n}{L_n}\right)} \frac{\sin\dfrac{\pi g_n}{L_n}}{\dfrac{\pi g_n}{L_n}}, \quad (32)$$

where I_0 is the Bessel function of zero order of the imaginary argument.

The additional factor is due to field penetration into the drift tubes, and is a factor producing coupling between the radial and axial motions.

A basic design parameter of the machine is the energy gain per wavelength in $M_0 c^2$ units

$$W_\lambda = \frac{eE_0 T\lambda}{M_0 c^2}\cos\Phi_s. \quad (33)$$

In terms of the parameter, Eq. (28) can be written in the nonrelativistic (nr) form as

$$\beta_{n,s} - \beta_{n-1,s} = W_\lambda. \quad (34,\text{nr})$$

In the relativistic range the fractional changes in velocity are small, and therefore if we write $W_{n+1,s} - W_{n,s} = \Delta W_{n,s}$ it follows that the momentum increment $\Delta P_{n,s} = \Delta W_{n,s}/V_{n,s}$, and therefore it follows that relativistically

$$P_{n,s} - P_{n-1,s} = M_0 c W_\lambda, \quad (34)$$

which includes Eq. (34,nr). From Eq. (34) it follows that *the momentum is a linear function of the number of drift tubes.* Hence, we can write for the synchronous particle

$$P_{n,s}/M_0 c = (n + n_0) W_\lambda, \quad (35)$$

$$W_{n,s}/M_0 c^2 = [1 + (n + n_0)^2 W_\lambda^2]^{\frac{1}{2}}. \quad (36)$$

Here n is taken to make $n=1$ the first gap of the machine and n_0 is the "effective number of gaps" corresponding to the injector.

3. General Stability Considerations for a "Long" Accelerator

(a) Types of Stability

To obtain satisfactory operation for a "long" linear accelerator, it is clearly necessary that the orbits be stable in phase and also stable radially. What length of such an accelerator would be considered "long" in this sense depends of course on the tolerances for injection conditions, voltage gradient, etc., that can be held. We show later that the periods of the various oscillations depend on the number $N = n + n_0$, i.e., the total effective number of drift tubes, including the injector. A linear accelerator of this type is "long" in the sense of requiring stability if it increases the momentum of the injected particles by a large factor. A large injection voltage thus tends to make an accelerator effectively "short."

Phase stability is produced in a linear accelerator if a late particle receives a larger degree of acceleration. This, in the case of a linear accelerator, means that the particle traverses the center of each gap at a time when the field is increasing.[20] Specifically, the condition for phase stability in a field $E_z(z, \omega t)$ is

$$\frac{\partial}{\partial\Phi}\left\{\int E_z\left(z, \frac{\omega z}{V} + \Phi\right) dz\right\} > 0. \quad (37)$$

The conditions for radial stability are more complicated. Focusing is obtained by the following mechanisms: (1) velocity focusing, sometimes called electrostatic or second-order focusing; (2) phase-dependent focusing; and (3) focusing produced by charges or currents contained within the beam.

[20] Note that this is the inverse of the condition pertaining to phase stability of a circular accelerator.

(b) Incompatibility of Simultaneous Radial and Phase Stability

If no charge is contained in the beam, a particle crossing a gap crosses as many lines directed towards the axis as away from the axis. A net radial momentum is thus produced if (a) the particle changes its velocity when crossing the gap, and (b) if the field varies in time. The former mechanism is the one that accounts for the focusing in electrostatic lenses. In the accelerator, this effect is important only in the first few gaps of a machine with low injection energy. The second effect rapidly becomes dominant in the later gaps. It is clear that the condition for phase focusing is that the field be decreasing during the passage of the particle across the gap. This condition appears incompatible with the phase stability condition expressed by Eq. (37). McMillan[21] has shown that this disagreement is a fundamental one, and cannot be removed by artifice in geometry. (McMillan's proof does not deal with the case wherein radial forces are introduced by fields other than the accelerating field, or the case involving periodic changes in the synchronous phase angle.)

(c) Radial Oscillations

The incompatibility between radial and phase stability can be removed if (1) the velocity of the particle changes appreciably in crossing the gap, or (2) the entrance to the next drift tube is closed by a grid or foil (Fig. 11).[22]

It can be shown that the change in velocity can account for a small region of phase stability without grids or foils. The analysis shows, however, that even if the particle crosses the center of the gap at the crest of the rf wave, the times of passing the entrance and exit of the gap are not symmetrical with respect to the time of passing the center, as the particle spends more time approaching the center than leaving the center. Owing to the time variation of the field, the focusing field at the entrance is effectively weaker than the defocusing field at the exit. This effect increases with the gap length and counteracts the effects of velocity focusing. In fact there is a critical gap length beyond which velocity focusing becomes negative (defocusing).

The second method of achieving radial and phase stability, using grids and foils, is more important in actual accelerators. The field configuration of Fig. 11 obviously gives a net inward momentum change to a particle accelerated across the gap.

It can be shown that this momentum change is

given by

$$\Delta P_r = -\frac{er}{2V^2}\left\{ VE_F - (1-\beta^2)\omega \right.$$

$$\left. \times \frac{\partial}{\partial\Phi}\int E_z\left(z, \frac{\omega z}{V}+\Phi\right)dz \right\}. \quad (38)$$

E_F is the electric field at the foil at the time of passage of the particle through the foil. This term can be evaluated only for a particular field. Calculations for the "square-wave" field show that foil or grid focusing gives rise to stable radial oscillation in the phase region

$$-\Phi < \pi/2\left(1-\frac{2g_n}{L_n}\right).$$

Figure 12 shows the region of stable focusing and phase stability for $g_n/L_n = 0.25$; the motion is completely stable for $-\pi/4 < \Phi < 0$. If we have $g_n/L_n > \frac{1}{2}$, foil or grid focusing is ineffective. An asymptotic solution for the radial oscillation can be obtained for large n

$$r_n \propto \beta_n^{\frac{1}{4}}\exp\left[\pm i\int^n \frac{K^{\frac{1}{2}}(\Phi)dn}{(n+n_0)^{\frac{1}{2}}[1+(n+n_0)^2|V\lambda^2]^{\frac{1}{4}}}\right], \quad (39)$$

$$K = \frac{\pi}{2\sin(\pi g_n/L_n)}$$

$$\times \frac{\cos\left(\frac{\pi g_n}{L_n}-\Phi_n\right)-2\beta_s^2\sin\frac{\pi g_n}{L_n}\sin\Phi_n}{\cos\Phi_n}. \quad (40)$$

Nonrelativistically one can obtain exactly, for synchronous orbits,

$$r_n = N^{\frac{1}{4}}\{AJ_1[2(K_sN)^{\frac{1}{2}}]+BY_1[2(K_sN)^{\frac{1}{2}}]\}, \quad (41)$$

where J_1 and Y_1 are Bessel functions, following the

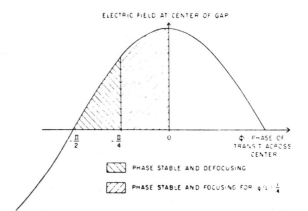

FIG. 12. Phase diagram indicating definition of Φ and regions of stability in grid-focused operation.

[21] E. M. McMillan, Phys. Rev. **80**, 493 (1950).

[22] To these conditions can be added: external focusing devices, such as strong-focusing lenses [J. P. Blewett, Phys. Rev. **88**, 1197 (1952)], and the use of a periodic modulation of the synchronous phase angle [M. L. Good, Phys. Rev. **92**, 538(A) (1953); L. B. Mullett, AERE-GP/M-147 (Harwell, England, 1953)]. The latter, when analyzed in detail, requires a very precise control of radio-frequency field amplitude.

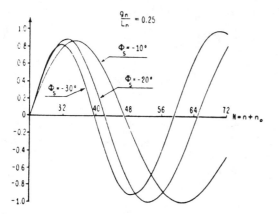

FIG. 13. Radial oscillations in grid- or foil-focused linear accelerator for various synchronous phase angles.

notation of Whittaker and Watson.[23] This has been plotted for $n_0 = 24$ and various values of Φ_s in Fig. 13.

The effects of small-angle multiple scattering on foil focusing have been treated by Serber.[8] The limit on the transparency of the grids that can be used is set by the field concentration on the grid wires.

(d) Phase Oscillations

The nature of the phase oscillations has been investigated analytically, and the following expressions have been obtained which describe the phase motion in the nonrelativistic and extremely relativistic (er) ranges:

$$\phi_n \propto \frac{1}{N^{\frac{1}{4}}} \cos[2(-2\pi \tan\Phi_s)^{\frac{1}{2}} N^{\frac{1}{4}} + \delta] \quad (42,\text{nr})$$

$$\phi_n \propto \frac{1}{N^{\frac{1}{4}}} \cos\{2[(-2\pi \tan\Phi_s)^{\frac{1}{2}}/W_\lambda{}^2] N^{-\frac{1}{4}} + \delta\}, \quad (42,\text{er})$$

where $\phi_n = \Phi_n - \Phi_s$. Thus in the extreme relativistic region the phase motion becomes nonoscillatory, as demanded by the asymptotic constancy of velocity.[24]

Whether or not a particle, injected into the machine, is accelerated through the machine depends on the phase and velocity that it has at the entrance to the machine. Calculations of the phase acceptance of a linear accelerator are summarized in Fig. 14. This is a plot of ϕ_n vs $n_0^{\frac{1}{4}} \Delta W/W_0$, the characteristic parameter for this phenomenon, for various values of ϕ_s. The fractional variation in injection energy is $\Delta W/W_0$. Any particle inside one of the closed curves will be phase stable but not necessarily radially stable. It can be seen that for lower injection energies (smaller n_0) the tolerances on the injection voltage become less critical.

(e) Unstable Operation

It is clear that if a linear accelerator is short enough it can be operated without grids. It will then be either phase unstable or radially unstable. The original rf linear accelerator of Sloan and Lawrence was operated

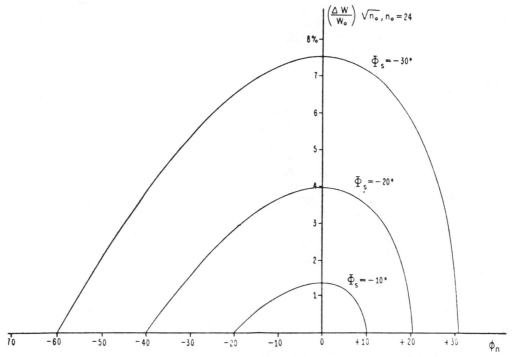

FIG. 14. Phase acceptance of a linear accelerator vs fractional variations in injection energy for various synchronous phase angles.

[23] Whittaker and Watson, *A Course in Modern Analysis* (The Cambridge University Press, New York, 1935).
[24] These equations differ from the form given in the University of California Radiation Laboratory Report No. UCRL-1216, in error, as was kindly pointed out by J. S. Bell (AERE, Harwell, England) to W. K. H. Panofsky.

without grids and was unstable. Experiments with the 40-foot linear accelerator without grids have shown that one can obtain an "unstable" beam of essentially the same magnitude as is obtainable with grids. However, the criticality of adjustment is greatly increased.

IV. MECHANICAL DESIGN

(1) Tank

The resonant cavity of the accelerator is housed in a vacuum tank 40 feet $6\frac{1}{2}\pm\frac{1}{4}$ inches long and $48\frac{1}{2}\pm\frac{1}{4}$ inches in diameter (inside dimensions), made of welded $\frac{1}{2}$-inch boiler plate, with flat steel ends $1\frac{1}{4}$ inches thick. The tank is divided longitudinally 4 inches above center, and the top lid is hinged at four points along the north side to permit the lid to be opened for access. Provisions for vacuum seal of the lid are described in Section IV (3). Opening of the lid is accomplished by a pair of hydraulic cylinders mounted near the middle of the lid and anchored to the pump manifold, which is a protuberance from the bottom half of the tank, on the north side. The entire tank is supported on two pads, each $\frac{1}{4}$ of the way from an end of the tank. The east-end pad rests on a 2-foot-long, 2-inch-diameter steel bar, which can roll east-west on a steel plate that in turn is fastened to the concrete floor. The west-end pad rests on two similar bars, which give some rotational stability to the tank. The 2-foot-long bars are sufficient to keep the tank from falling over, but additional stability against rotation is provided by resting the pump manifold on the floor. The tank is thus mounted so that it can expand freely with temperature changes.

The resonant cavity is also mounted so that it is free to expand and contract with temperature changes independently of the steel tank. In the southeast and southwest corners of the tank, groove pads with the groove pointed east-west are mounted at the height of the middle of the liner. In the north center of the tank there is a similar pad with its groove oriented north-south. The liner is not sufficiently rigid to be supported at just these three points, so 15 spring-loaded pads with hardened flat-ground surfaces are equally spaced around the sides of the tank to distribute the support points. On the liner there are $\frac{3}{8}$-inch bolts with hardened steel balls soldered to the ends by means of which the resonant cavity is supported at the 18 points, of which three constrain its position and motion. This elaborate mounting was installed after it was noticed that the steel vacuum tank warped in places as much as $\frac{1}{2}$ inch when the lid was raised and lowered. Methods of accommodating for this warp in the attachments between liner and tank, i.e., rf transmission lines, water-cooling lines, and end-tuning motors, are indicated elsewhere in this paper.

The tank, opened by the hydraulic lifts, is shown in Fig. 15.

FIG. 15. Tank opened for receiving liner.

(2) Liner Design

With the decision to separate the mechanical and electrical functions of the linear accelerator resonant cavity came the need for the design of an accurate, rigid, and light-weight tubular lining for the vacuum tank. This "liner" became a structure basically similar to a monocoque airplane fuselage of frame, stringer, and sheet construction, save that the sheet surfaces were on the inside. Because it was desired to avoid circumferential joints in the sheet surfaces, and because the cross section was constant, it became practicable to roll copper strip stock into panels of the length of the liner, with flanges and corrugations to serve as longitudinal stringers. Die-formed circumferential frames with a polygonal inside cross section were used, and thus the longitudinal panels could be flat and the necessary dimensional tolerances could be more easily maintained. The cross section of the liner as situated in the tank is diagrammed in Fig. 16. The liner is of dodecagon cross section, $38\frac{1}{8}$ inches across

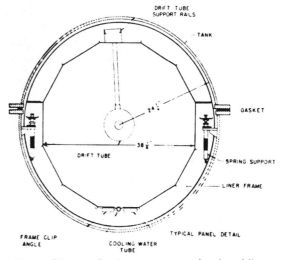

FIG. 16. Diagram showing arrangement of tank and liner.

flats and 480 inches in length, and split longitudinally into two unequal parts, of 150° and 210°, to permit support of the lower part along the horizontal center-lines.

(3) Liner Construction

The longitudinal copper panels were rolled from a copper strip 0.032 inch thick by 13 inches wide by 40 feet long to have a flange turned up at each edge and to have a semicircular channel in the center to receive a $\frac{5}{8}$-inch copper cooling-water tube. A total of 40 "pump out" slots one-half inch wide by 12 inches long were used, with the slots strapped every 4 inches. Thirty-three circumferential frames were made in two parts from 0.072-inch 24 ST aluminum alloy. The internal flange, to which the copper was attached, was made of short lengths of extruded angle riveted to the frame. Clearance cutouts for the cooling tubes and drift-tube supports were provided and reinforced where necessary by attached angles. The frames were finished with an aircraft zinc chromate primer.

Support rails for the drift tubes were run from end to end of the liner through cutouts in the frames. These rails were interrupted every 30 inches to allow for differential thermal expansion. Openings for the drift-tube stems could then be cut through the copper panel at any point without interference with frame position. Support of each drift tube was accomplished by two clamp plates that were placed on each side of the salient flange of the angles that formed the rail (Fig. 17). Three-inch-diameter holes were punched in the end and the side of the liner, opposite glass viewing ports in the vacuum tank. These allow one to observe sparking and other related phenomena in the liner.

Vertical aluminum braces are attached between the end of the liner and the "end-tuner" structure, to provide rigidity to the tuner. An end tuner consists of a drift tube extending into the resonant·cavity from the end of the liner, with length controllable by a worm gear driven by a flexible shaft leading to a motor outside the vacuum tank. Electrical contact is made between this drift tube and the end of the liner by means of a tight-fitting slotted collar of silver-plated steel, fitting around the drift tube and bolted to the liner. The

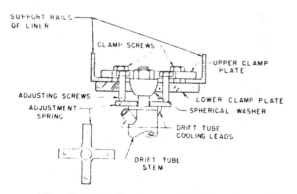

FIG. 17. Drift-tube clamps and arrangements.

west-end (entrance) tuner drift tube is 4.750 inches in diameter, and adjustable in length from 2 inches to 4 inches. The east-end tuner drift tube is 2.750 inches in diameter and is adjustable from 4 inches to 7 inches in length.

(4) Drift-Tube Construction and Support

The drift tubes are basically cylinders of varying length and diameter, each supported by a single stem perpendicular to the liner axis at the center line of each drift tube. The drift-tube diameter varies from $4\frac{3}{4}$ inches to $2\frac{3}{4}$ inches, the first eleven drift tubes being constant at $4\frac{3}{4}$ inches, and the remaining 35 drift tubes diminishing to $2\frac{3}{4}$ inches, in steps of approximately 0.060 inch. The drift-tube lengths vary from about $4\frac{3}{8}$ inches for the first drift tube to 11 inches for the last drift tube. The drift-tube body is made of a copper tube, with the end at the beam exit made from a copper plate hard-soldered into the tube, and with a threaded ring in the opposite end. Into this threaded ring is screwed a cap, which in turn receives a grid holder. The exit end of the drift tube has a re-entrant opening, formed by a brass tail tube about 3 inches in length, and varying from 1 inch inside diameter to $1\frac{1}{2}$ inches inside diameter, for the range of drift-tube sizes. All external edges are uniformly rounded with a radius of $\frac{3}{8}$ inch. The threaded cap was originally designed to be screwed into the drift tube body after the grid holder had been inserted from the inside. On the initial runs, serious sparking occurred across the contact surface between the drift-tube cap and body, even though special care had been taken to insure high contact pressures at this point. Thus it became necessary to solder the drift-tube cap to the body with a low-temperature eutectic alloy, and to redesign the grid holder so that it could be inserted from the front of the drift tube. The threaded construction, however, did permit change of the drift-tube lengths by shims after the voltage distribution measurements had been completed.

The drift-tube stems are made of one-inch-diameter brass tubing, soft-soldered into a reamed boss on the transverse center line of the drift tube. Through this stem is passed a quarter-inch-diameter copper tube, which makes a loop around the inside of the drift-tube body, to which it is soft-soldered, and then returns through the stem for circulation of cooling water. The use of brass stems was found to be a mistake, and it was subsequently found necessary to silver-plate the drift-tube body and stem, to reduce rf losses in the brass stem and tail tube. The drift-tube stem is closed by a plug which carries a threaded extension. The drift-tube components are diagrammed in Fig. 18.

Each drift tube is supported (see Fig. 17) by a pair of plates clamped on the salient flanges of the drift-tube support rails supported by the linear frames. One of these clamp plates has a large clearance hole, while the other is provided with a seat for a spherical washer and with four tapped holes uniformly spaced about the seat. The threaded extension of the drift-tube stem passes through a cross of steel, heat-treated to a spring temper, and then through a hemispherical washer, beyond which it is terminated with a nut. With the hemispherical washer resting in the seat provided in the clamp plate, and with four cap screws through the threaded holes bearing against the arms of the cross, the drift tube is held firmly in position, and can be adjusted along the three coordinate axes by means of the two pairs of screws on opposite sides of the stem, and by means of the nut on the threaded extension.

(5) Liner and Drift-Tube Cooling

A semicircular distribution manifold is soldered to the end plates of the liner on both top and bottom parts. From this manifold radial tubes lead to fittings on the ends of the 40-foot tubes soldered to each liner panel; thus the panels are cooled by water flowing through these long tubes in parallel, in circuits that come

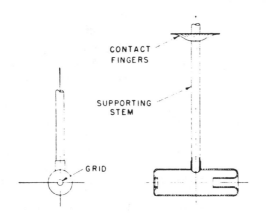

FIG. 18. Outline drawing of a typical drift tube.

through the tank at one end, through the distribution manifold, along the panels into the collection manifold and out through a discharge lead at the opposite end of the tank. Separate circuits are maintained for the upper and lower parts of the liner.

The drift tubes are also cooled in parallel by a third water-cooling circuit. Two tubes are supported in openings in the liner frames, with one tube serving as supply header and the other as collection header. These tubes have nipples hard-soldered to them, one adjacent to each drift tube, into which the $\frac{1}{4}$-inch copper tubes passing through the drift-tube stems are soldered. With the water introduced at one end of the tank and removed from the other, the cooling-water pressure drop through each parallel flow path is maintained the same. Checks on the operation of the parallel flow system are made by putting hot water through the lines, and feeling all the tubes to determine that they are receiving their quotas of water and that no obstructions exist in the individual circuits.

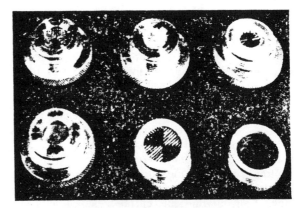

FIG. 19. Various types of slat grids and beryllium foils mounted in holders.

(6) Grids

As was shown in Section III, radial focusing and phase stability can be achieved in the machine by introducing charge within the beam; i.e., by arranging the entrance ends of drift tubes so that electric field lines terminate within the beam. This was first attempted by putting 3×10^{-5}-inch-thick beryllium foils across the entrance of each drift tube. Sparking in the tank destroyed them, however, and grids were used instead. There is, of course, greater field concentration on grids than on flat foils. To a first approximation, if one considers a grid to be merely a foil with many holes punched out, the field is increased by a factor equal to the ratio of the total area to the area occupied by conductors, since the same number of lines of force end on the conductor, but on a smaller net area. The ratio of aperture to field strength can be improved by using a slat grid of structure similar to a klystron grid, instead of a perforated plate, since some of the field lines terminate on the flat sides of the slats. Figure 19 shows several grid shapes that have been used, and also the beryllium foils.

Grids were fabricated from $0.002 \times \frac{1}{16}$-inch tungsten ribbon. A 95° bend was put in short sections of this strip by means of modified vise-grip pliers. Brass rings mounted on a carbon mandrel were slit by a multiband saw to a depth of 72 mils. The bent strips were set into these slots and hard-soldered around the rim, which was then given a finishing lathe cut. The grids were polished to approximate the optimum "rounded" shape, by immersing in 2-molar NaOH and passing 5 amps ac for 5 seconds between the finished grid and a tungsten rod. They were next soldered with pure tin into copper holders, which hid the brass ring, and could be screwed into the entrance end of the drift tubes. Grids were given a thorough visual inspection through a 30× stereoscopic microscope, and sharp points were removed. Then they were tested by photometering the transmission of a beam of parallel light from a tungsten-strip-filament lamp. The optical path was: pinhole over strip filament lamp, collimating lens, grid, focusing lens, 0.010-inch-diameter pinhole, photronic cell. It is interesting to note that this was a "good geometry" experiment, so that an aperture of, say, 90 percent of the total opening gave a photocell reading of 80 percent.[25]

The final test of the grids was made by investigating their behavior in vacuum under dc field equal to the rf field of the linear accelerator. If the cold emission was less than 50 μa after the grids had been given up to 5 minutes run-in time, they were accepted. A grid in its mounting is shown in Fig. 20. At the present time, the grid aperture has been considerably increased by removing all but the four central L-shaped slats. No significant increase in the x-ray background from the tank accompanied this change, which shows that field emission from other parts of the drift tube is still responsible for most of the electron drain in the machine. We do not believe that the etching procedure was necessary in practice. We tested electrically several grids that had not been etched, and could not find any significant difference from the normal ones.

The 15-cycle rf pulses produce fields of 10^7 volts/meter between the drift tubes. This causes pulsed forces of 4 g/cm^2 on the drift-tube ends. This pulsed force is sufficient to loosen the grid holders, so set screws are used to lock them into the drift tubes.

(7) Radiation Shielding

The stray radiation around the linear accelerator has been investigated to determine its sources and energies. Two major radiations are found, x-rays and fast neutrons. The x-rays come almost entirely from electron bombardment of the drift-tube ends. This was determined by exposing x-ray films behind an iron-slat collimator "telescope" laid on top of the accelerator. By absorption measurements it was established that the x-ray energies ranged up to $\sim$2 Mev near the exit end of the accelerator. In order to provide sufficient attenuation for the x-rays to permit personnel to occupy the building during operation, a total shielding of $1\frac{1}{2}$ inches of lead is used. This is divided into a $\frac{1}{2}$-inch thickness directly mounted on the tank, and a 1-inch-thick vertical wall extending the length of the tank but independent of it. This shield reduced the x-ray level in the building to acceptable levels for personnel safety.

Fig. 20. Grids of the type now used in the linear accelerator.

[25] That is, the diffracted component of the beam is *not* recorded by the photocell.

It was later found that the fast-neutron flux was above the tolerances currently accepted. In order to reduce the neutron component two walls of basalite concrete, 10 feet in height, were installed along the length of the machine. The walls are 18 inches thick on the north side, and 12 inches thick on the south side of the machine. This shield also provides appreciable attenuation for the x-rays. The radiation level in the shop area of the building ($\sim$30 feet from the accelerator) is of the order of 5 mr/hr at full voltage.

For a short time after the tank has been let down to air and re-evacuated, the x-ray radiation level is higher by a factor of 2 to 5, but improves quickly with running of the rf, which serves to outgas the system. Outgassing can also be speeded by running hot water through the liner cooling lines, though that has very seldom been done. After the accelerator has been run for several months, the x-ray level has slowly risen by a factor of five, owing to oil deposits on the drift-tube surfaces (Fig. 21). The tank is usually opened about twice a year to clean off the oil deposits.

(8) Vacuum System

The basic requirements of the linear accelerator vacuum system are, first, a base pressure 10^{-5} mm or less, and second, a pumpdown and bake-out time of reasonable length, say, 8 hours. To fulfill these conditions in a steel tank 40 feet long, of 15 000-liter capacity, containing about 500 vacuum seals and joints, and several hundred feet of polyethylene cable, it was obviously necessary to provide a fast pump, and to keep leaks and outgassing to a minimum. That this has been done successfully can be seen from the fact that at present, the base pressure is about 2×10^{-6} mm, the rate of pressure rise with the pumps closed off is as low as 10^{-7} mm/sec, and bake-out times as short as 4 hours have been recorded.

(a) Pumps

The pumps used are: a 30-inch three-stage diffusion pump with a pumping speed of 7000 liter/sec, and an 8-inch two-stage diffusion pump, in series, backed by two 43-c.f.m. Kinney rotary mechanical pumps in parallel. The diffusion pumps use Litton oil, which is effectively kept out of the tank by a refrigerated baffling system. The pumping speed measured inside the liner is 2500 liters/sec, which is considered a reasonable fraction of the speed of the pump alone.

(b) Vacuum Seals and Joints

A very wide variety of seals is used on the linear accelerator. Besides the usual types, such as soft- and hard-solder joints, gaskets, and arc-welded seams, there are Wilson seals, sylphon seals, aircraft-type spark plugs, and even certain standard rf connectors that have been found to be vacuum-tight.

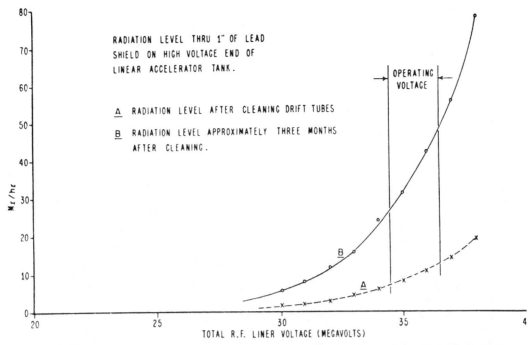

FIG. 21. X-ray level near output end of the linear accelerator as a function of operating radio-frequency voltage.

Perhaps the most noteworthy of the gasket seals is the main tank gasket, which runs completely around the tank and is almost 90 feet long. To avoid the expense of machining a flange 40 feet long we decided to use a molded rubber gasket held to the flange with special screws and retained by a $\frac{1}{4}$-inch-square strip of steel tack-welded to the flange along the vacuum side (Fig. 22). Despite the unmachined flange, the performance of this gasket has exceeded our expectation. Three bolts on each side suffice to hold the flanges together, and when the pumps are started the external air pressure exerts ample force to complete the seal. The heads of the screws act to separate the flanges and prevent them from damaging the gasket.

The vacuum seal to the outside of the radio-frequency transmission line is made with a standard 4-inch-diameter rubber "O-ring." This seal and the transmission line are installed and removed entirely from the outside of the vacuum tank. This seal seldom leaks, although there is some motion of the transmission line in it owing to the differential thermal expansion between the liner and the vacuum tank and the warping of the tank due to the change in pressure during pumpdown. The vacuum seal to the inner conductor of the transmission line is made with flat teflon gaskets in compression between the insulator and the copper conductors that make up the transmission line.

V. OSCILLATORS

The present machine was designed to operate with an average voltage gradient of 0.90 megavolt per foot, or a total end-to-end voltage of 36 megavolts (peak value). This differs from the energy gain (28 Mev) of the particles due to the operating phase angle and transit-time loss. Since the shunt impedance of the liner on the fundamental mode is 311 megohms, the radio-frequency power required by the accelerator is approximately 2.1 megawatts. As the machine is pulsed "on" for 600 microseconds fifteen times per second, i.e., a duty cycle of 0.9 percent, the average power is approximately 20 kilowatts.

The accelerator was first put into operation in October 1948, using eighteen and later twenty-six

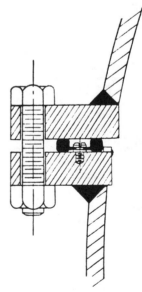

FIG. 22. Cross section of main tank gasket.

power oscillators, each of which developed approximately 85 kw from four GL-434 triode tubes in parallel connection. The GL-434 tubes, which were obtained on war surplus, had such a high casualty rate while in operation, with many actually becoming gassy on the shelf, that it was decided to build an oscillator system using a more rugged tube. The oscillator system now in use was installed in February 1950, and consists of three pre-exciter oscillators and nine single-tube power oscillators, each connected to the liner through a separate transmission line and coupling loop.

The present pre-exciters consist of some slightly modified radar-type "BC-677" oscillators, which are loosely coupled to the liner. These exciters perform three important functions for the machine: they excite the liner through the "multipactor region" (see below), select the correct mode of oscillation, and supply the low level of radio-frequency energy to the liner necessary to start the power oscillators.

Multipactor action[26] is almost universally found in devices using radio-frequency voltages in a vacuum. It is a secondary-electron multiplication process which can produce severe loading at low power levels, but once the system is above a given level the effect disappears. To illustrate this process, consider two parallel metal plates spaced some small distance apart in a vacuum. In general the secondary-electron emission ratio for metals (with the usual surface contamination) is greater than unity. If an increasing radio-frequency voltage is applied between the plates, there is a minimum value of voltage such that an electron can just cross the gap in exactly one-half cycle. If the energy of the electron is of the correct magnitude, more than one secondary electron will be released, and these electrons will see a voltage such as to accelerate them back across the gap again where they will make still more electrons. This process builds up very rapidly, and can drastically reduce the shunt impedance of the resonator at low power levels. With the geometry found in the

accelerator the multipactoring action limits the voltage to very low amplitudes if it is allowed to occur.

The straightforward cures are to make the electron transit time different in the two directions by a dc bias, or to raise the radio-frequency voltage so rapidly that there are not enough rf cycles in the critical region for the discharge to become large. The low energy end of the machine provides extremely favorable geometry for such a discharge between drift tubes. The spacings between the drift tubes in this region are of the order of an inch, and it can be shown that this gap is resonant for multipactor action at around 2000 volts. In addition there are many such gaps, any one of which can be responsible for the discharge.

The first cure attempted was to isolate every other drift tube from ground for dc and to apply a bias such as to make the electron transit time different in one direction from that in the other. This method worked, but brought with it many difficulties in providing a suitable radio-frequency by-pass condenser from the drift-tube stems to the liner. The problem was subsequently solved by the experimental discovery that three pre-exciters coupled in at the high-energy end of the liner (where the drift-tube spacings are the longest and the most unsuitable for multipactoring) could deliver sufficient energy (taking advantage of the very low group velocity of propagation in the liner as a wave guide at cutoff) to drive the rest of the liner up through the multipactor voltage region so rapidly that the multipactoring did not have time to build up.

The correct mode is selected by careful manual tuning of the three loosely coupled pre-exciter oscillators. The main oscillators will not start without the low-level pre-excitation because the energy storage or "flywheel effect" of the liner is large and their power gain for small signals is low. The pre-exciters establish

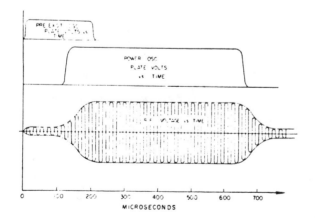

Fig. 23. Plate and rf voltage *vs* time in microseconds.

[26] P. T. Farnsworth, J. Franklin Inst. 218, 411 (1934).

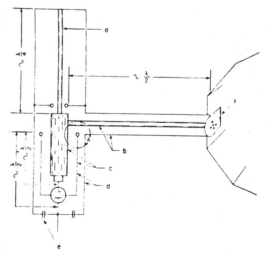

Fig. 24. Linac oscillator schematic. (a) Transmission line stub $\lambda/4$ used as mechanical support and to bring in cooling air and cathode power to osc tube; (b) Output transmission line; (c) Drive circuit (grid to cathode); (d) Output circuit (grid to plate); (e) dc plate blocking condenser; (f) Coupling loop.

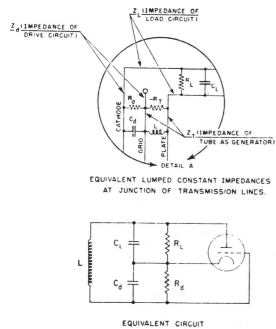

FIG. 25. Linac oscillator schematic. Detail *A* refers to impedances seen in Region *A* of Fig. 24.

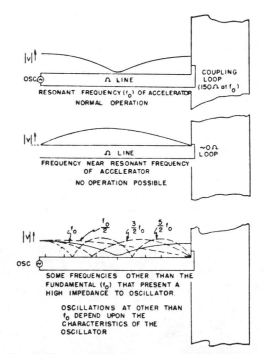

FIG. 26. Voltage distribution along transmission line for correct and incorrect modes of operation.

the correct mode at low level (approximately 1 percent of final power) by being pulsed on 30 to 100 microseconds before the main oscillators are turned on. This operation cycle is indicated in Fig. 23.

Each of the nine power oscillators uses an Eimac tube type 3W10000A3 in a modified Colpitts circuit. The oscillator is made of coaxial transmission lines. The inner (drive circuit) and the outer (plate circuit) coaxial lines connect the tube to the load (coupling loop in the liner) through the output transmission line. The relative characteristic impedances and the lengths of these lines control the amplitudes and phases of the radio-frequency voltages appearing on the tube and load circuits. Figures 24 and 25 are schematic and equivalent oscillator circuit diagrams.

The magnitude of the load impedance is adjusted by varying the size of the coupling loop in the liner. The self-inductance of this coupling loop is purposely designed to be so small that the coupled impedance of the liner at resonance dominates. These loops are copper plates 6 in. long and 6 in. wide and spaced approximately $\frac{3}{8}$ inch from the wall of the liner. The plates are connected to the liner at one end and to the transmission line center conductor at the other end. The impedance of the self-inductance of these loops at 200 Mc is approximately 15 ohms, and the coupled impedance from the liner is approximately 150 ohms. The importance of this impedance ratio is qualitatively illustrated in Fig. 26. The coupling loop presents only the impedance of its low self-inductance to the oscillator for frequencies which are near but not on the liner resonant frequency. This low impedance upsets the magnitudes and phases of the voltages within the

oscillator and discourages it from oscillating on frequencies to which the liner is not resonant. In addition, the magnitudes and phases of the voltages appearing at the tube vary so rapidly with frequency that oscillation is discouraged for frequencies appreciably removed from the liner resonant frequency.

Each oscillator unit rests on, and is partially enclosed by, a metal cabinet with interlocked doors for personnel protection [see Figs. 27(a) and 27(b)]. All high-voltage, air-cooling, and water-cooling connections are made inside this cabinet and are designed to be quickly disconnected. The "quick disconnect" idea was also applied in designing the "plug in" connection between the oscillator and the coaxial transmission that feeds the radio-frequency power from the oscillator to the liner. This "quick disconnect" feature allows a faulty oscillator to be replaced in approximately five minutes.

Each oscillator is pretested at a peak power output of more than 350 kw (duty cycle 0.9 percent, plate voltage 18.5 kv, over-all efficiency 50 percent) although in actual operation it delivers 250 kw (plate voltage 15 kv).

VI. THE POWER SYSTEM

1. General Description

The electrical components available, the cooling requirements, and general power requirements dictate that the accelerator be operated as a pulsed machine. The pulse length (600 μsec maximum) is a compromise between cavity build-up time and available energy-storage capacity. As high a duty as is permitted by power considerations is desirable.

For the parameters involved here, energy storage in rotating machinery is not practical. Accordingly, the system adopted is a pulse-forming synthetic transmission line continuously charged through a reactor.

The approximate design specifications for the power supply are given in Table II.

For fulfillment of these specifications it was decided to use a standard three-phase mercury-vapor rectifier circuit charging a pulse-forming network through a reactor. The network is discharged into the load by a triggered spark gap through a pulse transformer.

The power oscillators used in the present design require that the cavity be pre-excited before they can generate rf power. This requires a separate power

system to feed the pre-exciter oscillators; this system is identical in design to the "main" supply outlined above except that the pulse-forming network is charged through an emission-limited diode in place of the reactor, and that no pulse transformer is used. The block diagram of the equipment is shown in Fig. 28. Figure 29 gives the design parameters for the pulse-forming lines.

2. Spark Gaps and Pulse Transformer

Current is switched by a set of spark gaps. These gaps operate in air and are made with copper electrodes. An air jet de-ionizes the arc mechanically. The gaps are mounted in a soundproof box.

The main oscillators are matched to the pulse-forming lines by a pulse transformer operating with a step-up ratio of 2:1. The transformer is relatively small; it weighs approximately 2000 pounds.

The spark gaps are triggered by a 100-kv pulse of approximately 1.0-μsec duration. This pulse is provided by discharging a condenser through the primary of a pulse transformer of the design of Kerns and Baker[27] by means of a hydrogen thyratron. The grids of the thyratrons are driven by a central pulse generator.

This circuit delivers properly timed pulses to the pre-exciter and main oscillator thyratrons. The repetition rate can be set to any desired frequency up to the limit imposed by the allowable duty cycle. The pulse generator supplies a trigger pulse to the Van de Graaff generator which keys the ion source on after the cavity is built up, and also supplies miscellaneous trigger pulses to actuate scope sweeps, counter gates, etc.

(a)

(b)

FIG. 27. Oscillators installed on the linear accelerator.

TABLE II. Design specifications for power supply for linear accelerator.

| | Main supply | | | |
| | Per oscillator | | Total | |
	Peak	Average	Peak	Average
dc input power	450 kw	3.9 kw	4000 kw	35. kw
voltage	14.0 kv	14.0 kv (peak)	14.0 kv	14.0 kv (peak)
current	32 a	0.28 a	290 a	2.5 a
rf output power	250 kw	2.2 kw	2200 kw	20 kw
impedance	$R_{dc} = \frac{V_p}{I_p} = 440\Omega$		$R_{dc} = \frac{V_p}{I_p} = 49\Omega$	
cathode power	3.4 kw		30.6 kw	
repetition rate	15 cps			
pulse length	600 μsec			

| | Pre-exciter supply | | | |
| | Per oscillator | | Total | |
	Peak	Average	Peak	Average
dc input power	180 kw	1.1 kw	530 kw	3.5 kw
voltage	10 kv	10 kv (peak)	10 kv	10 kv (peak)
current	18 a	0.11 a	53 a	0.35
rf output power	7 kw	0.05 kw	21 kw	0.15 kw
impedance	$R_{dc} = \frac{V_p}{I_p} = 560\Omega$		$R_{dc} = \frac{V_p}{I_p} = 190\Omega$	
cathode power	1.2 kw		3.6 kw	
repetition rate	15 cps			
pulse length	440 μsec			

[27] Baker, Edwards, Farly, and Kerns, Rev. Sci. Instr. 19, 899 (1948).

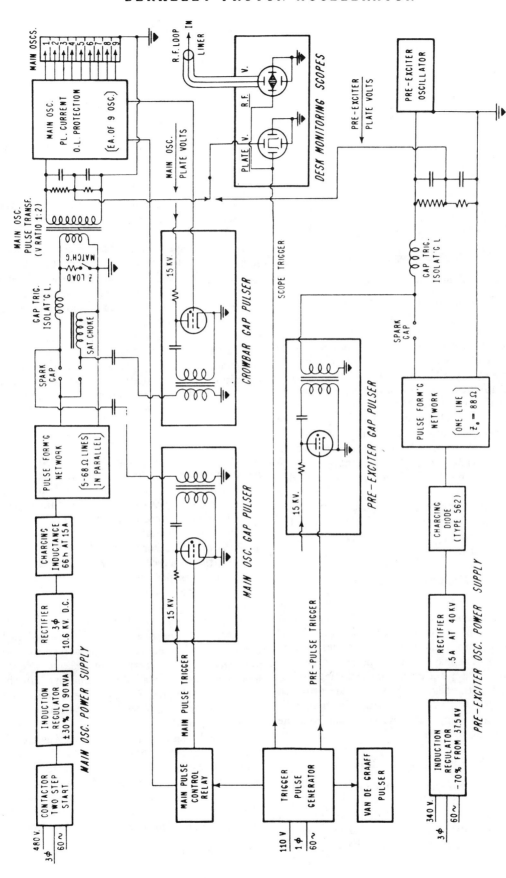

FIG. 28. Block diagram, power supply system.

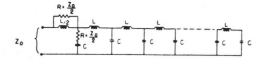

		MAIN OSC FIVE (LINES IN PARALLEL)	PRE-EXCIT (ONE LINE)
L =	INDUCTANCE PER SECTION -----	1 1 4 mh/line	2 0 mh
C =	CAPACITANCE PER SECTION ----	25 μfd/line	25 μfd
$Z_0 = \sqrt{\frac{L}{C}}$ =	CHARACTERISTIC IMPEDANCE	68 Ω/line	90 Ω
$t = \sqrt{LC}$ =	DELAY/SECTION -------	1.6 7 μ sec	2 2 μsec
N =	NUMBER OF SECTIONS -----	18	10
$T = 2N\sqrt{LC}$ =	PULSE LENGTH -------	600 μ sec	440 μ sec

FIG. 29. Schematic diagram of pulse-forming line and electrical characteristics for main and pre-pulse lines.

VII. ACCESSORY EQUIPMENT

1. Electron Catcher Magnet

The high axial rf electric fields existing in the resonant cavity can accelerate any free electrons (e.g., formed between drift tubes or on drift tube surfaces) to considerable energies, corresponding to approximately 60 percent of the voltage across the gap. Electrons formed near the exit end of the No. 46 drift tube by secondary electron multiplication or gas ionization can be accelerated to an energy of 1.1 Mev. These electrons can produce x-rays, which constitute an appreciable personnel hazard, as well as interfering with certain experimental equipment. A small electromagnet, producing approximately 3000 gauss between pole pieces $2\frac{1}{2}$ in. in diameter and 2 in. apart, is provided to deflect these electrons vertically, so that they strike a carbon cylinder, thus producing only soft x-rays which are absorbed in a three-inch lead shield around the cylinder. This magnet can also be used for small-angle vertical steering of the proton beam.

2. Deflecting Magnet

An 8000-gauss electromagnet is used to deflect the 32-Mev proton beam horizontally, and thus separate it from the lower components of 4 and 8 Mev. The latter are produced by protons spending 2 rf cycles in each drift-tube space. The magnet also removes the 16-Mev H_2^+ component if a $\frac{1}{4}$-mil aluminum stripping foil is inserted ahead of the field. Output ports are provided at angles of $-10°$, $0°$, $10°$, $20°$, and $30°$. The magnet is used primarily as a switch, to move the beam from one experimental setup to another.

3. Bombardment Facilities

After the beam has passed through the deflecting magnet, it is available for the bombardment of targets. Apparatus at this end of the machine is similar to that found in cyclotron installations and need not be discussed here. Two useful items are a Faraday cup in vacuum, for beam integration, and a rotating foil

changer, which is employed for varying the energy of the beam. Such pieces of research equipment as β-ray spectrographs, cloud chambers, and scattering chambers have been installed in the bombardment room at various times.

VIII. PERFORMANCE

1. Energy

The output energy of the linear accelerator has been determined by range measurements in aluminum, using the range-energy relation as computed by Aron et al.[28]; the energy can be varied over a range of ±150 kev by moving the "end tuners" (half drift tubes) at the ends of the machine. The measured energy is close to 31.7 Mev.

The energy spread of the beam can only be inferred at present from the sharpness of the threshold of the reaction[29] $C^{12}(p,n)N^{12}$. When the resulting excitation curve is corrected for absorber straggling, the rms energy width of the beam can be shown to be $\Delta E < 100$ kev on the high-energy side. The shape of the energy spectrum on the low-energy side is not susceptible to measurement by this technique. If, however, one were to define the beam by slits, and subject it to a magnetic deflection, the inhomogeneity introduced by the passage of protons through the slit pairs would certainly be greater than that already present.

2. Current

For most purposes, the time-average current is of most importance. During the past six months, an experimenter could be sure of having $\frac{1}{4}$ microampere of average current, if he could use it. In many experiments, the beam is purposely cut down by a large factor. Occasionally, the beam is somewhat higher, and 0.37 μa is the present record value of $\langle I \rangle_{Av}$. These values are for operation at a recurrence rate of 15 cycles per second.

At 15 cycles per second, the beam pulse length is about 400 μsec, so that the duty cycle is 1 in 166. The peak current at $\langle I \rangle = 0.37$ μa is therefore ~60 μa. If the grids are assumed to cause a loss of about $\frac{1}{3}$ of the beam and the phase angle is assumed to be 36°, the calculated injection current agrees well with the measured injection current of 1.5 ma. Neither grid loss nor phase angle is well known, but these are reasonable estimates. The agreement indicates that no large gains in output current are to be had unless major changes are made.

Two such changes have been investigated. A radio-frequency "buncher" has been built and tested. Protons from the Van de Graaff generator are velocity-modulated by the buncher, and after a drift space of 15 feet, are formed into small groups occupying a narrow phase angle as they enter the accelerator. The peak output current from the accelerator has been tripled by

[28] W. A. Aron et al., 'Range-energy curves,' AECU-663 (1949).
[29] L. W. Alvarez, Phys. Rev. 75, 1815 (1949).

this buncher. The energy tolerance on the injected beam must be decreased severalfold to make the bunching of practical importance. (A small change in average velocity of the injected beam makes the bunch arrive at the wrong phase, so the output current is thereby decreased.) A program to develop an energy-stabilizing circuit has therefore been undertaken.

The second improvement is the use of electrostatic strong-focusing lenses of the Christofilos-Brookhaven[22] type. A quadrupole electrode system has been installed in each drift tube, and when it is used the grids are removed. Under these conditions, the external average beam has been increased to $\frac{1}{2}$ μa. If the focusing voltage could have been raised to its designed value, the current would have been about 1 μa. The increase in current with focusing voltage comes from the larger phase angle at which the defocusing forces are counterbalanced by the electrostatic focusing forces. Unfortunately, sparking in the lenses limited the voltage to the lower value. Deterioration of the cables has forced us to abandon the strong-focusing feature and return to grids, but these difficulties probably would not have occurred if the machine had been designed for strong focusing at the start. It is remarkable that enough space was available inside the drift tubes to make the test as successful as it was.

IX. ACKNOWLEDGMENTS

The design and construction of the linear accelerator was in every sense a cooperative affair, and contributions from a large number of men are involved. Only those who had the responsibility for the major design features, and who were members of the group for the period of two years during which the most intensive work was done, are listed as authors, but many others helped. Much of the work was done by graduate students in the Department of Physics, and major contributions were made by the electrical and mechanical design groups of this laboratory.

We are indebted to Professor E. O. Lawrence for encouraging us to embark on the project. The Manhattan District of the U. S. Army Engineers gave us strong backing in the early days, and the Atomic Energy Commission has continued this support. The Signal Corps generously donated large quantities of radar equipment, without which we probably would not have been able to build the accelerator.

Mr. William Baker is responsible for the redesign of the radar oscillators. There is a good possibility that without this redesign the accelerator would never have produced a beam. Among the graduate students, special acknowledgements must be made of the valuable electronic contributions of Lawrence H. Johnston, Robert Mozley, Bruce Cork, Ernest Martinelli, Robert Phillips, Lee Aamodt, Thomas Parkin, Richard Shuey, William Toulis, Val J. Ashby, and Donald R. Cone. In the field of mechanical design, important contributions were made by A. W. Chesterman, A. E. Kaehler, E. A. Day, R. L. Olson, D. A. Vance, Virginia McClain, and Florence Mosher. The laboratory Electrical Design Group was of the greatest assistance, and we would like to acknowledge the work of Saul Lissauer, J. C. Kilpatrick, Walter Sessions, Porter Evans, and C. A. Harris.

A group of former radar technicians played an important role in the building of the machine, and some of these men are operating and servicing it at the present time. We wish to thank especially Philip Carnahan, Albert J. Bartlett, Alva Ray Davis, Jr., Wilfred P. Kimlinger, James A. McFaden, Wendell W. Olson, and Frank Grobelch. Finally, we wish to acknowledge the important contributions in various fields of Dr. Robert Serber, who developed the mathematical theory of beam stability, Richard Crawford, David Garbellano, Velma Turner, Leonard Deckard, and Craig Nunan. For the past few years, Robert Watt has been in charge of the operation of the accelerator. He and Craig Nunan supervised the installation of electrostatic "strong focusing" lenses, which was reported at the Brookhaven conference in December of 1952.

11

The Tandem Accelerator: Workhorse of Nuclear Physics

Peter H. Rose

I first met Luis W. Alvarez in August 1958, about the time our High Voltage Engineering's (HVE) first tandem accelerator was shipped to the Chalk River Laboratories in Canada. I was tired from two hectic years of development and testing and planned a vacation at my wife's parents' home in Idaho. Since I was going to be close to the West Coast and because I was a senior HVE physicist, Denis Robinson, the HVE president, suggested I visit Luie. Luie was disappointed that during the excitement and flurry of the first tandem accelerator's success, his contribution and pioneering work had not been acknowledged.

As a fairly recent employee, the origins of the tandem project at HVE were unknown to me. I had come across Bennett's patent[1] on energy doubling in direct current accelerators using negative ions while I was still studying in England. I had appreciated the value of the idea, so I assumed that a company like HVE would extend their technology and build such a machine. I was naive, as even entrepreneurial companies need to know that there will be customers for a new product before building it! Further, Bennett's idea never caught the attention of his physics peers who might have appreciated it. It was not until the concept was reinvented by Luie, who was in the physics mainstream, that the idea received the recognition it deserved.

The events leading to the first tandem accelerator go back to the close professional and personal relationships developed between the talented scientists and engineers working at the MIT Radiation Laboratory on radar and at Los Alamos on the Manhattan Project. Luie was one of these, and it was during this period that he and Robinson got to know each other.

When HVE came into being in 1946, Robinson and Luie continued to meet at American Physical Society meetings and in Berkeley. Robinson remembers first learning of the negative ion accelerator on March 12, 1951. Luie called him at home and suggested the idea of an energy-doubling accelerator using negative ions. Robinson's notes describe a horizontal machine with serial voltage columns and acceleration tubes powered by a Cockroft Walton that forms the stem of a T. This is the accelerator Luie has described in the accompanying article.

Luie tells of how he thought of the idea that changing an accelerated ion's charge from negative to positive inside an electrode would double its energy.[2] However, Luie may actually have had the same idea earlier. One of the principal instruments used by the team at Los Alamos was the 2.4 million volts pressurized Van de Graaff accelerator developed by Ray Herb and his University of Wisconsin colleagues. Herb remembers that in the late forties when he was visiting Los Alamos, Luie was excited about an experiment by Jim Tuck, who had obtained high yields of negative ions by passing protons through thin foils. Herb's discussion with Luie resulted in Herb's developing an effective negative ion source which was the basis for HVE's original tandem design.

In the years that followed the initial phone call to Robinson, Luie encouraged Robinson to develop this machine as a new HVE product. Unfortunately, little money was then available for nuclear research. Further, there was the expectation that the electrostatic accelerators being developed at MIT and Los Alamos would easily achieve their 10 megavolt design voltages. HVE meanwhile developed a simple, effective accelerator with an ion source in the terminal capable of running up to 6.5 megavolts. This machine was shipped to the Oak Ridge National Laboratory in April 1951. With the success of this Van de Graaff accelerator, HVE stole the initiative from the university and government laboratories and for twenty years was the world leader in direct current accelerators. Luie continued to advocate energy doubling when he met Robinson, who was willing to proceed but needed the support of his colleagues, Robert Van de Graaff and John Trump, and a customer. In a busy world these things take time.

By January 1955, Van de Graaff and Trump had become excited about building this new type of accelerator, which Van rechristened the "tandem." Luie had previously referred to it as a "swindletron," scarcely a suitable name for a commercial product! The conceptual design looked good, and the project only awaited a customer. In February 1955, W. B. Lewis, Robinson's World War II boss and then Chalk River Laboratory Director, made a verbal commitment to buy the first machine. This started the Tandem pro-

1. W. H. Bennett, U.S. Patent 2206556 (1940).

2. L. W. Alvarez, Rev. Sci. Instrum. **22**, 705 (1951).

gram, although the actual purchase order was not received until January 1957.

None of this history was known to me when I visited Luie in 1958. My job had been to develop a negative ion injector and make the first tandem accelerator work. I went to Luie's office in Berkeley eager to meet this famous physicist and convince him that we at HVE were grateful for the opportunities his invention had created for us. Naturally, he was interested in all the details I was able to give him. I remember he was intrigued that we saw no sign of any electron loading in the terminal region. He had feared the possibility of electron oscillation through the stripper canal. Luie could not have been more generous and congratulated me on the success of HVE's first tandem accelerator. He was clearly delighted with the outcome. He now must be even prouder of the large body of significant and elegant physics that has been made possible by this family of machines.

Changing the charge of energetic particles appeals to physicists. As Luie points out, it makes it possible to accelerate particles from ground potential and back again without a time-varying field. Large direct current voltages are difficult to obtain. So there was considerable excitement in 1929 when Davis and Barnes at Columbia University found a method of attaching electrons to fast alpha-particles, thereby neutralizing their positive charge. These neutral particles could drift to a high-voltage terminal, where the electrons would be stripped, causing the bare alpha-particles to accelerate to "ground" voltage, where they would be reneutralized by adding electrons and so on until the desired energy was reached. Unfortunately, the experimental results of this study were in error, and "charge-changing" as a means of accelerating particles was abandoned for several years.

In 1932, 140 keV protons were obtained by charge-exchanging from positive to neutral to positive in gas targets,[3] while in 1936 negative hydrogen ions and neutral atoms were formed by charge exchange processes to obtain three, four, and even five stages of acceleration.[4] However, the measured currents of nanoamperes were at energies too low to attract the attention of nuclear physicists. The same fate met W. H. Bennett and P. F. Darby's negative ion source, which was capable of 10 microamperes.[5] Because nuclear cross sections were small, much larger currents were needed to make measurements, so Bennett's patent and the rest of this early work was forgotten. Without Luie's reinvented idea of using negative ions for energy doubling when the necessary technology was becoming available and the enthusiasm and skills brought to it by Van

de Graaff, the tandem accelerator might never have been built. Even so, it took years of active promotion before work started on building an accelerator.

Luie realized that a tandem accelerator could take two forms. The machine could have two acceleration tubes side by side in a column connected by a 180° analyzing magnet. If the negative ions are stripped prior to analysis, the magnet can be modest in size and the positive ion tube need only accelerate the selected ion species. The simpler tube used in the first tandems had a negative and a positive ion tube mounted in line and separated by a stripper target. Here the positive ion column accelerates all stripped ions, thus limiting the maximum current. This limitation became increasingly important as research moved to heavier ions and has only recently been removed by the addition of a charge filter in the terminal.

When I joined HVE in October 1956, the design of the first tandem was well advanced. I soon discovered that Robinson, Trump, and Van de Graaff personally knew everyone having any interest in accelerators. So I had information I could cull from the literature or, better still, from the experts themselves. I eagerly embraced this extraordinary opportunity. HVE was then an exciting place, and my transition from an English university to industrial life in the United States was easier and more enjoyable than I expected. Although the physical parameters and specifications of the machine had been decided, a 5 megavolts terminal and a 5 microampere beam, there were many uncertainties. The task was a remarkable one for a small company using its own resources.

A significant cost in pressurized vertical Van de Graaff accelerators was the tower that housed the machine. This tower had to be high and sturdy enough to permit the tank to be lifted off the column to service the terminal and high-voltage structure. Moreover, with an in-line tandem system the pressure vessel became about twice as long as for a conventional machine. To prevent the cost of a tall tower pricing the tandem out of the market, Van de Graaff and Jack Danforth, a talented engineer, designed a horizontal accelerator with glass-to-metal bonded columns coupling the tank to the columns by a strong spring to keep them under compression. This 30 foot long glass bridge was stiff and had only small deflections due to the terminal weight and operational stresses.

This horizontal tandem accelerator could be installed on any large flat floor, and arranging the analyzing and beam-switching magnets, as well as experimental apparatus, was easy and flexible. This was a factor in the tandem accelerator's acceptance and resulted in 62 of these machines being built and installed in laboratories all over the world. Many are still in use in the forefront of nuclear physics and research.

3. C. Gerthsen, Naturwissenschaften **20**, 743 (1932).

4. O. Peter, Ann. Phys. (Paris) **27**, 299 (1936).

5. W. H. Bennett and P. F. Darby, Phys. Rev. **49**, 97 (1936); **49**, 442 (1936); and **49**, 881 (1936).

We had no doubt that we could accelerate negative hydrogen ions to the terminal, or that these high-velocity ions would be completely stripped of their electrons and become protons. It was known that even the thinnest carbon or beryllium foils would provide 100% stripping efficiency, but the lifetime of these foils would be comparatively short. Van de Graaff wanted to produce a machine that would not have this problem and recognized that gas targets would be indestructible. Some cross-section measurements on a 2 megavolts machine confirmed that there was just enough room in our terminal to fit a long half-centimeter diameter tube with sufficient gas thickness to strip the negative ions. The gas from the stripper was pumped down the positive ion acceleration tube, just as in a conventional accelerator. A short canal at the entrance of the terminal acted as a baffle to prevent stripper gas from going into the negative ion acceleration tube and prematurely stripping the fragile negative particles. The target gas pressure was controlled by a needle valve, but apart from this control the only dynamic parts in the terminal were the drive motor and the charging belt. The high-voltage section was thus reduced to elegant simplicity. Previously, the complications and short lifetime of the terminal ion source was the major cause of down time. When this was eliminated, the machine ran for thousands of hours without any need to enter the pressure vessel for servicing.

The problem that concerned me most was focusing the negative ion beam through the long small-diameter stripper canal in the terminal. Calculations indicated that high efficiencies were possible at high energies, but misalignments as small as 1 mm or beam aberrations could cause the beam to be lost. Since a variable energy accelerator like the tandem is expected to produce a beam of continuously variable energy, the ion optical system must match the injected beam to the tandem if the accelerated beam intensity is to remain constant. Van de Graaff and I liked the idea of generating negative ions by a proton source at ground potential that formed negative ions in a gas canal at 40 keV, an energy at which the positive-to-negative ion ratio is still just over 1%. The injected negative ions would, by energy-doubling in this inverted tandem, have an energy of 80 keV, a suitable injection energy, and we could still use einzel lenses to match the beam to the acceleration tube. At that time we had had no experience with strong focusing lenses at low energies, and we were afraid of space charge blowup or that unexpected aberrations might cause construction delays. However, at the high-energy end of the pressure vessel, the einzel lens was impractical, and a quadrupole lens was gratefully employed. Had strong focusing not been discovered, the tandem accelerator would not have been practical, since the proton beam leaving the accelerator was diverging and large, unlike the focused beam that could be obtained from the ion source in a terminal machine. I now see how lucky we were that this technology became available shortly before we needed it.

However, strong focusing was only one of the novelties. Just as important were the seeds planted in Ray Herb's mind by Luie in 1946, which bore fruit when the measurements of negative ion yields from proton beams in gases showed it was possible to build a high-current negative ion source without the scattering introduced by a foil.[6] The source developed by Ray Herb's students[7] was used as a model for the HVE negative ion injector.

By the spring of 1958, the first fully assembled tandem stood in grandeur on the factory floor. It seemed impossible that we could thread the beam from the ion source to the distant target through the various lenses, acceleration tubes, apertures, magnets, and so on. Our fears were well founded, as the beam from the ion source vanished, and only the ability of the machine to hold voltage gave us no trouble. Luckily, in a linear machine debugging is relatively easy, since adjustments are more or less independent. Even so, it took months of detailed work to get the machine operating satisfactorily. The tests culminated with an arduous week of experiments conducted by the Chalk River physicists, which clearly demonstrated energy doubling could produce high-energy, intense, well-focused beams. The machine was also flexible, with all the same ease of energy variation and precision of the previous Van de Graaff accelerators.[8] Acceptance of this machine by the nuclear physics community was instantaneous, and considerable commercial success for the company followed.

Critical to the success of the HVE machines were the acceleration tube designs of Trump and Van de Graaff, and the invention by Van de Graaff at about this time of an accelerating field inclined first in one direction and then in the other, making possible even higher voltages.[9] The accelerated particles followed a weaving path through the tube, while secondary particles were quickly swept away, eliminating high energy x-rays and reducing tube loading. Taking advantage of these newly invented inclined field tubes, HVE extended the voltage capability of the original 6 megavolts tandem by lengthening the column, and increasing the tank diameter to 10 feet, which immediately gave a machine capable of 7.5 megavolts. Other improve-

6. A. C. Whittier, Can. J. Phys. **32,** 275 (1954).

7. J. A. Weinman and J. R. Cameron, Rev. Sci. Instrum. **27,** 288 (1956).

8. H. E. Gove, J. A. Kuehner, A. E. Litherland, E. Almqvist, D. A. Bromley, A. J. Ferguson, P. H. Rose, R. P. Bastide, N. Brooks, and R. J. Connor, Phys. Rev. Lett. **1,** 251 (1958).

9. R. J. Van de Graaff, P. H. Rose, and A. B. Wittkower, Nature (London) **195,** 1292 (1962).

ments allowed this machine to operate with 9 or even 10 million volts on the terminal. It was soon realized because of these higher energies helium, lithium, and heavier ions could be accelerated to energies above the Coulomb barrier of target nuclei, thus widening the choice of projectile for nuclear interactions. Only atoms which do not form negative ions, like nitrogen and some of the noble gases, could not be used.

With the tandem energy approaching those of linacs and cyclotrons, the advantages of direct-current acceleration encouraged laboratories to invest in even more expensive, higher voltage machines. Thus the HVE 13 and 20 megavolt machines were built, using a new horizontal column capable of supporting the much larger structures needed to reach these higher voltages. The investment was heavy, particularly in the 20 megavolt machine, the development of which coincided with the decrease in government funding in the late sixties. The death of Van de Graaff in 1967 was another blow to the company that further reduced HVE's ability to support accelerator development. While it ended an era for HVE, and also the lavish support for low-energy nuclear physics, it did not mean the end of the tandem industry. However, with the breakup of the team that made HVE successful, I left in 1971 to start a company which built low-energy ion implantation accelerators for the semiconductor industry.

This story of the tandem accelerator would not be complete without mentioning the use, particularly of small machines, in carbon-14 dating. Here again it intersects Luie's life. Luie had decided to leave the exploiting of the tandem accelerator to Denis Robinson because, among other reasons, the Berkeley accelerators were then capable of much higher energies. However, one of Luie's major triumphs was to use a cyclotron as a sensitive mass spectrometer to discover that helium-3 is stable and occurs naturally with an abundance of about one part in a million.

As described in chapter 19, Luie and Richard Muller returned to this technique in 1977 to search for integral charged quarks. Their high discrimination comes from the resolving power of the cyclotron, considerably enhanced by the unambiguous information obtained from the detection of individual energetic ions. It was in this climate of scientific discovery that Rich Muller realized the same technique could be used for radioisotope dating using hydrogen-3, carbon-14 and other ions. At about the same time, it was realized independently in several places, notably at Rochester by Ken Purser,[10] that the tandem accelerator could be used for dating with the advantage of eliminating the nonacceleratable nitrogen-14 background. Applying this method, it is now possible to date material as old as 45,000 years with sample sizes less than one-thousandth as large as those needed by the classical radioactivation method.

It is fitting to conclude my remarks by noting that a great physicist not only has original ideas but also stimulates the same in those around him. While Luie has given full credit to Rich Muller for those first experiments on accelerator radioisotope dating, Luie's genius was certainly a catalyst. If this thought needs reinforcement, examine the last paragraph of his accompanying paper, where he almost casually identifies a solution to a problem that was plaguing synchrotrons. This was another brilliant idea that found acceptance once intense ion sources had been developed. Negative ions have now improved the injection efficiency of synchrotrons from a previously low value to one close to unity.

10. K. H. Purser, A. E. Litherland, and H. E. Gove, Nucl. Instrum. Methods **162**, 637 (1979).

Reprinted from THE REVIEW OF SCIENTIFIC INSTRUMENTS, Vol. 22, No. 9, 705–706, September, 1951
Printed in U. S. A.

Energy Doubling in dc Accelerators

LUIS W. ALVAREZ
*Department of Physics, Radiation Laboratory, University of California,
Berkeley, California*
(Received May 4, 1951)

IT is generally believed that charged particles cannot be accelerated from ground potential to ground potential unless they pass through a system which has associated with it a time varying magnetic field. Dc electric fields must satisfy the equation $\oint E ds = 0$, while the time varying fields used in radiofrequency accelerators and betatrons are freed from this restriction of scalar potential theory. In 1932, A. J. Dempster[1] produced protons with an energy of 45 kev, by passing them from an electrode at $+22.5$ kv dc to ground. The protons were first accelerated to ground potential, with an energy gain of 22.5 kev. A small fraction of the protons then picked up an electron from a residual gas molecule, and "coasted" to a second electrode at $+22.5$ kv. Then a small fraction of these neutral hydrogen atoms lost their electrons, and were accelerated to ground with a second gain in energy equal to 22.5 kev. An accelerator of this type is obviously impractical for several reasons. The probability of neutralizing a proton varies inversely with a high power of the particle velocity, so the scheme would not work at energies of interest to nuclear physicists. Even at the low energies where neutralization is not negligible, the energy spread of the beam would be wide because charge exchange could take place at all points along the beam trajectory.

It does appear, however, that charge variation can be utilized in a practical manner to circumvent the apparent restrictions of potential theory. If one accelerates negative hydrogen ions (H^- or D^-) to an electrode at $+V$ volts, strips off the two electrons by a thin foil at that point, and accelerates the protons or deuterons to ground potential, he will have doubled the effective voltage of the accelerator. The stripping foil can be of thin collodion, for example, so the energy loss and scattering would be negligible. The stripping cross sections are of order 10^{-16} cm², so a foil with 10^{16} atoms per cm² will strip more than one-half of the beam. Such a foil weighs less than 1 microgram per cm². Therefore any physically realizable foil will give good stripping; its thickness can be so small as to give no appreciable straggling or scattering. (The energy loss can be a few hundred electron volts.)

An accelerator of this type is now being designed to give 4-Mev protons. It is to be constructed by L. C. Marshall and J. Woodyard of the Electrical Engineering Department of the University of California. The acceleration column is a conventional one, as presently used in Van de Graaff generators. The over-all length is 6 feet, so the gradient is about 0.7 MV per foot. The whole system is immersed in oil, so no high pressure tank is required. The voltage is supplied by a Cockcroft-Walton circuit, which operates at 500 kc. Such a circuit has been brought to a high state of development by J. Woodyard, and is used as the primary ion source power supply for the Bevatron. The individual sections of the circuit are constructed as "plug-in units," each of which contains two rectifiers, two transformers, and two capacitors. In the event of the failure of some component, one can replace the whole unit which was giving trouble.

One of the most attractive features of the machine is that the ion source is near ground potential, so no electronic or electromechanical devices are required at high potential. One can compensate automatically for variations in the high potential, by changing the injection voltage; the output energy can thus be kept exceedingly constant in time.

Many existing Van de Graaff generators are equipped with duplicate accelerating columns. One column is used as a "differential pumping tube," to handle the gas from the ion source at high potential. It should be possible to accelerate negative ions up the pumping column, which would no longer be needed for its original purpose. In the high voltage electrode, the ions could be bent through 180° in a magnet, then stripped, and finally accelerated back to ground through the usual acceleration column. The magnet would be of the annular type, to save weight, and would require adjustment in its field whenever the energy was charged.

In a magnetic field of 18,000 gauss, the diameter of a proton orbit is approximately:

$$D = 6.3(E)^{\frac{1}{2}} \text{ inches,}$$

where E is the proton energy in Mev. Since the spacing between centers of the two columns is normally greater than 12 inches, it is seen that the doubling method is practical, in a mechanical sense, for all presently operating Van de Graaff generators.

The literature on ion sources for negative hydrogen ions is rather limited, but Bennett[2] describes a source which gave 0.02 μamp of H^- ions. Dr. James Tuck has recently informed the author of some experimental results on negative hydrogen ion production at Los Alamos. The work was done by Arnold, Phillips, Sawyer, Stovall, and Tuck. It was found that if low energy protons or deuterons were sent through a thin foil of Al or SiO, that up to 20 percent of the incident ions emerged with negative charge. For deuterons of 30 kev incident on a foil 10 μg per cm² thick, the ratio $[D^-/D^+$ incident] is 0.20. The ratio $[D^-/D^+]$ emergent, has been measured at several incident energies, and has the following values: 50 kev, 4.5 percent; 30 kev, 7 percent; 20 kev, 10 percent; 10 kev, 26 percent. The currents of negative hydrogen ions are limited only by the heating of the foil. One could obtain high currents by using a rotating foil, if heating turned out to be serious. A rotating foil would not be troublesome mechanically, since no pressure difference exists across the foil. Ion sources giving proton currents in excess of one milliampere are now available, so the accelerator proposed in this note should be capable of giving a few hundred microamperes of high energy protons.

The problem of injecting protons into a synchrotron might be simplified by using the charge exchange mechanism described above. There is a well-known theorem which states that it is impossible to inject charged particles into a stable orbit in a steady magnetic field. H^- ions could be directed from outside a synchrotron field, onto a thin foil placed at the midpoint of the aperture of the "doughnut." If the ions were tangent to the equilibrium orbit as they passed through the foil, they would reverse their curvature and follow stable circular paths. This method of injection might be very useful in the initial testing of a proton synchrotron, where small currents would be satisfactory. One could explore the orbits in a low, steady field, and assure himself that the magnetic design was correct.

The author is indebted to Dr. Tuck for permission to quote his data before their publication. This work was done under the auspices of the Atomic Energy Commission.

[1] A. J. Dempster, Phys. Rev. **42**, 901 (1932).
[2] W. Bennett, Phys. Rev. **49**, 91 (1936).

12

The Development of the Hydrogen Bubble Chamber

M. Lynn Stevenson

Luis W. Alvarez and Frank Crawford returned from the 1953 American Physical Society meeting in Washington with news of an exciting invention of a young physicist from the University of Michigan. Don Glaser had developed a "bubble" chamber in which diethyl ether bubbles worked the way water droplets do in a cloud chamber. We easily extended Van der Waal's equation of state for a real gas, which explained the cloud chamber operation, to the bubble chamber. At that time, Frank and I were using liquid hydrogen targets for our thesis research, which we bombarded with protons to produce a deuteron and a positively charged pion. As Panofsky's graduate students, we had assisted in his experiments on the capture of negative pions in hydrogen and deuterium. Thus, we were interested in developing a hydrogen bubble chamber to do similar experiments with the heavier new particles.

The first entry in my bubble chamber logbook, May 5, 1953, was the pressure versus volume (P-V) diagram of Van der Waal's equation in reduced variables that applied to any material. Our strategy had three steps: reproduce Glaser's results, develop a liquid nitrogen bubble chamber to work out the cryogenic bugs, and build a liquid hydrogen chamber.

Within a week of Luie's return, his capable technician Pete Schwemin had assembled the necessary equipment to reproduce Glaser's ether results. But we learned quickly that Glaser was a much better "chemist" than we were. Our glass bulbs and tubing were never sufficiently clean for the liquid to remain superheated long enough to allow a cosmic ray to trigger the system. We had to use a radioactive source to show sensitivity. By February 24, 1954, Pete had built his sixteenth chamber, "Pete's 16."

On May 17, 1953, we prepared the P-V diagram for nitrogen, and five months later our first try proved unsuccessful. After we first saw radiation-induced boiling in nitrogen, we expected to soon be seeing tracks. This never happened, and to this day no one has ever seen tracks in nitrogen. Meanwhile, John Wood had joined our group and, at Luie's suggestion, started a parallel effort using liquid hydrogen based on Hildebrand and Nagel's design. By January 25, 1954, John had photographed the first tracks in a liquid hydrogen chamber, and his results were published in *Physical Review* on May 1, 1954. My logbook on that day lists

the next things to be done: develop stereoscopic photography, build larger chambers, and study bubble growth with time and temperature using pulsed particle sources.

The most important event in this period was our observation of tracks in our accidentally dirty chamber. On May 5, 1954, we made pictures of tracks in a composite chamber with brass walls and glass windows using both a pulsed neutron and a polonium-beryllium source.

We were getting close to doing physics. By August 24, 1954, when we created the bubble chamber logbook, we had two versions of a 4 inch diameter chamber, Doug Parmentier's "Model 4A" and Pete Schwemin's "Model 4B." Safety was constantly on our minds. Our checklist for the 4 inch hydrogen bubble chamber had 43 items that had to be completed before liquid hydrogen could even be brought into the building, and ten items for an operating chamber.

During this time the Alvarez group was busy with several experiments. Bud Good joined us in May 1954, just as Frank and I were exploiting the polarized proton beam from the 184 inch cyclotron to redo our thesis experiment. Our interest was turning to the Bevatron, and during the first part of 1955 we made a fruitless search for particles with lifetimes of a fraction of a second. We placed a scintillator telescope close to the Bevatron beam to catch the decay fragments and gated the telescope on a few hundred milliseconds after the beam slammed into a nearby target.

I recall Leprince-Ringuet's visit with Luie in the room we called the "bullpen," which housed our entire group. He presented the latest evidence on the masses of the different heavy mesons. Luie's remark that they must all be the same particle began our efforts to make precision lifetime measurements of both the positive τ and θ mesons.

Meanwhile, as the bubble chambers grew larger, Luie's associates from the proton linac group, headed by Don Gow, began joining our project. One of these was Bob Watt, who provides his own commentary on his years with Luie.

By the spring of 1959, Philippe Eberhard and I had produced a beam of antiprotons to use with the 72 inch bubble chamber. I can still hear Bob Watt's voice ringing throughout the Bevatron as he dashed in shouting, "Six foot long tracks!'' The growth rate of bubble chambers exceeded that of cyclotrons, as is vividly seen in Figure 5 in the accompanying article (Luie's Nobel lecture), which shows those most responsible for this incredible achievement: Paul Hernandez, Pete Schwemin, Ron Rinta, Bob Watt, Luie, and Glen Eckman.

Equally impressive was the development of the data analysis system, achieved by Art Rosenfeld and Frank Solmitz, with the help of Gerry Lynch and Margaret Alston Garnjost.

Reproduced here is the partial text of my 1965 nomination of Luie for the Nobel Prize for Physics. When the honor did come, in 1968, Luie insisted on sharing his prize by taking his colleagues and their wives to Stockholm. All of us are deeply grateful to him for so enriching our lives.

To the Nobel Committee for Physics
January 22, 1965

Dear Sirs:

Thank you for the honor of being allowed to make a nomination for the Nobel Prize for Physics for 1965.

I should like to nominate Luis W. Alvarez for his work on the elementary particles. The pioneering work of Alvarez in the construction of the large hydrogen bubble chamber and the development of the high speed data processing system at the Lawrence Radiation Laboratory has made possible the discovery of many of the known resonant states and the determination of the properties of many others.

Alvarez stands with respect to elementary particle physics and the Glaser bubble chamber as Blackett stood with respect to nuclear physics and the Wilson cloud chamber. Both Alvarez and Blackett had visions of how to best utilize the devices of the originators. Seaborg played the same role with the original discovery of Plutonium by McMillan when the latter was called away to other duties.

Once the bubble chamber principle was known (1952), Alvarez had the vision of a large liquid hydrogen chamber operating as an integral part of the high energy accelerator. He recognized that hydrogen was the ideal liquid for understanding high energy particle physics. While others were demonstrating how their "clean", all glass, chambers could remain sensitive (superheated) for long periods of time, Alvarez was conceiving large, composite chambers of metal walls and glass windows that could utilize the pulsed characteristics of the accelerators. There was no need to make chambers "clean".

He moved decisively to achieve that goal. He recognized that a large scale industrial effort was necessary. The ultimate cost was in the millions of dollars. He argued that sums of money comparable to the cost of the accelerator were justified for detection equipment when the detection device would measurably increase the output of physics from the accelerator. Here he took a stand that was to bring him into conflict with some of the best experimental physicists of his time. He argued that the large hydrogen chamber was far superior to any other detection device for the purpose of studying elementary particle physics at high energy. The Atomic Energy Commission was convinced by his arguments and gave him the money. By 1958, only six years from the time of Glaser's original work, the 72-inch liquid hydrogen chamber was an operating device. The growth rate (in linear dimension) of the bubble chambers from the small "thimble" size of the original one to the 72-inch chamber size exceeded the growth rate of cyclotrons by more than a factor of two. To illustrate the vigor with which Alvarez pursued his objective and of the difficulty of building large chambers, we cite the fact that it was five years later before another chamber of comparable size was operating elsewhere. From the beginning, Alvarez made available all the engineering studies, (which constitutes a six-foot long shelf of books) to any group interested in constructing chambers. It is noteworthy that in the excellent review paper on bubble chambers by Hilding Slatis (Nuclear Instruments and Methods **5**,1, 1959) none of the 56 articles cited bears Alvarez's name, although, throughout the text of the article numerous references are made to him. But here it is clear that Alvarez's primary objective was to do particle physics and not to publish articles in the open literature on bubble chambers.

Along with the chamber development went the necessary development of a high speed data processing system with its exploitation of the high speed digital computer.

Alvarez's zeal for doing physics in this *big* way bordered on evangelism and probably, more than any other factor, brought him many critics. But he was right. Of the 36 known particles, 13 have been found in the Alvarez chambers; of these 5 were announced in publications bearing his name, 5 in publications of his group that did not bear his name, and 3 in publications of other groups to whom he freely gave film. No other single person has been associated in such a decisive way with the discovery of so many particles. . . .

It is for the development of the large hydrogen bubble chamber program and the discovery of 10 particles in high energy physics that I recommend Luis W. Alvarez be awarded the Nobel Prize for Physics for 1965. . . .

Thank you again for the privilege of making this nomination.

Sincerely yours,

M. L. Stevenson
Professor of Physics
University of California
Berkeley, California

Life with Luie

Robert D. Watt

My association with Luis W. Alvarez began in 1949 and lasted until I took his 82-inch bubble chamber to the Stanford Linear Accelerator Center, seventeen exciting years later. During this time I came to understand that, although Luie might have 100 ideas each day, 50 were probably useless, another 25 too difficult to do, and among the remaining 25 one or two would be worth a Nobel Prize. It was then left to us to drag our feet on all but the best of these ideas. Luie's really good suggestions are well known, while the others are too numerous to mention.

When I first joined Luie's group, it was to supervise the operation of the 40 foot Linear Accelerator. This machine had been designed and constructed by a group of young physicists among whom were doctors Hugh Bradner and W. K. H. Panofsky. The group soon became bored with the routine care and maintenance of the accelerator and decided it was time to turn this task over to professional operators. Ollie Olson, Wilfred Kimlinger, Jack Franck, and a number of us slowly improved the reliability of the equipment by redesigning and replacing the weak components.

Our greatest problem was the reliability of the 202 megacycle power sources used to excite the cavity. For the first few years, we had one man, full time, rebuilding oscillators and replacing their small radar power tubes. On a normal day of operation, the loud call, "New head!" could be heard every twenty minutes. Although the practice made the crew expert at replacing these power sources, 25% of the machine time was lost to this problem. The oscillator problem was solved by Jack Franck, who designed and built a half-dozen large oscillator amplifiers to feed the power transmission lines into the cavity. After this, radio-frequency problems were very rare indeed.

Another recurrent problem was breakage of the Textolite supports that held the Van de Graaff high-voltage head in position. Luie suggested that I visit Van de Graaff installations around the country to find out how they solved the problem. It quickly became apparent that most places had no problem—because their Van de Graaffs were vertical machines. Ours was horizontal, with the high-voltage sections cantilevered out some 15 feet. We solved our textolite problem with diagonal supports in tension made of glass fibers embedded in a matrix of epoxy. The Textolite supports did not break after that.

One exciting adventure I shared with Luie was building the world's first strong-focusing accelerator. Luie had heard about a new mathematical scheme for transporting charged particles. He knew that it would be discussed at the next American Physical Society meeting in several months and so decided to test the theory on real hardware. We all worked around the clock converting the 40 foot proton linac to electrostatic strong focusing, by inserting hyperbolic electrodes in the drift tubes. To install the hyperbolic electrodes, each tube had to be taken out of the machine and opened up. The tube stems were hollow and carried the high-voltage leads down to the electrodes. After fitting the electrodes, we had to reinstall the tube endcaps and seal their circumferential seam to the drift tubes. Since very large currents circulated there, it was necessary to bridge the gap with silver, which we hand-plated on. Each drift tube took five or six hours to modify, and there were 46 of them. It was not unusual for us to draft help from any available person, including Luie.

After all this work, we tested the linac and found that indeed the strong-focusing principle worked. We had increased the beam intensity threefold. Unfortunately, the high-voltage leads inside the drift tube stems degraded with the several-thousand volts we applied and slowly shorted out. Our great success only lasted a few weeks, but we proved the principle.

Conversation with Luie in those days was always a pleasure, because I never needed to say more than the first three words of a sentence to convey an idea to him. Within the space of these few words, he always understood the total content of my sentence, and any further utterance on my part would have been redundant. What Luie never understood, however, was that although his answer to my words was always correct, what I would have said, had I continued, usually was not. In this way, I was credited with much more knowledge than I truly had.

The birth and training of the 72 inch bubble chamber was one of our most trying periods. It was horrible to start, make sensitive, and operate. Once, when I was very discouraged, Luie backed me into a corner in the control room and said, "Bob, you're doing this all wrong. Don't think of it as a large 72 inch chamber. Treat it as a group of small 10 inch chambers in series and in parallel." And as soon as I changed my thinking to that mode, the bubble chamber, obligingly, began operating better. Unfortunately, it was a few months before all these "10 inch" chambers realized that they should be sensitive simultaneously and have the same bubble density.

When the 72-inch bubble chamber was working well, we often did small experiments for Luie. In this way, I became involved in his first monopole search. Robbie Smits and Pete Schwemin had several coils wound, each with many thousands of turns, and through these coils they hoped to repeatedly circulate samples of various materials. The two coils were arranged so that the fluctuating magnetic field of the Bevatron located a hundred yards away would be bucked out, but the coils were in series for a monopole passing through one and then the other coil. I became interested when it became obvious that some sort of very high speed circulation device was needed to get enough monopole passes per second to produce a detectable signal. Bill Richards quickly produced a wheel about six inches in diameter which became the rotor in a gas-driven, gas-supported sample carrier. This rotor held about a hundred small samples and could be separated into two parts for insertion into the coils. The rotation speed was easily held at 100 revolutions per second, adequate to give a strong 0.5 microvolt signal for a monopole with a single Dirac magnetic charge.

After the system was finished and reasonably reliable, we started pushing various samples through it: Bevatron and AGS (alternating gradient synchrotron) targets and most elements. Luie soon realized that the best place to find monopoles was at the north or south magnetic poles of Earth, to which its magnetic field would guide them. He also knew that he should get the samples himself so that he would know the locations from which they came. In short order, he went to the South Pole, picked up samples, and brought them to us for testing. His trip was for naught, of course, but as he remarked "A negative result is as important as a positive result."

After failing to find monopoles on Earth, it became clear that we would have to examine the Moon. At that time, he was not able to qualify as an astronaut, but he was able to get permission to use the samples brought back by the astronauts. I went to NASA near Houston to arrange for the use of one of their quarantine areas to install our monopole search equipment. Our 100 passes per second rotary equipment lacked the reliability and sensitivity for a scientific project like the Space Program. We needed something better. It was Philippe Eberhard's inspiration to use superconducting coils for the detection device, where now each pass would increase the current stored in the coil and after 100 or so passes would give a good signal for a single monopole. With Ron Ross and John Taylor we built the device and tested it at Berkeley. When the lunar samples arrived at NASA, we all went to Houston. On a three-shift-a-day basis, we ran a large amount of lunar material through the detector. Unfortunately, there were still no monopoles.

I observed during my years with Luie that he always knew the capability of each of us and expected us to perform at that level. He was quick to congratulate a good performance and rarely failed to notice a bad one.

The most striking example of his ability to share credit with others happened several years after I moved to Stanford University. At 4:00 in the morning, I was awakened by the telephone and greeted with this message: "Hi Bob, this is Luie. We just won the Nobel Prize!" He didn't say "I"; he said "we," and this is typical of his assumption that a team of people who work together deserve credit together. His Nobel lecture is more of a tribute to the accomplishments of his associates than it is a description of his contributions. This trait of Luie's has made it very satisfying for those of us who have worked with him.

RECENT DEVELOPMENTS IN PARTICLE PHYSICS

by

Luis W. Alvarez

The Lawrence Radiation Laboratory Berkeley, California

Nobel Lecture, December 11, 1968

When I received my B. S. degree in 1932, only two of the fundamental particles of physics were known. Every bit of matter in the universe was thought to consist solely of protons and electrons. But in that same year, the number of particles was suddenly doubled. In two beautiful experiments, Chadwick showed that the neutron existed, (1) and Anderson photographed the first unmistakable positron track. (2) In the years since 1932, the list of known particles has increased rapidly, but not steadily. The growth has instead been concentrated into a series of spurts of activity.

Following the traditions of this occasion, my task this afternoon is to describe the latest of these periods of discovery, and to tell you of the development of the tools and techniques that made it possible. Most of us who become experimental physicists do so for two reasons; we love the tools of physics because to us they have intrinsic beauty, and we dream of finding new secrets of nature as important and as exciting as those uncovered by our scientific heroes. But we walk a narrow path with pitfalls on either side. If we spend all our time developing equipment, we risk the appellation of "plumber", and if we merely use the tools developed by others, we risk the censure of our peers for being parasitic. For these reasons, my colleagues and I are grateful to the Royal Swedish Academy of Science for citing both aspects of our work at the Lawrence Radiation Laboratory at the University of California—the observations of a new group of particles and the creation of the means for making those observations.

As a personal opinion, I would suggest that modern particle physics started in the last days of World War II, when a group of young Italians, Conversi, Pancini, and Piccioni, who were hiding from the German occupying forces, initiated a remarkable experiment. In 1946, they showed (3) that the "mesotron", which had been discovered in 1937 by Neddermeyer and Anderson (4) and by Street and Stevenson (5), was not the particle predicted by Yukawa (6) as the mediator of nuclear forces, but was instead almost completely unreactive in a nuclear sense. Most nuclear physicists had spent the war years in military-related activities, secure in the belief that the Yukawa meson was available for study as soon as hostilities ceased. But they were wrong.

The physics community had to endure less than a year of this nightmarish state; Powell and his collaborators (7) discovered in 1947 a singly charged particle (now known as the pion) that fulfilled the Yukawa prediction, and that decayed into the "mesotron", now known as the muon. Sanity was restored to particle physics, and the pion was found to be copiously produced in Ernest Lawrence's 184-inch cyclotron, by Gardner and Lattes (8) in 1948. The cosmic ray studies of Powell's group were made possible by the elegant nuclear emulsion technique they developed in collaboration with the Ilford laboratories under the direction of C. Waller.

In 1950, the pion family was filled out with its neutral component by three independent experiments. In Berkeley, at the 184-inch cyclotron, Moyer, York, et al. (9) measured a Doppler-shifted γ-ray spectrum that could only be explained as arising from the decay of a neutral pion, and Steinberger, Panofsky, and Steller (10) made the case for this particle even more convincing by a beautiful experiment using McMillan's new 300-MeV synchrotron. And independently at Bristol, Ekspong, Hooper, and King (11) observed the two-γ-ray decay of the π^0 in nuclear emulsion, and showed that its lifetime was less than 5×10^{-14} second.

In 1952 Anderson, Fermi, and their collaborators (12) at Chicago started their classic experiments on the pion-nucleon interaction at what we would now call low energy. They used the external pion beams from the Chicago synchrocyclotron as a source of particles, and discovered what was for a long time called the pion-nucleon resonance. The isotopic spin formalism, which had been discussed for years by theorists since its enunciation in 1936 by Cassen and Condon (13), suddenly struck a responsive chord in the experimental physics community. The were impressed by the way Brueckner (14) showed that "I-spin" invariance could explain certain ratios of reaction cross sections, if the resonance, which had been predicted many years earlier by Pauli and Dancoff (15) were in the 3/2 isotopic spin state, and had an angular momentum of 3/2.

By any test we can now apply, the "3,3 resonance" of Anderson, Fermi, et al. was the first of the "new particles" to be discovered. But since the rules for determining what constitutes a discovery in physics have never been codified—as they have been in patent law—it is probably fair to say that it was not customary, in the days when the properties of the 3,3 resonance were of paramount importance to the high energy physics community, to regard that resonance as a "particle". Neutron spectroscopists study hundreds of resonances in neutron-nucleus system which they do not regard as separate entities, even though their lives are billions of times as long. I don't believe that an early and general recognition that the 3,3 resonance should be listed in the "table of particles" would in any way have speeded up the development of high energy physics.

Although the study of the production and the interaction of pions had passed in a decisive way from the cosmic ray groups to the accelerator laboratories in the late 1940's, the cosmic-ray-oriented physicists soon found two new families of "strange particles"—the K mesons and the hyperons. The existence of the strange particles has had an enormous inpact on the work done by our group at Berkeley. It is ironic that the parameters of the Bevatron were fixed and the decision to build that accelerator had been made before a single physicist in Berkeley really believed in the existence of strange particles. But as we look back on the evidence, it is obvious that the observations were well made, and the conclusions were properly drawn. Even if we had accepted the existence—and more pertinently the importance—of these particles, we would not have known what energy the Bevatron needed to produce strange particles; the associated production mechanism of Pais (16) and its experimental proof by Fowler, Shutt, et al. (17) were still in the future. So the fact that, with a few notable exceptions, the Bevatron has made its greatest contributions to physics in the field of strange particles must be attributed to a very fortunate set of accidents.

The Bevatron's proton energy of 6.3 GeV was chosen so that it would be able to produce antiprotons, if such particles could be produced. Since, in the interest of keeping the "list of particles" tractable, we no longer count antiparticles nor individual members of I-spin multiplets, it is becoming fashionable to regard the discovery of the antiproton as an "obvious exercise for the student". (If we were to apply the "new rules" to the classical work of Chadwick and Anderson, we would conclude that they hadn't done anything either—the neutron is simply another I-spin state of the proton, and Anderson's positron is simply the obvious antielectron!) In support of the non-obvious nature of the Segrè group's discovery of the antiproton (18) I need only recall that one of the most distinguished high energy physicists I know, who didn't believe that antiprotons could be produced, was obliged to settle a 500-dollar bet with a colleague who held the now universally accepted belief that all particles can exist in an antistate.

I have just discussed in a very brief way the discovery of some particles that have been of importance in our bubble chamber studies, and I will continue the discussion throughout my lecture. This account should not be taken to be authoritative—there is no authority in this area—but simply as a narrative to indicate the impact that certain experimental work had on my own thinking and on that of my colleagues.

I will now return to the story of the very important strange particles. In contrast to the discovery of the pion, which was accepted immediately by almost everyone—one apparent exception will be related later in this talk—the discovery and the eventual acceptance of the existence of the strange particles stretched out over a period of a few years. Heavy, unstable particles were first seen in 1947, by Rochester and Butler (19), who photographed and properly interpreted the first two "V particles" in a cosmic-ray-triggered cloud chamber. One of the V's was charged, and was probably a K meson. The other was neutral, and was probably a K^0. For having made these observations, Rochester and Butler are generally credited with the discovery of strange particles. There was a disturbing period of two years in which Rochester and Butler operated their chamber and no more V particles were found. But in 1950 Anderson, Leighton, et al. (20) took a cloud chamber to a mountain top and showed that it was possible to observe approximately one V

†1—693039. Alvarez

particle per day under such conditions. They reported, "To interpret these photographs, one must come to the same remarkable conclusion as that drawn by Rochester and Butler on the basis of these two photographs, viz., that these two types of events represent, respectively, the spontaneous decay of neutral and charged unstable particles of a new type."

Butler and his collaborators then took their chamber to the Pic-du-Midi and confirmed the high event rate seen by the CalTech group on White Mountain. In 1952 they reported the first cascade decay (21)—now known as the Ξ^- hyperon.

While the cloud chamber physicists were slowly making progress in understanding the strange particles, a parallel effort was under way in the nuclear emulsion-oriented laboratories. Although the first K meson was undoubtedly observed in Leprince-Ringuet's cloud chamber (22) in 1944, Bethe (23) cast sufficient doubt on its authenticity that it had no influence on the physics community and on the work that followed. The first overpowering evidence for a K meson appeared in nuclear emulsion, in an experiment by Brown and most of the Bristol group (24), in 1949. This so-called τ^+ meson decayed at rest into three coplanar pions. The measured ranges of the three pions gave a very accurate mass value for the τ meson of 493.6 MeV. Again there was a disturbing period of more than a year and a half before another τ meson showed up.

In 1951, the year after the τ meson and the V particles were finally seen again, O'Ceallaigh (25) observed the first of his kappa mesons in nuclear emulsion. Each such event involved the decay at rest of a heavy meson into a muon with a different energy. We now know these particles as K^+ mesons decaying into $\mu^+ + \pi^0 + \nu$, so the explanation of the broad muon energy spectrum is now obvious. But it took some time to understand this in the early 1950's, when these particles appeared one by one in different laboratories. In 1953, Menon and O'Ceallaigh (26) found the first $K_{\pi 2}$ or θ meson, with a decay into $\pi^+ + \pi^0$. The identification of the θ and τ mesons as different decay modes of the same K mesons is one of the great stories of particle physics, and it will be mentioned later in this lecture.

The identification of the neutral Λ emerged from the combined efforts of the cosmic ray cloud chamber groups, so I won't attempt to assign credit for its discovery. But it does seem clear that Thompson et al. (27) were the first to establish the decay scheme of what we now know as the K_1^0 meson: $K_1^0 \rightarrow \pi^+ + \pi^-$. The first example of a charged Σ hyperon was seen in emulsion by the Genoa and Milan groups (28), in 1953. And after that, the study of strange particles passed, to a large extent, from the cosmic ray groups to the accelerator laboratories.

So by the time the Bevatron first operated, in 1954, a number of different strange particles had been identified; several charged particles and a neutral one all with masses in the neighborhood of 500 MeV, and three kinds of particles heavier than the proton. In order of increasing mass, these were the neutral Λ, the two charged Σ's (plus and minus), and the negative cascade (Ξ^-), which decayed into a Λ and a negative pion.

The strange particles all had lifetimes shorter than any known particles except the neutral pion. The hyperons all had lifetimes of approximately 10^{-10} second, or less than 1 % of the charged pion lifetime. When I say that they were called strange particles because their observed lifetimes presented such a puzzle for theoretical physicists to explain, I can imagine the lay members in this audience saying to themselves, "Yes, I can't see how anything could come apart so fast." But the strangeness of the strange particles is not that they decay so rapidly, but that they last almost a million million times longer than they should—physicists couldn't explain why they didn't come apart in about 10^{-21} second.

I won't go into the details of the dilemma, but we can note that a similar problem faced the physics community when the muon was found to be so inert, nuclearly. The suggestion by Marshak and Bethe (29) that it was the daughter of a strongly interacting particle was published almost simultaneously with the independent experimental demonstration by Powell et al. mentioned earlier. Although invoking a similar mechanism to bring order into the strange-particle arena was tempting, Pais (16) made his suggestion that strange particles were produced "strongly" in pairs, but decayed "weakly" when separated from each other.

Gell-Mann (30) (and independently Nishijima (31) then made the first of his several major contributions to particle physics by correctly guessing the rules that govern the production and decay of all the strange particles. I use the word "guessing" with the same sense of awe I feel when I say that Champollion guessed the meanings of the hieroglyphs on the Rosetta Stone. Gell-Mann had first to assume that the K meson was not an I-spin triplet, as it certainly appeared to be, but an I-spin doublet plus its antiparticles, and he had further to assume the existence of the neutral Σ and of the neutral Ξ. And finally, when he assigned appropriate values of his new quantum number, strangeness, to each family, his rules explained the one observed production reaction and predicted a score of others. And of course it explained all the known decays, and predicted another. My research group eventually confirmed all of Gell-Mann's and Nishijima's early predictions, many of them for the first time, and we continue to be impressed by their simple elegance.

This was the state of the art in particle physics in 1954, when William Brobeck turned his brainchild, the Bevatron, over to his Radiation Laboratory associates to use as a source of high energy protons. I had been using the Berkeley proton linear accelerator in some studies of short-lived radioactive species, and I was pleased at the chance to switch to a field that appeared to be more interesting. My first Bevatron experiment was done in collaboration with Sula Goldhaber; (32) it gave the first real measurement of the τ meson lifetime. My next experiment was done with three talented young post-doctoral fellows, Frank S. Crawford, Jr., Myron L. Good, and M. Lynn Stevenson. An early puzzle in K-meson physics was that two of the particles (the θ and τ) had similar, but poorly determined lifetimes and masses. That story has been told in this auditorium by Lee (33) and Yang, (34) so I won't

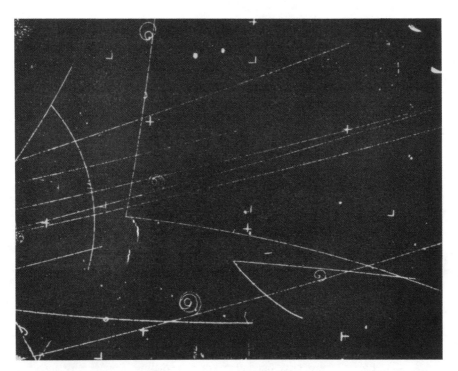

Fig. 1.
Caption: $\pi^- + p \rightarrow K^0 + \Lambda$.

repeat it now. But I do like to think that our demonstration (35), simultaneously with and independently from one by Fitch and Motley (36), that the two lifetimes were not measurably different, plus similar small limits on possible mass differences set by von Friesen et al. (37) and by Birge et al. (38), nudged Lee and Yang a bit toward their revolutionary conclusion.

Our experiences with what was then a very complicated array of scintillation counters led me and my colleagues to despair of making meaningful measurements of what we perceived to be the basic reactions of strange particle physics:

$$\pi^- + p \rightarrow \Lambda \quad + \quad K^0$$
$$\downarrow \qquad\quad \downarrow$$
$$p + \pi^- \quad \pi^- + \pi^+$$

the production reaction is indicated by the horizontal arrows, the subsequent decays by the vertical arrows. Figure 1 shows a typical example of this reaction, as we saw it later in the 10-inch bubble chamber. We concluded, correctly I believe, that none of the then known techniques was well suited to study this reaction. Counters appeared hopelessly inadequate to the task, and the spark chamber had not yet been invented. The Brookhaven diffusion cloud chamber group (17) had photographed only a few events like that shown in Fig. 1, in a period of two years. It seemed to us that a track-recording technique was called for, but each of the three known track devices had drawbacks that ruled it out as a serious contender for the role we envisaged. Nuclear emulsion, which had been so spectacularly successful in the hands of Powell's group, depended on the contiguous nature of the successive tracks at a production or decay vertex. The presence of neutral and therefore nonionizing particles between related charged particles, plus lack of even a rudimentary time resolution, made nuclear emulsion techniques virtually unusable in this new field. The two known types of cloud chambers appeared to have equally insurmountable difficulties. The older Wilson expansion chamber had two difficulties that rendered it unsuitable for the job: if used at atmospheric pressure, its cycling period was measured in minutes, and if one increased its pressure to compensate for the long mean free path of nuclear interactions, its cycling period increased at least as fast as the pressure was increased. Therefore the number of observed reactions per day started at an almost impossibly low value, and dropped as "corrective action" was taken. The diffusion cloud chamber was plagued by "background problems", and had an additional disadvantage—its sensitive volume was confined in the vertical direction to a height of only a few centimeters. What we concluded from all this was simply that particle physicists needed a track-recording device with solid or liquid density (to increase the rate of production of nuclear events by a factor of 100), with uniform sensitivity (to avoid the problems of the sensitive layer in the diffusion chamber), and with fast cycling time (to avoid the Wilson chamber problems). And of course, any cycling detector would permit the association of charged tracks joined by neutral tracks, which was denied to the user of nuclear emulsion.

In late April of 1953 I paid my annual visit to Washington, to attend the meeting of the American Physical Society. At lunch on the first day, I found myself seated at a large table in the garden of the Shoreham Hotel. All the seats but one were occupied by old friends from World War II days, and we reminisced about our experiences at the MIT radar laboratory and at Los Alamos. A young chap who had not experienced those exciting days was seated at my left, and we were soon talking of our interests in physics. He expressed concern that no one would hear his 10-minute contributed paper, because it was scheduled as the final paper of the Saturday afternoon session, and therefore the last talk to be presented at the meeting. In those days of slow airplanes, there were even fewer people in the audience for the last paper of the meeting than there are now—if that is possible. I admitted that I wouldn't be there, and asked him to tell me what he would be reporting. And that is how I heard first hand from Donald Glaser how he had invented the bubble chamber, and to what state he had brought its development. And of course he has since described those achievements from this platform (39). He showed me photographs of bubble tracks in a small glass bulb, about 1 centimeter in diameter and 2 centimeters long, filled with diethyl ether. He stressed the need for absolute cleanliness of the glass bulb, and said that

†2—693039. Alvarez

he could maintain the ether in a superheated state for an average of many seconds before spontaneous boiling took place. I was greatly impressed by his work, and it immediately occurred to me that this could be the "big idea" I felt was needed in particle physics.

That night in my hotel room I discussed what I had learned with my colleague from Berkeley, Frank Crawford. I told Frank that I hoped we could get started on the development of a liquid hydrogen chamber, much larger than anything Don Glaser was thinking about, as soon as I returned to Berkeley. He volunteered to stop off in Michigan on the way back to Berkeley, which he did, and learned everything he could about Glaser's technique.

I returned to Berkeley on Sunday, May 1, and on the next day Lynn Stevenson started to keep a new notebook on bubble chambers. The other day, when he saw me writing this talk, he showed me that old notebook with its first entry dated May 2, 1953, with Van der Waal's equation on the first page, and the isotherms hydrogen traced by hand onto the second page. Frank Crawford came home a few days later, and he and Lynn moved into the "student shop" in the synchrotron building, to build their first bubble chamber. They were fortunate in enlisting the help of John Wood, who was an accelerator technician at the synchrotron. The three of them put their first efforts into a duplication of Glaser's work with hydrocarbons. When they had demonstrated radiation sensitivity in ether, they built a glass chamber in a Dewar flask to try first with liquid nitrogen and then with liquid hydrogen.

I remember that on several occasions I telephoned to the late Earl Long at the University of Chicago, for advice on cryogenic problems. Dr. Long gave active support to the liquid hydrogen bubble chamber that was being built at that time by Roger Hildebrand and Darragh Nagle at the Fermi Institute in Chicago. In August of 1953 Hildebrand and Nagle (40) showed that superheated hydrogen boiled faster in the presence of a gamma-ray source than it did when the source was removed. This is a necessary (though not sufficient) condition for successful operation of a liquid hydrogen bubble chamber, and the Chicago work was therefore an important step in the development of such chambers. The important unanswered question concerned the bubble density—was it sufficient to see tracks of "minimum ionizing" particles, or did liquid hydrogen—as my colleagues had just shown that liquid nitrogen did—produce bubbles but no visible tracks?

John Wood saw the first tracks in a 1.5-inch-diameter liquid hydrogen bubble chamber in February of 1954 (41). The Chicago group could certainly have done so earlier, by rebuilding their apparatus, but they switched their efforts to hydrocarbon chambers, and were rewarded by being the first physicists to publish experimental results obtained by bubble chamber techniques. Figure 2 is a photograph of Wood's first tracks.

At the Lawrence Radiation Laboratory, we have long had a tradition of close cooperation between physicists and technicians. The resulting atmosphere, which contributed so markedly to the rapid development of the liquid hydrogen bubble chamber, led to an unusual phenomenon: none of the scientific papers on the development of bubble chamber techniques in my

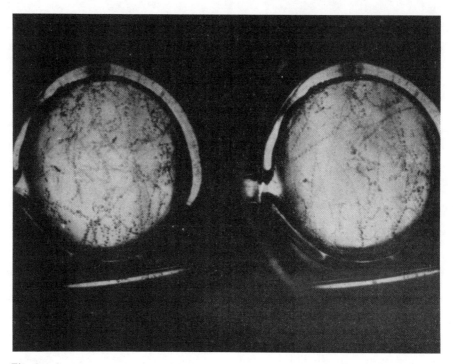

Fig. 2.
First tracks in Hydrogen.

research group were signed by experimenters who were trained as physicists or who had had previous cryogenic experience. The papers all had authors who were listed on the Laboratory records as technicians, but of course the physicists concerned knew what was going on, and offered many suggestions. Nonetheless, our technical associates carried the main responsibility, and published their findings in the scientific literature. I believe this is a healthy change from practices that were common a generation ago; we all remember papers signed by a single physicist that ended with a paragraph saying, "I wish to thank Mr. _____ , who built the apparatus and took much of the data."

And speaking of acknowledgments, John Wood's first publication, in addition to thanking Crawford, Stevenson, and me for our advice and help, said, "I am indebted to A. J. Schwemin for help with the electronic circuits." "Pete" Schwemin, the most versatile technician I have ever known, became so excited by his initial contact with John Wood's 1.5-inch-diameter all-glass chamber that he immediately started the construction of the first metal bubble chamber with glass windows. All earlier chambers had been made completely of smooth glass, without joints, to prevent accidental boiling at sharp points; such boiling of course destroyed the superheat and made the chamber insensitive to radiation. Both Glaser and Hildebrand stressed the long times their liquids could be held in the superheated condition; Hildebrand and Nagle averaged 22 seconds, and observed one superheat period of 70 seconds. John Wood reported (41), "We were discouraged by our inability to attain the long times of superheat, until the track photographs showed that it was not important in the successful operation of a large bubble chamber." I have always felt that second to Glaser's discovery of tracks this was the key observation in the whole development of bubble chamber technique. As long as one "expanded the chamber" rapidly, bubbles forming on the wall didn't destroy the superheated condition of the main volume of the liquid, and it remained sensitive as a track-recording medium.

Pete Schwemin, with the help of Douglas Parmentier, built the 2.5-inch-diameter hydrogen chamber in record time, as the world's first "dirty chamber". I've never liked that expression, but it was used for a while to distinguish chambers with windows gasketed to metal bodies from all-glass chambers. Because of it "dirtiness", the 2.5-inch chamber boiled at its walls, but still showed good tracks throughout its volume. Now that "clean" chambers are of historical interest only, we can be pleased that the modern chambers need no longer be stigmatized by the adjective "dirty".

Lynn Stevenson's notebook shows a diagram of John Wood's chamber dated January 25, 1954, with Polaroid pictures of tracks in hydrogen. A month later he recorded details of Schwemin's 2.5-inch chamber, and drew a complete diagram dated March 5. (That was the day after the Physical Review received Wood's letter announcing the first observation of tracks.) On April 29, Schwemin and Parmentier photographed their first tracks; these

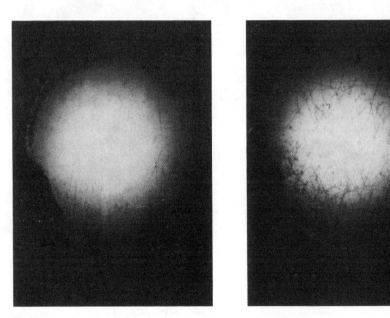

Fig. 3.
Tracks in 2 1/2 inch chamber: Neutrons (left); Gamma Rays (right).

Fig. 4.
4 inch chamber. D. Parmentier (left), A. J. Schwemin (right).

are shown in Fig. 3. (Things were happening so fast at this time that the 2.5-inch system was never photographed as a whole before it ended up on the scrap pile.)

In August, Schwemin and Parmentier separately built two different 4-inch-diameter chambers. Both were originally expanded by internal bellows, and Parmentier's 4-inch chamber gave tracks on October 6. Schwemin's chamber produced tracks three weeks later, and survived as *the* 4-inch chamber. See Fig. 4. The bellows systems in both chambers failed, but it turned out to be

Fig. 5.
Display of chambers, November 1968. From left to right, 1 1/2, 4, 6, 10, 15 and 72 inch chambers; Hernandez, Schwemin, Rinta, Watt, Alvarez and Eckman.

easier to convert Schwemin's chamber to the vapor expansion system that was used in all our subsequent chambers until 1962. (In that year, the 25-inch chamber introduced the "Ω-bellows" that is now standard for large chambers.)

Figure 5 shows all our chambers displayed together a few weeks ago, at the request of Swedish Television. As you can see, we all look pretty pleased to see so many of our "old friends" side by side for the first time.

Figure 6 shows an early picture of multiple meson production in the 4-inch chamber. This chamber was soon equipped with a pulsed magnetic field, and in that configuration it was the first bubble chamber of any kind to show magnetically curved tracks. It was then set aside by our group as we pushed on to larger chambers. But it ended its career as a useful research tool at the Berkeley electron synchrotron, after almost two million photographs of 300-MeV bremsstrahlung passing through it had been taken and analyzed by Bob Kenney et al. (43).

In the year 1954, as I have just recounted, various members of my research group had been responsible for the successful operation of four separate liquid hydrogen bubble chambers, increasing in diameter from 1.5 inches to 4 inches. By the end of that eventful year, it was clear that it would take a more concerted engineering-type approach to the problem if we were to progress to the larger chambers we felt were essential to the solution of high energy physics problems. I therefore enlisted the assistance of three close associates, J. Donald Gow, Robert Watt, and Richard Blumberg. Don Gow and Bob Watt had taken over full responsibility for the development and operation of the 32-MeV linear accelerator that had occupied all my attention from its inception late in 1945 until it first operated in late 1947. Neither of them had any ex-

Fig. 6.
Multiple Meson Production in 4 inch chamber.

perience with cryogenic techniques, but they learned rapidly, and were soon leaders in the new technology of hydrogen bubble chambers. Dick Blumberg had been trained as a mechnical engineer, and he had designed the equipment used by Crawford, Stevenson, and me in our experiments, then in progress, on the Compton scattering of γ rays by protons (44).

Wilson Powell had built two large magnets to accommodate his Wilson Cloud Chambers, pictures from which adorned the walls of every cyclotron laboratory in the world. He very generously placed one of these magnets at our disposal, and Dick Blumberg immediately started the mechanical design of the 10-inch chamber—the largest size we felt could be accommodated in the well of Powell's magnet. Blumberg's drafting table was in the middle of the single room that contained the desks of all the members of my research group. Not many engineers will tolerate such working conditions, but Blumberg was able to do so and he produced a design that was quickly built in the main machine shop. All earlier chambers had been built by the experimenters themselves. The design of the 10-inch chamber turner out to be a much larger job than we had foreseen. By the time it was completed, eleven members of the Laboratory's Mechanical Engineering Department had worked on it, including Rod Byrns and John Mark. The electrical engineering aspects of all our large chambers were formidable, and we are indebted to Jim Shand for his leadership in this work for many years.

Great difficulty was experienced with the first operation of the 10-inch chamber; too much hydrogen was vaporized at each "expansion". Pete Schwemin quickly diagnosed the trouble and butilt a fast-acting valve that permitted the chamber to be pulsed every 6 seconds, to match the Bevatron's cycling time.

It would be appropriate to interrupt this description of the bubble chamber development program to describe the important observations made possible by the operation of the 10-inch chamber early in 1956, but instead, I will preserve the continuity by describing the further development of the hardware. In December of 1954, shortly after the 4-inch chamber had been operated in the cyclotron building for the first time, it became evident to me that the 10-inch chamber we had just started to design wouldn't be nearly large enough to tell us what we wanted to know about the strange particles. The tracks of these objects had been photographed at Brookhaven (17), and we knew they were produced copiously by the Bevatron.

The size of the "big chamber" was set by several different criteria, and fortunately all of them could be satisfied by one design. (Too often, a designer of new equipment finds that one essential criterion can be met only if the objects is very large, while an equally important criterion demands that it be very small.) All "dirty chambers" so far built throughout the world had been cylindrical in shape, and were characterized by their diameter measurement. By studying the relativistic kinematics of strange particles produced by Bevatron beams, and more particularly by studying the decay of these particles, I convinced myself that the big chamber should be rectangular, with a length of at least 30 inches. This length was next increased to 50 inches in order that there would be adequate amounts of hydrogen upstream from the required decay region, in which production reactions could take place. Later the length was changed to 72 inches, when it was realized that the depth of the chamber could properly be less than its width and that the change could be made without altering the volume. The production region corresponded to about 10 % of a typical pion-proton mean free path, and the size of the decay region was set by the relativistic time-dilated decay lengths of the strange particles, plus the requirement that there be a sufficient track length available in which to measure magnetic curvature in a "practical magnetic field" of 15,000 gauss. In summary, then, the width and depth of the chamber came rather simply from an examination of the shape of the ellipses that characterize relativistic transformations at Bevatron energies, plus the fact that the magnetic field spreads the particles across the width but not along the depth of the chamber.

The result of this straightforward analysis was a rather frightening set of numbers: The chamber length was 72 inches; its width was 20 inches, and its depth was 15 inches. It had to be pervaded by a magnetic field of 15,000 gauss, so its magnet would weigh at least 100 tons and would require 2 or 3 megawatts to energize it. It would require a window 75 inches long by 23 inches wide and 5 inches thick to withstand the (deuterium) operating pressure of 8 atmospheres, exerting force of 100 tons on the glass. No one had any experience with such large volumes of liquid hydrogen; the hydrogen-oxygen rocket engines that now power the upper stages of the Saturn boosters were still gleams in the eyes of their designers—these were pre-Sputnik days. The safety aspects of the big chamber were particularly worrisome. Low temperature laboratories had a reputation for being dangerous places in which to work, and they didn't deal with such large quantities of liquid hydrogen, and what supplies they did use were kept at atmospheric pressure.

For some time, the glass window problem seemed insurmountable—no one had ever cast and polished such a large piece of optical glass. Fortunately for the eventual success of the project, I was able to persuade myself that the chamber body could be constructed of a transparent plastic cylinder with metallic end plates. This notion was later demolished by my engineering colleagues, but it played an important role in keeping the project alive in my own mind until I was convinced that the glass window could be built. As an indication of the cryogenic "state of the art" at the time we worried about the big window, I can recall the following anecdote. One day, while looking through a list of titles of talks at a recent cryogenic conference, I spotted one that read, "Large glass window for viewing liqued hydrogen." Eagerly I turned to the paper—but it described a metallic Dewar vessel equipped with a glass window 1 inch in diameter!

Don Gow was now devoting all his time to hydrogen bubble chambers, and in January of 1955 we interested Paul Hernandez in taking a good hard engineering look at the problems involved in building and housing the 72-inch bubble chamber. We were also extremely fortunate in being able to interest the cryogenic engineers at the Boulder, Colorado, branch of the National Bureau of Standards in the project. Dudley Chelton, Bascomb Birmingham and Doug Mann spent a great deal of time with us, first educating us in large-scale liquid hydrogen techniques, and later cooperating with us in the design and initial operation of the big chamber.

In April of 1955, after several months of discussion of the large chamber, I wrote a document entitled "The Bubble Chamber Program at UCRL." This paper showed in some detail why it was important to build the large chamber, and outlined a whole new way of doing high energy physics with such a device. It stressed the need for semiautomatic measuring devices (which had not previously been proposed), and described how electronic computers would reconstruct tracks in space, compute momenta, and solve problems in relativistic mechanics. All these techniques are now part of the "standard bubble chamber method", but in April of 1955 no one had yet applied them. Of all the papers I have written in my life, none gives me so much satisfaction on rereading as does this unpublished prospectus.

After Paul Hernandez and Don Gow had estimated that the big chamber,

†3—693039, Alvarez

including its building and power supplies, would cost about 2.5 million dollars, it was clear that a special AEC appropriation was required; we could no longer build our chambers out of ordinary laboratory operating money. In fact, the document I've just described was written as a sort of proposal to the AEC for financial support—but without mentioning money! I asked Ernest Lawrence if he would help me in requesting extra funds from the AEC. He read the document, and agreed with the points I had made. He then asked me to remind him of the size of the world's largest hydrogen chamber. When I replied that it was 4 inches in diameter, he said he thought I was making too large an extrapolation in one step, to 72 inches. I told him that the 10-inch chamber was on the drawing board, and if we could make it work, the operation of the 72-inch chamber was assured. (And if we couldn't make it work, we could refund most of the 2.5 million.) This wasn't obvious until I explained the hydraulic aspects of the expansion system of the 72-inch chamber; it was arranged so that the 20-inch wide, 72-inch long chamber could be considered to be a large collection of essentially independently expanded 10-inch square chambers. He wasn't convinced of the wisdom of the program, but in a characteristic gesture, he said, "I don't believe in your big chamber, but I do believe in you, and I'll help you to obtain the money." I therefore accompanied him on his next trip to Washington, and we talked in one day to three of the five Commissioners: Lewis Strauss, Willard Libby (who later spoke from this podium), and the late John Von Neumann, the greatest mathematical physicist then living. That evening, at a cocktail party at Johnny Von Neumann's home, I was told that the Commission had voted that afternoon to give the laboratory the 2.5 million dollars we had requested. All we had to do now was build the thing and make it work!

Design work had of course been under way for some time, but it was now rapidly accelerated. Don Gow assumed a new role that is not common in physics laboratories, but is well known in military organizations; he became my "chief of staff". In this position, he coordinated the efforts of the physicists and engineers; he had full responsibility for the careful spending of our precious 2.5 million dollars, and he undertook to become an expert second to none in all the technical phases of the operation, from low temperature thermodynamics to safety engineering. His success in this difficult task can be recognized most easily in the success of the whole program, culminating in the fact that I am speaking here this afternoon. I am sorry that Don Gow can't be here today; he died several years ago, but I am reminded of him every day—my three-year-old son is named Donald in his memory.

The engineering team under Paul Hernandez's direction proceeded rapidly with the design, and in the process solved a number of difficult problems in ways that have become standard "in the industry". A typical problem involved the very considerable differential expansion between the stainless steel chamber and the glass window. This could be lived with in the 10-inch chamber, but not in the 72-inch. Jack Franck's "inflatable gasket" allowed the glass to be seated against the chamber body only after both had been cooled to liquid hydrogen temperature.

Just before leaving for Stockholm, I attended a ceremony at which Paul Hernandez was presented with a trophy honoring him as a "Master Designer" for his achievements in the engineering of the 72-inch chamber. I had the pleasure of telling in more detail than I can today of his many contributions to the success of our program. One of his associates recalled a special service that he rendered not only to our group but to all those who followed us in building liquid hydrogen bubble chambers. Hernandez and his associates wrote a series of "Engineering Notes", on matters of interest to designers of hydrogen bubble chambers, that soon filled a series of notebooks that spanned 3 feet of shelf space. Copies of theses were sent to all interested parties on both sides of the Atlantic, and I am sure that they resulted in a cumulative savings to all bubble chamber builders of several million dollars; had not all this information been readily available, the test programs and calculations of our engineering group would have required duplication at many laboratories, at a large expense of money and time. Our program moved so rapidly that there was never time to put the Engineering Notes into finished form for publication in the regular literature. For this reason, one can now read review articles on bubble chamber technology, and be quite unaware of the part that our Laboratory played in its development. There are no references to papers by members of our group, since those papers were never written—the data that would have been in them had been made available to everyone who needed them at a much earlier date.

And just to show that I was also deeply involved in the chamber design, I might recount how I purposely "designed myself into a corner" because I thought the results were important, and I thought I could invent a way out of a severe difficulty, if given the time. All previous chambers had had two windows, with "straight through" illumination. Such a configuration reduces the attainable magnetic field, because the existence of a rear pole piece would interfere with the light-projection system. I made the decision that the 72-inch chamber would have only a top window, thereby permitting the magnetic field to be increased by a lower pole piece and at the same time saving the cost of the extra glass window, and also providing added safety by eliminating the possibility that liquid hydrogen could spill through a broken lower window. The only difficulty was that for more than a year, as the design was firmed up and the parts were fabricated, none of us could invent a way both to illuminate and to photograph the bubbles through the same window. Duane Norgren, who has been responsible for the design of all our bubble chamber cameras, discussed the matter with me at least once a week in that critical year, and we tried dozens of schemes that didn't quite do the job. But as a result of our many failures, we finally came to understand all the problems, and we eventually hit on the retrodirecting system known as coat hangers. This solution came none too soon; if it had been delayed by a month or more, the initial operation of the 72-inch chamber would have been correspondingly delayed. We took many other calculated risks in designing the system; if we had postponed the fabrication of the major hardware until we had solved all the problems on paper, the project might still not be completed. Engineers

Fig. 7.
72 inch bubble chamber in its building.

are conservative people by nature; it is the ultimate disgrace to have a boiler explode or a bridge collapse. We were therefore fortunate to have Paul Hernandez as our chief engineer; he would seriously consider anything his physics colleagues might suggest, no matter how outlandish it might seem at first sight. He would firmly reject it if it couldn't be made safe, but before rejecting any idea for lack of safety he would use all the ingenuity he possessed to make it safe.

Fig. 8.
"Franckenstein."

We felt that we needed to build a test chamber to gain experience with a single-window system, and to learn to operate with a hydrogen refrigerator, our earlier chambers had all used liquid hydrogen as a coolant. We therefore built and operated the 15-inch chamber in the Powell magnet, in place of the 10-inch chamber that had served us so well.

The 72-inch chamber operated for the first time on March 24, 1959, very nearly four years from the time it was first seriously proposed. Figure 7 shows it at about that time. The "start-up team" consisted of Don Gow, Paul Hernandez, and Bob Watt, all of whom had played key roles in the initial operation of the 15-inch chamber. Bob Watt and Glenn Eckman have been responsible for the operation of all our chambers from the earliest days of the 10-inch chamber, and the success of the whole program has most often rested in their hands. They have maintained an absolutely safe operating record in the face of very severe hazards, and they have supplied their colleagues in the physics community with approximately ten million high-quality stereo photographs. And most recently, they have shown that they can design chambers as well as they have operated them. The 72-inch chamber was recently enlarged to an 82-inch inze, incorporating to a large extent the design concepts of Watt and Eckman.

Although I haven't done justice to the contributions of many close friends and associates who shared in our bubble chamber development program, I must now turn to another important phase of our activities—the data-analysis program. Soon after my 1955 prospectus was finished, Hugh Bradner undertook to implement the semiautomatic measuring machine proposal. He first made an exhaustive study of commercially available measuring machines, encoding techniques, etc., and then, with Jack Franck, designed the first "Franckenstein". This rather revolutionary device has been widely copied, to such an extent that objects of its kind are now called "conventional" measuring machines (Fig. 8). Our first Franckenstein was operating reliably in 1957, and in the summer of 1958 a duplicate was installed in the U.S. exhibit at the "Atoms for Peace" exposition in Geneva. It excited a great deal of interest in the high energy physics community, and a number of groups set out to make similar machines based on its design. Almost everyone thought at first that our provision for automatic track following was a needless waste of money, but over the years, that feature has also come to be "conventional".

Jack Franck then went on to design the Mark II Franckenstein, to measure 72-inch bubble chamber film. He had the first one ready to operate just in time to match the rapid turn-on of the big chamber, and he eventually built three more of the Mark II's. Other members of our group then designed and perfected the faster and less expensive SMP system, which added significantly to our "measuring power". The moving forces in this development were Pete Schwemin, Bob Hulsizer, Peter Davey, Ron Ross, and Bill Humphrey (45). Our final and most rewarding effort to improve our measuring ability was fulfilled several years ago, when our first Spiral Reader became operational. This single machine has now measured more than one and a half million high energy interactions, and has, together with its almost identical twin, measured one and a quarter million events in the last year. The SAAB Company here in Sweden is now building and selling Spiral Readers to European laboratories.

The Spiral Reader had a rather checkered career, and it was on several occasions believed by most workers in the field to have been abandoned by our group. The basic concept of the spiral scan was supplied by Bruce McCormick, in 1956. Our attempts to reduce his ideas to practice resulted in failure, and shortly after that, McCormick moved to Illinois, where he has since been engaged in computer development. As the cost transistorized circuits dropped rapidly in the next years, we tried a second time to implement the Spiral Reader concept, using digital techniques to replace the analog devices of the earlier machine. The second device showed promise, but its "hard-wired logic" made it too inflexible, and the unreliability of its electronic components kept it in repair most of the time. The mechanical and optical components of the second Spiral Reader were excellent, and we hated to drop the whole project simply because the circuitry didn't come up to the same standard. In 1963, Jack Lloyd suggested that we use one of the new breed of small high-speed, inexpensive computers to supply the logic and the control circuits for the Spiral Reader. He then demonstrated great qualities of leadership by delivering to our research group a machine that has performed even better than he had promised it would. In addition to his development of the hardware, he initiated POOH, the Spiral Reader filtering program, which was brought to a high degree of perfection by Jim Burkhard. The smooth and rapid transition of the Spiral Reader from a developmental stage into a useful

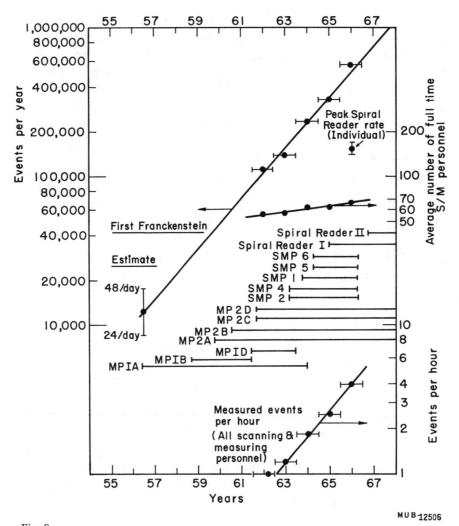

Fig. 9.
Measuring Rates.

MUB 12506

operational tool was largely the result of several years of hard work on the part of Gerry Lynch and Frank Solmitz. Figure 9, from a talk I gave two and a half years ago (46), shows how the measuring power of our group has increased over the years, with only a modest increase in personnel.

According to a simple extrapolation of the exponential curve we had been on from 1957 through 1966, we would expect to be measuring 1.5 million events per year some time in 1969. But we have already reached that rate and we will soon be leveling off about there because we have stopped our development work in this area.

The third key ingredient of our development program has been the continually increasing sophistication in our utilization of computers, as they have increased in computational speed and memory capacity. While I can speak from a direct involvement in the development of bubble chambers and measuring machines, and in the physics done with those tools, my relationship to our computer programming efforts is largely that of an amazed spectator. We were most fortunate that in 1956 Frank Solmitz elected to join our group. Although the rest of the group thought of themselves as experimental physicists, Solmitz had been trained as a theorist, and had shown great aptitude in the development of statistical methods of evaluating experimental data. When he saw that our first Franckenstein was about to operate, and no computer programs were ready to handle the data it would generate, he immediately set out to remedy the situation. He wrote HYDRO, our first system program for use on the IBM 650 computer. In the succeeding twelve years he has continued to carry the heavy responsibility for all our programming efforts. A major breakthrough in the analysis of bubble chamber events was made in the years 1957 through 1959. In this period, Solmitz and Art Rosenfeld, together with Horace Taft from Yale University and Jim Snyder from Illinois, wrote the first "fitting routine", GUTS, which was the core of our first "kinematics program, KICK". To explain what KICK did, it is easiest to describe what physicists had to do before it was written. HYDRO and its successor, PANG, listed for each vertex the momentum and space angles of the tracks entering or leaving that vertex, together with the calculated errors in these measurements. A physicist would plot the angular coordinates on a stereographic projection of a unit sphere known as a Wolff-plot. If he was dealing with a three-track vertex—and that was all we could handle in those days—he would move the points on the sphere, within their errors, if possible, to make them coplanar. And of course he would simultaneously change the momentum values, within their errors, to insure that the momentum vector triangle closed, and energy was conserved. Since momentum is a vector quantity, the various conditions could be simultaneously satisfied only after the angles and the absolute values of the momenta had been changed a number of times in an iterative procedure. The end result was a more reliable set of momenta and angles, constrained to fit the conservation laws of energy and momentum. In a typical case, an experienced physicist could solve only a few Wolff-plot problems in a day. (Lynn Stevenson had written a specific program, COPLAN, that solved a particular problem of interest to him that was later handled by the more versatile GUTS.)

GUTS was being written at a time when one highly respected visitor to the group saw the large pile of PANG printout that had gone unanalyzed because so many of our group members were writing GUTS—a program that was planned to do the job automatically. Our visitor was very upset at what he told me was a "foolish deployment of our forces". He said, "If you would only get all those people away from their program writing, and put them to work on Wolff-plots, we'd have the answer to some really important physics in a month or two". I said I was sure we'd end up with a lot more physics in the next years if my colleagues continued to write GUTS and KICK. I'm sure that those who wrote these pioneering "fitting and kinematics programs" were subjected to similar pressures. Everyone in the high energy physics community has long been indebted to these farsighted men because they knew that what they were doing was right. KICK was soon developed so that it gave an overall fit to several interconnected vertices, with various hypothetical identities of the several tracks assumed in a series of attempts at a fit. The relationship beteen energy and momentum depends on mass, so a highly constrained fit can be obtained only if the particle responsible for each track is properly identified. If the degree of constraint is not so high, more than one "hypothesis" (set of track identifications) may give a fit, and the physicist must use his judgment in making the identification.

As another example in this all-too-brief sketch of the computational aspects of our work, I will mention an important program, initiated by Art Rosenfeld and Ron Ross, that has removed much of the remaining drudgery from the bubble chamber physicists' life. SUMX is a program that can easily by instructed to search quickly through large volumes of "kinematics program output," printing out summaries and tabulations of interesting data. (Like all our pioneering programs, SUMX was replaced by an improved and more versatile program—in this case, KIOWA. But I will continue to talk as though SUMX were still used.) A typical SUMX printout will be a computer-printed document 3 inches thick, with hundreds of histograms, scatter plots, etc.

Hundreds of histograms are similarly printed showing numbers of events with effective masses for many different combinations of particles, with various "cuts" on momentum transfer, etc. What all this amounts to is simply that a physicist is no longer rewarded for his ability in deciding what histograms he should tediously plot and then examine. He simply tells the computer to plot all histograms of any possible significance, and then flips the pages to see which ones have interesting features.

One of my few real interactions with our programming effort came when I suggested to Gerry Lynch the need for a program he wrote that is known as GAME. In my work as a nuclear physicist before World War II, I had often been skeptical of the significance of the "bumps" in histograms, to which importance was attached by their authors. I developed my own criteria for judging statistical significance, by plotting simulated histograms, assuming the curves to be smooth; I drew several samples of "Monte Carlo distributions", using a table of random numbers as the generator of the samples. I usually found that my skepticism was well founded because the "faked" histograms showed as much structure as the published ones. There are of course many statistical tests designed to help one evaluate the reality of bumps in histograms, but im my experience nothing is more convincing than an examination of a set of simulated histograms from an assumed smooth distribution.

GAME made it possible, with the aid of a few control cards, to generate a hundred histograms similar to those produced in any particular experiment. All would contain the same number of events as the real experiment, and would be based on a smooth curve through the experimental data. The standard procedure is to ask a group of physicists to leaf through the 100 histograms—with the experimental histogram somewhere in the pile—and vote on the apparent significance of the statistical fluctuations that appear. The first time this was tried, the experimenter—who had felt confident that his bump was significant—didn't know that his own histogram was in the pile, and didn't pick it out as convincing; he picked out two of the computer-generated histograms as looking significant, and pronounced all others—including his own—as of no significance! In view of this example, one can appreciate how many retractions of discovery claims have been avoided in our group by the liberal use of the GAME program.

As a final example from our program library, I'll mention FAKE, which, like SUMX, has been widely used by bubble chamber groups all over the world. FAKE, written by Gerry Lynch, generates simulated measurements of bubble chamber events to provide a method of testing the analysis programs to determine how frequently they arrive at an incorrect answer.

Now that I have brought you up to date on our parallel developments of hardware and software (computer programs), I can tell you what rewards we have reaped, as physicists, from their use. The work we did with the 4-inch chamber at the 184-inch cyclotron and at the Bevatron cannot be dignified by the designation "experiments", but it did show examples of π-μ-e decay and neutral strange-particle decay. The experiences we had in scanning the 4-inch film merely whetted our appetite for the exciting physics we felt sure would be manifest in the 10-inch chamber, when it came into operation in Wilson Powell's big magnet.

Robert Tripp joined the group in 1955, and as his first contribution to our program he designed a "separated beam" of negative K mesons that would stop in the 10-inch chamber. We had two different reasons for starting our bubble chamber physics program with observations of the behavior of K^- mesons stopping in hydrogen. The first reason involved physics: The behavior of stopping π^- mesons in hydrogen had been shown by Panofsky (47) and his co-workers to be a most fruitful cource of fundamental knowledge concerning particle physics. The second reason was of an engineering nature: Only one Bevatron "straight section" was available for use by physicists, and it was in constant use. In order not to interfere with other users, we decided to set the 10-inch chamber close to a curved section of the Bevatron, and use secondary particles, from an internal target, that penetrated the wall of the vacuum chamber and passed between neighboring iron blocks in the return yoke of the Bevatron magnet. This physical arrangement gave us negative particles (K^- and π^- mesons) of a well-defined low momentum. By introducing an absorber into the beam, we brought the K^- mesons almost to rest, but allowed the lighter π^- mesons to retain a major fraction of their original momentum. The Powell magnet provided a second bending that brought the K^- mesons into the chamber, but kept the π^- mesons out. That was the theory of this first separated beam for bubble chamber use. But in practice, the chamber was filled with tracks of pions and muons, and we ended up with only one stopped K^- per roll of 400 stereo pairs. It is now common for experimenters to stop one million K^- mesons in hydrogen, in a single experimental run, but the 137 K^- mesons we stopped in 1956 (48) gave us a remarkable preview of what has now been learned in the much longer exposures. We measured the relative branching of $K^- + p$ into

$$\Sigma^- + \pi^+ : \Sigma^+ \pi^- : \Sigma^0 + \pi^0 : \Lambda + \pi^0.$$

And in the process, we made a good measurement of the Σ^0 mass. We plotted the first decay curves for the Σ^+ and Σ^- hyperons, and we observed for the first time the interactions of Σ^- hyperons and protons at rest. We felt amply rewarded for our years of developmental work on bubble chambers by the very interesting observations we were now privileged to make.

We had a most exciting experience at this time, that was the result of two circumstances that no longer obtain in bubble chamber physics. In the first place, we did all our own scanning of the photographic film. Such tasks are now carried out by professional scanners, who are carefully trained to recognize and record "interesting events". We had no professional scanners at the time, because we wouldn't have known how to train them before this first film became available. And even if they had been trained, we would not have let them look at the film—we found it so completely absorbing that there was always someone standing behind a person using one of our few film viewers, ready to take over when the first person's eyes tired. The second circumstance that made possible the accidental discovery I am about to describe was the very poor quality of our separated K^- beam—by modern standards. Most of the tracks we observed were made by negative pions or muons, but we also saw many positively charged particles—protons, pions, and muons.

At first we kept no records of any events except those involving strange particles; we would look quickly at each frame in turn, and shift to the next one if no "interesting event" showed up. In doing this scanning, we saw many examples of π^+-μ^+-e^+ decays, usually from a pion at rest, and we soon learned about how long to expect the μ^+ track to be—about 1 centimeter. I did my scanning on a stereo viewer, so I probably had a better feeling for the length of a μ^+ track in space than did my colleagues, who looked at two projections of the stereo views, sequentially. Don Gow, Hugh Bradner, and I often scanned at the same time, and we showed each other whatever interesting events came into view. Each of us showed the others examples of what we thought was an unusual decay scheme: $\pi^- \rightarrow \mu^- \rightarrow e^-$. The decay of a μ^- at rest into an e^-, in hydrogen, was expected from the early observations by Conversi et al. (3), but Panofsky (47) had shown that a π^- meson couldn't decay at rest in hydrogen. Our first explanation for our observations was simply that the pion had decayed just before stopping. But we gradually became convinced that this explanation really didn't fit the facts. There were too many muon tracks of about the same length, and none that were appreciably longer or shorter, as the decay-in-flight hypothesis would predict. We now began to keep records of these "anomalous decays", as we still called them, and we found occasional examples in which the muon was horizontal in the chamber, so its length could be measured. (We had as yet no way of reconstructing tracks in space from two stereo views.) By comparing the measured length of the negative muon

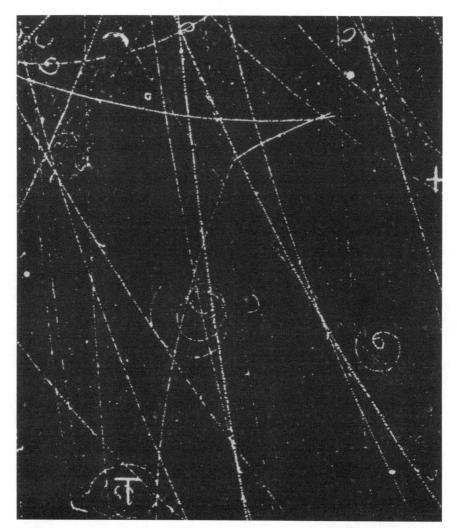

Fig. 10.
Muon Catalysis (with gap).

track with that of its more normal positive counterpart, we estimated that the negative muons had an energy of 5.4 MeV, rather than the well-known positive muon energy (from positive pion decay at rest) of 4.1 MeV. This confirmed our earlier suspicion that the long primary negative track couldn't be that of a pion, but it left us just as much in the dark as to the nature of the primary.

After these observations had been made, I gave a seminar describing what we had observed, and suggesting that the primary might be a previously unknown weakly interacting particle, heavier than the pion, that decayed into a muon and a neutral particle, either neutrino or photon. We had just made the surprising observation, shown in Fig. 10, that there was often a gap, measured in millimeters, between the end of the primary and the beginning of the secondary. This finding suggested diffusion by a rather long-lived negative particle that orbited around and neutralized one of the protons in the liquid hydrogen. We had missed many tracks with these "gaps" because no one had seen such a thing before; we simply ignored such track configurations by subconsciously assuming that they were unassociated events in a badly cluttered bubble chamber.

One evening, one of the members of our research team, Harold Ticho from our Los Angeles campus, was dining with Jack Crawford, a Berkeley astrophysicist he had known when they were students together. They discussed our observations at some length, and Crawford suggested the possibility that a fusion reaction might somehow be reponsible for the phenomenon. They calculated the energy released in several such reactions, and found that it agreed with experiment if a stopped muon were to be binding together a proton and a deuteron into an HD μ^--molecular ion. In such a "mulecule" the proton and deuteron would be brought into such close proximity for such a long time that they would fuse into ^{3}He, and could deliver their fusion energy to the muon by the process of internal conversion. However, they couldn't think of any mechanism that would make the reaction happen so often—the fraction of deuterons in liquid hydrogen is only 1 in 5000. They had, however, correctly identified the reaction, but a key ingredient in the theoretical explanation was still missing.

The next day, when we had all accepted the idea that stopped muons were catalyzing the fusion of protons and deuterons, our whole group paid a visit to Edward Teller, at his home. After a short period of introduction to the observations and to the proposed fusion reaction, he explained the high probability of the reaction as follows: the stopped muon radiated its way into the lowest Bohr orbit around a proton. The resulting muonic hydrogen atom, $p\mu^-$, then had many of the properties of a neutron, and could diffuse freely through the liquid hydrogen. When it came close to the deuteron in an HD molecule, the muon would transfer to the deuteron, because the ground state of the μ^-d atom is lower than that of the μ^-p atom, in consequence of "reduced mass" effect. The new "heavy neutron" $d\mu^-$ might then recoil some distance as a result of the exchange reaction, thus explaining the "gap." The final stage of capture of a proton into a $pd\mu^-$ molecular ion was also energetically favorable, so a proton and deuteron could now be confined close enough together by the heavy negative muon to fuse into a ^{3}He nucleus plus the energy given to the internally converted muon.

We had a short but exhilarating experience when we thought we had solved all of the fuel problems of mankind for the rest of time. A few hasty calculations indicated that in liquid HD a single negative muon would catalyze enough fusion reactions before it decayed to supply the energy to operate an accelerator to produce more muons, with energy left over after making the liquid HD from sea water. While everyone else had been trying to solve this problem by heating hydrogen plasmas to millions of degrees, we had apparently stumbled on the solution, involving very low temperatures instead. But soon, more

Fig. 11.
Double Muon Catalysis.

realistic estimates showed that we were off the mark by several orders of magnitude—a "near miss" in this kind of physics!

Just before we published our results (49), we learned that the "μ-catalysis" reaction had been proposed in 1947 by Frank (50) as an alternative explanation of what Powell et al. had assumed (correctly) to be the decay of π^+ to μ^+. Frank suggested that it might be the reaction we had just seen in liquid hydrogen, starting with a μ^-, rather than with a π^+. Zeldovitch (51) had extended the ideas of Frank concerning this reaction, but because their

Fig. 12.

No spectrometers on

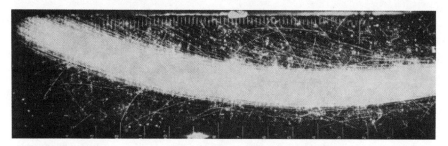

One spectrometer on

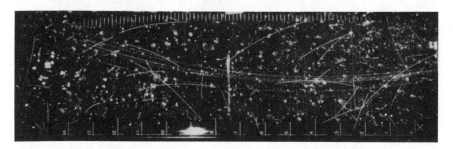

Two spectrometers on

K⁻ BEAM IN 72-INCH BUBBLE CHAMBER

papers were not known to anyone in Berkeley, we had a great deal of personal pleasure that we otherwise would have missed.

I will conclude this episode by noting that we immediately increased the deuterium concentration in our liquid hydrogen and observed the expected increase in fusion reactions, and saw two examples of successive catalyses by a single muon (Fig. 11). We also observed the catalysis of $D + D \rightarrow {}^3H + {}^1H$ in pure liquid deuterium.

A few months after we had announced our μ-catalysis results, the world of particle physics was shaken by the discovery that parity was not conserved in beta decay. Madame Wu and her collaborators (52), acting on a suggestion by Lee and Yang (53), showed that the β rays from the decay of oriented ^{60}Co nuclei were emitted preferentially in a direction opposite to that of the spin. Lee and Yang suggested that parity nonconservation might also manifest itself in the weak decay of the Λ hyperon into a proton plus a negative pion. Crawford et al. had moved the 10-inch chamber into a negative pion beam, and were analyzing a large sample of Λ's from associated production events. They looked for an "up-down asymmetry" in the emission of pions from Λ's, relative to the "normal to the production plane," as suggested by Lee and Yang. As a result, they had the pleasure of being the first to observe parity nonconservation in the decay of hyperons (54).

In the winter of 1958, the 15-inch chamber had completed its engineering test run as a prototype for the 72-inch chamber, and was operating for the first time as a physics instrument. Harold Ticho, Bud Good, and Philippe Eberhard (55) had designed and built the first separated beam of K^- mesons with a momentum of more than 1 GeV/c. Figure 12 shows the appearance of a bubble chamber when such a beam is passed through it, and when one or both of the electrostatic separators are turned off. The ingenuity which has been brought to bear on the problem of beam separation, largely by Ticho and Murray, is difficult to imagine, and its importance to the success of our program cannot be overestimated (55). Joe Murray has recently joined the Stanford Linear Accelerator Center, where he has in a short period of time built a very successful radiofrequency-separated K beam and a backscattered laser beam.

The first problem we attacked with the 15-inch chamber was that of the Ξ^0. Gell-Mann had predicted that the Ξ^- was one member of an I-spin doublet, with strangeness minus 2. The predicted partner of the Ξ^- would be a neutral hyperon that decayed into a Λ and a π^0—both neutral particles that would, like the Ξ^0, leave no track in the bubble chamber. A few years earlier, as an after-dinner speaker at a physics conference, Victor Weisskopf had "brought down the house" by exhibiting an absolutely blank cloud chamber photograph, and saying that it represented proof of the decay of a new neutral particle into two other neutral particles! And now we were seriously planning to do what had been considered patently ridiculous only a few years earlier.

According to the Gell-Mann and Nishijima strangeness rules, the Ξ^0 should be seen in the reaction

$$
\begin{array}{ccc}
K^- + p \rightarrow \Xi^0 & + & K^0 \\
\downarrow & & \downarrow \\
\Lambda + \pi^0 & & \pi^- + \pi^+ \\
\downarrow & & \\
\pi^- + p & &
\end{array}
$$

In the one example of this reaction that we observed, Fig. 13, the charged pions from the decay of the neutral K^0 yielded a measurement of the energy and direction of the unobserved K^0. Through the conservation laws of energy and momentum (plus a measurement of the momentum of the interacting K^-

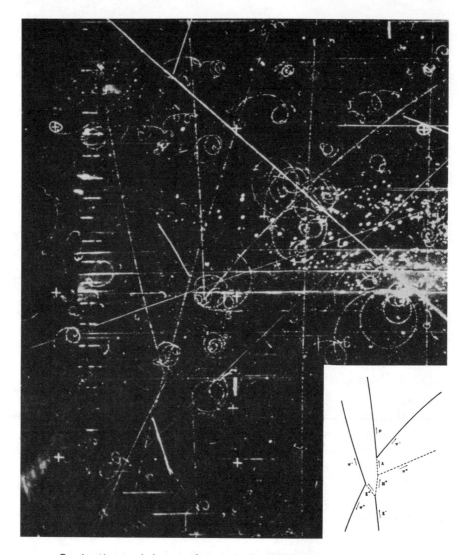

Production and decay of a neutral cascade hyperon (Xi zero).

Fig. 13.

track) we could calculate the mass of the coproduced Ξ^0 hyperon plus its velocity and direction of motion. Similarly, measurements of the π^- and proton gave the energy and direction of motion of the unobserved Λ, and proved that it did not come directly from the point at which the K^- meson interacted with the proton. The calculated flight path of the Λ intersected the calculated flight path of the Ξ^0, and the angle of intersection of the two unobserved but calculated tracks gave a confirming measurement of the mass of the Ξ^0 hyperon, and proved that it decayed into a Λ plus a π^-. This single hard-won event was a sort of tour de force that demonstrated clearly the power of the liquid hydrogen bubble chamber plus its associated data-analysis techniques.

Although only one Ξ^0 was observed in the short time the 15-inch chamber was in the separated K^- beam, large numbers of events showing strange-particle production were available for study. The Franckensteins were kept busy around the clock measuring these events, and those of us who had helped to build and maintain the beam now concentrated our attention on the analysis of these reactions. The most copious of the simple "topologies" was $K^- p \rightarrow$ two charged prongs plus a neutral V-particle. According to the strangeness rules, this topology could represent either

$$K^- + p \rightarrow \Lambda + \pi^+ + \pi^-$$
$$\downarrow$$
$$\pi^- + p$$

or

$$K^- \rightarrow p \rightarrow \overline{K}{}^0 + p + \pi^-$$
$$\downarrow$$
$$\pi^- + \pi^+$$

The kinematics program KICK was now available to distinguish between these two reactions, and to eliminate those examples of the same topology in which an unobserved π^0 was produced at the first vertex. SUMX had not yet been written, so the labor of plotting histograms was assumed by the two very able graduate students who had been associated with the K^- beam and its exposure to the 15-inch chamber since its planning stages: Stanley Wojcicki and Bill Graziano. They first concentrated their attention on the energies of the charged pions from the production vertex in the first of the two reactions listed above. Since there were three particles produced at the vertex—a charged pion of each sign plus a Λ—one expected to find the energies of each of the three particles distributed in a smooth and calculable way from a minimum value to a maximum value. The calculated curve is known in particle physics as the "phase-space distribution". The decay of a τ meson into three charged pions was a well known "three-particle reaction" in which the dictates of phase space were rather precisely followed.

But when Wojcicki and Graziano finished transcribing their data from

KICK printout into histograms, they found that phase-space distributions were poor approximations to what they observed. Figure 15 shows the distribution of energy of both positive and negative mesons, together with the corresponding "Dalitz plot", which Richard Dalitz (56) had originated to elucidate the "τ-θ puzzle", which had in turn led to Lee and Yang's parity-nonconservation hypothesis.

The peaked departure from a phase-space distribution had been observed only once before in particle physics, where it had distinguished the reaction

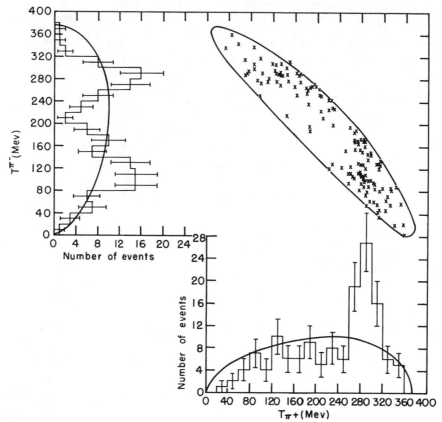

Fig. 14.
Discovery of the Y₁* (1385).

$p + p \rightarrow \pi^+ + d$ from the "three-body reaction" $p + p \rightarrow \pi^+ + p + n$. (Although no new particles were discovered in these reactions, they did contribute to our knowledge of the spin of the pion [57]). But such a peaking had been observed in the earliest days of experimentation in the artificial disintegration of nuclei, and its explanation was known from that time. Oliphant and Rutherford (58) observed the reaction $p + {}^{11}B \rightarrow 3\ {}^4He$. This is a three-body reaction, and the energies of the α particles had a phase-space-like distribution except for the fact that there was a sharp spike in the energy distribution at the highest α-particle energy. This was quickly and properly attributed (58) to the reaction

$$p + {}^{11}B \rightarrow {}^8Be + {}^4He$$
$$\downarrow$$
$${}^4He + {}^4He$$

In other words, some of the reactions proceeded via a two-body reaction, in which one α particle recoiled with unique energy against a quasistable ${}^8Be$ nucleus. But the ${}^8Be$ nucleus was itself unstable, coming apart in 10^{-16}

second into two α particles of low relative energy. The proof of the fleeting existence of ^{8}Be was the peak in the high energy α-particle distribution, showing that initially only two particles, ^{8}Be and ^{4}He, participated in the reaction.

The peaks seen in Fig. 14 were thus a proof that the $\pi^\pm$ recoiled against a combination of $\Lambda + \pi^\mp$ that had a unique mass, broadened by the effects of the uncertainty principle. The mass of the $\Lambda\pi$ combination was easily calculable as 1385 MeV, and the I-spin of the system was obviously 1, since the I-spin of the Λ is 0, and the I-spin of the π is 1. This was then the discovery of the first "strange resonance," the Y_1^* (1385). Although the famous Fermi 3, 3 resonance had been known for years, and although other resonances in the π^- nucleon system had since shown up in total cross section experiments at Brookhaven and Berkeley, CalTech and Cornell (59) the impact of the Y_1^* resonance on the thinking of particle physicists was quite different—the Y_1^* really acted like a new particle, and not simply as a resonance in a cross section.

We announced the Y_1^* at the 1960 Rochester High Energy Physics Conference (60), and the hunt for more short-lived particles began in earnest. The same team from our bubble chamber group that had found the Y_1^*(1385) now found two other strange resonances before the end of 1960—the K^*(890) (61), and the Y_0^*(1405) (62).

Although the authors of these three papers have for years been referred to as "Alston et al.", I think that on this occasion it is proper that the full list be named explicitly. In addition to Margaret Alston (now Margaret Garnjost)

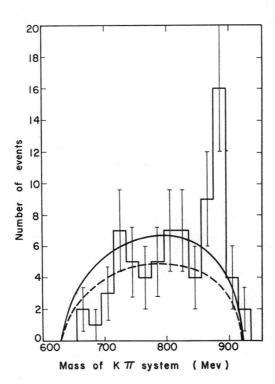

Fig. 15.
Discovery of the K* (890).

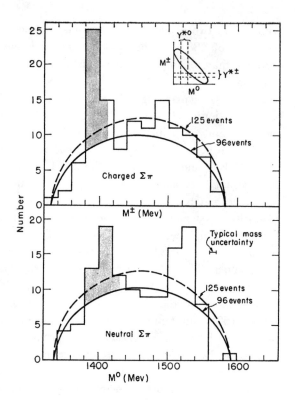

Fig. 16.
Discovery of the Y₀* (1405).

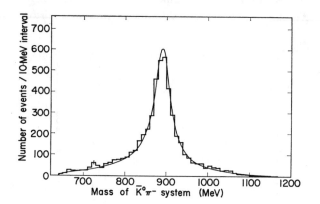

Fig. 17.
Present day K* (890).

and Luis W. Alvarez, and still in alphabetical order, the authors are:

Philippe Eberhard, Myron L. Good, William Graziano, Harold K. Ticho, and Stanley G. Wojcicki.

Figures 15 and 16 show the histograms from the papers announcing these two new particles; the K^* was the first example of a "boson resonance" found by any technique. Instead of plotting these histograms against the energy of one particle, we introduced the now universally accepted technique of plotting them against the effective mass of the composite system: $\Sigma + \pi$ for the $Y_0^*(1405)$ and $K + \pi$ for the $K^*(890)$. Figure 17 shows the present state of

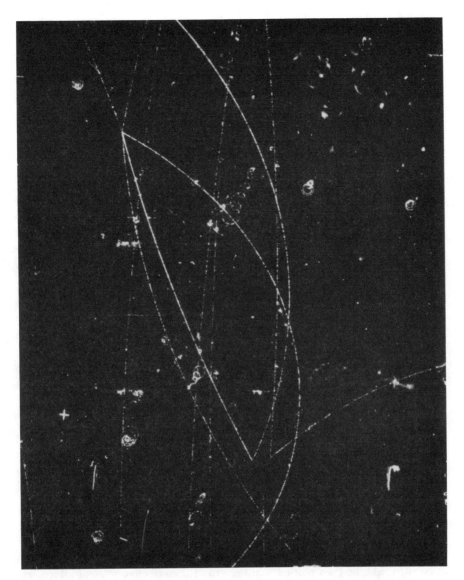

Fig. 18.
First production of Anti-Lambda.

the art relative to the $K^*(890)$; there is essentially no phase-space background in this histogram, and the width of the resonance is clearly measurable to give the lifetime of the resonant state via the uncertainty principle.

These three earliest examples of strange-particle resonances all had lifetimes of the order of 10^{-23} second, so the particles all decayed before they could traverse more than a few nuclear radii. No one had foreseen that the bubble chamber could be used to investigate particles with such short lives; our chambers had been designed to investigate the strange particles with lifetimes of 10^{-10} second—10^{13} times as long.

In the summer of 1959, the 72-inch chamber was used in its first planned physics experiment. Lynn Stevenson and Philippe Eberhard designed and constructed a separated beam of about 1.6-GeV/c antiprotons, and a quick scan of the pictures showed the now famous first example of antilambda production, via the reaction

$$\bar{p} + p \rightarrow \overline{\Lambda} + \Lambda$$
$$\downarrow \qquad \downarrow$$
$$\pi^+ + \bar{p} \quad \pi^- + p$$

Figure 18 shows this photograph, with the antiproton from the antilambda decay annihilating in a four-pion event. I believe that everyone who attended the 1959 High Energy Physics Conference in Kiev will remember the showing of this photograph—the first interesting event from the newly operating 72-inch chamber.

Hofstadter's classic experiments on the scattering of high energy electrons by protons and neutrons (63) showed for the first time how the electric charge was distributed throughout the nucleons. The theoretical interpretation of the experimental results (64) required the existence of two new particles, the vector mesons now known as the ω and the ϱ. The adjective "vector" simply means that these two mesons have one unit of spin, rather than zero, as the ordinary π and K mesons have. The ω was postulated to have I-spin = 0, and the ϱ to have I-spin = 1; the ω would therefore exist only in the neutral state, while the ϱ would occur in the +, —, and 0 charged states.

Many experimentalists, using a number of techniques, set out to find these important particles, whose masses were only roughly predicted. The first success came to Bogdan Maglić, a visitor to our group, who analyzed film from the 72-inch chamber's antiproton exposure. He made the important decision to concentrate his attention on proton-antiproton annihilations into five pions—two negative, two positive, and one neutral. KICK gave him a selected sample of such events; the tracks of the π^0 couldn't be seen, of course, but the constraints of the conservation laws permitted its energy and direction to be computed. Maglić then plotted a histogram of the effective mass of all neutral three-pion combinations. There were four such neutral combinations for each event; the neutral pion was taken each time together with all four possible pairs of oppositely charged pions. SUMX was just beginning to work, and still had bugs in it, so the preparation of the histogram was a very tedious and time-consuming chore, but as it slowly emerged, Maglić had the thrill of seeing a bump appear in the side of his phase-space distribution. Figure 19 shows a small portion of the whole distributions, with the peak that signaled the discovery of the very important ω meson.

Although Bogdan Maglić originated the plan for this search, and pushed through the measurements by himself, he graciously insisted that the paper announcing his discovery (65) should be co-authored by three of us who had developed the chamber, the beam, and the analysis program that made it possible.

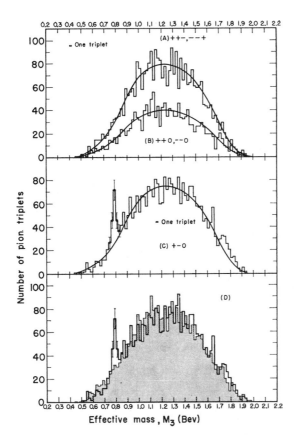

Fig. 19.
Discovery of the ω Meson.

The ϱ meson is the only one from this exciting period in the development of particle physics whose discovery cannot be assigned uniquely. In our group, the two Franckensteins were being used full time on problems that the senior members felt had higher priority. But a team of junior physicists and graduate students, Anderson et al. (66), found that they could make accurate enough measurements directly on the scanning tables to accomplish a "Chew—Low extrapolation." Chew and Low had described a rather complicated procedure to look for the predicted dipion resonance now known as the ϱ meson. Figure 20 shows the results of this work, which convinced me that the ϱ existed and had its predicted spin of 1. The mass of the ϱ was given as about 650 MeV, rather than its now accepted value of 765 MeV. (This low value is now explained in terms of the extreme width of the ϱ resonance.) The evidence for the ϱ seemed to me even more convincing than the early evidence Fermi and his co-workers produced in favor of the famous 3, 3 pion-nucleon resonance.

But one of the unwritten laws of physics is that one really hasn't made a discovery until he has convinced his peers that he has done so. We had just persuaded high energy physicists that the way to find new particles was to look for bumps on effective-mass histograms, and some of them were therefore

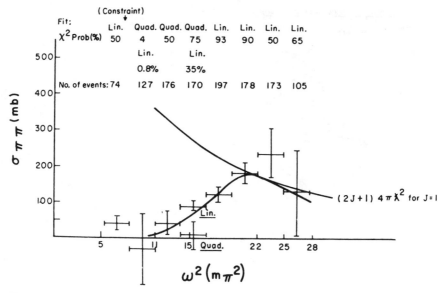

Fig. 20.
First evidence for the ρ Meson.

unimpressed by the Chew—Low demonstration of the ϱ. Fortunately, Walker and his collaborators (67) at Wisconsin soon produced an effective-mass ideogram with a convincing bump at 765 MeV, and they are therefore most often listed as the discoverers of the ϱ.

Ernest Lawrence very early established the tradition that his laboratory would share its resources with others outside its walls. He supplied short-lived radioactive materials to scientists in all departments at Berkeley, and he sent longer-lived samples to laboratories throughout the world. The first artificially created element, technetium, was found by Perrier and Segrè (68), who did their work in Palermo, Sicily. They analyzed the radioactivity in a molybdenum deflector strip from the Berkeley 28-inch cyclotron that had been bombarded for many months by 6-MeV deuterons.

We followed Ernest Lawrence's example, and thus participated vicariously in a number of important discoveries of new particles. The first was the η found at Johns Hopkins, by a group headed by Aihud Pevsner (69). They analyzed film from the 72-inch chamber, and found the η with a mass of 550 MeV, decaying into $\pi^+\pi^-\pi^0$. Within a few weeks of the discovery of the η, Rosenfeld and his co-workers (70) at Berkeley, who had independently observed the η, showed quite unexpectedly that I spin was not conserved in its decay. Figure 21 shows the present state of the art with respect to the ω and η mesons; the strengths of their signatures in this single histogram is in marked contrast to their first appearances in 72-inch bubble chamber experiments.

In the short interval of time between the first and second publications on the η, the discovery of the $Y_0^*(1520)$ was announced by Ferro—Luzzi, Tripp,

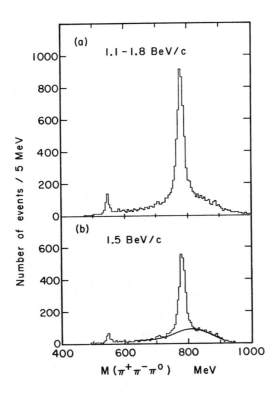

Fig. 21.

Present day histogram showing ω and η Mesons.

and Watson (71), using a new and elegant method. Bob Tripp has continued to be a leader in the application of powerful methods of analysis to the study of the new particles.

The discovery of the $\mathit{\Xi}^*(1530)$ hyperon was accomplished in Los Angeles by Ticho and his associates (72), using 72-inch bubble chamber film. Harold Ticho had spent most of his time in Berkeley for several years, woking tirelessly on every phase of our work, and many of his colleagues had helped prepare the high energy separated K^- beam for what came to be known as the K72 experiment. The UCLA group analyzed the two highest-momentum K^- exposures in the 72-inch chamber, and found the $\mathit{\Xi}^*$ (1530) just in time to report it at the 1962 High Energy Physics Conference in Geneva. (Confirming evidence for this resonance soon came from Brookhaven [73]).

Murray Gell-Mann had recently enunciated his important ideas concerning the "Eightfold Way" (74), but his paper had not generated the interest it deserved. It was soon learned that Ne'eman had published the same suggestions, independently (75).

The announcement of the $\mathit{\Xi}^*$ (1530) fitted exactly with their predictions of the mass and other properties of that particle. One of their suggestions was that four I-spin multiplets, all with the same spin and parity, would exist in a "decuplet" with a mass spectrum of "lines" showing an equal spacing. They put the Fermi 3, 3 resonance as the lowest mass member, at 1238 MeV. The second member was the Y_1^* (1385), so the third member should have a mass of $(1385) + (1385 — 1238) = 1532$. The strangeness and the multiplicity of each member of the spectrum was predicted to drop 1 unit per member, so the Ξ^* (1530) fitted their predictions completely. It was then a matter of simple arithmetic to set the mass, the strangeness, and the charge of the final member —the Ω^-. The realization that there was now a workable theory in particle physics was probably the high point of the 1962 International Conference on High Energy Physics.

Since the second and third members of the series—the ones that permitted the prediction of the properties of the Ω^- to be made—had come out of our bubble chamber, it was a matter of great disappointment to us that the Bevatron energy was insufficient to permit us to look for the Ω^-. Its widely acclaimed discovery (76) had to wait almost two years, until the 80-inch chamber at Brookhaven came into operation.

Since the name of the Ω had been picked to indicate that it was the last of the particles, the mention of its discovery is a logical point at which to conclude this lecture. I will do so, but not because the discovery of the Ω signaled the end of what is sometimes called the population explosion in particle physics—the latest list (77) contains between 70 and 100 particle multiplets, depending upon the degree of certainty one demands before "certification". My reason for stopping at this point is simply that I have discussed most of the particles found by 1962—the ones that were used by Gell-Mann and Ne'eman to formulate their SU(3) theories—and things became much too involved after that time. So many groups were then in the "bump-hunting business" that most discoveries of new resonances were made simultaneously in two or more laboratories.

I am sorry that I have neither the time nor the ability to tell you of the great beauty and the power that has been brought to particle physics by our theoretical friends. But I hope that before long, you will hear it directly from them.

In conclusion, I would like to apologize to those of my colleagues and my friends in other laboratories, whose important work could not be mentioned because of time limitations. By making my published lecture longer than the oral presentation, I have reduced the number of apologies that are necessary, but unfortunately I could not completely eliminate such debts.

REFERENCES

1. J. Chadwick, Proc. Roy. Soc. (London) 136A, 692 (1932).
2. C. D. Anderson, Science 76, 238 (1932).
3. M. Conversi, E. Pancini, and O. Piccioni, Phys. Rev. 71, 209 (1947).
4. S. H. Neddermeyer and C. D. Anderson, Phys. Rev. 51, 884 (1937).
5. J. C. Street and E. C. Stevenson, Phys. Rev. 51, 1005 (1937).
6. H. Yukawa, Proc. Phys.-Math. Soc. Japan 17, 48 (1935).
7. C. M. G. Lattes, H. Muirhead, G. P. S. Occhialini, and C. F. Powell, Nature 159, 694 (1947).
8. E. Gardner and C. M. G. Lattes, Science 107, 270 (1948).
9. R. Bjorklund, W. E. Crandall, B. J. Moyer, and H. F. York, Phys. Rev. 77, 213 (1950)
10. J. Steinberger, W. K. H. Panofsky, and J. Steller, Phys. Rev. 78, 802 (1950).
11. A. G. Carlson (now A. G. Ekspong), J. E. Hopper, and D. T. King, Phil. Mag. 41, 701 (1950).
12. H. L. Anderson, E. Fermi, E. A. Long, R. Martin, and D. E. Nagle, Phys. Rev. 85, 934 (1952).
13. B. Cassen and E. U. Condon, Phys. Rev. 50, 846 (1936).
14. K. A. Brueckner, Phys. Rev. 86, 106 (1952).
15. W. Pauli and S. M. Dancoff, Phys. Rev. 62, 85 (1942).
16. A. Pais, Phys. Rev. 86, 663 (1952).
17. W. B. Fowler, R. P. Shutt, A. M. Thorndike, and W. L. Whittemore, Phys. Rev. 91, 1287 (1953); 93, 861 (1954); 98, 121 (1955).
18. O. Chamberlain, E. Segrè, C. Wiegand, and T. Ypsilantis, Phys. Rev. 100, 947 (1955).
19. G. D. Rochester and C. C. Butler, Nature 160, 855 (1947).
20. A. J. Seriff, R. B. Leighton, C. Hsiao, E. D. Cowan, and C. D. Anderson, Phys. Rev. 78, 290 (1950).
21. R. Armenteros, K. H. Barker, C. C. Butler, A. Cachon, and C. M. York, Phil. Mag. 43, 597 (1952).
22. L. Leprince-Ringuet and M. L'Héritier, Compt. Rend. 219, 618 (1944).
23. H. A. Bethe, Phys. Rev. 70, 821 (1946).
24. R. M. Brown, U. Camerini, P. H. Fowler, H. Muirhead, C. F. Powell, and D. M. Ritson, Nature 163, 47 (1949).
25. C. O'Ceallaigh, Phil. Mag. 42, 1032 (1951).
26. M. G. K. Menon and C. O'Ceallaigh, Proc. Roy. Soc. (London) A221, 292 (1954).
27. R. W. Thompson, A. V. Buskirk, L. R. Etter, C. J. Karzmark, and R. H. Rediker, Phys. Rev. 90, 329 (1953).
28. A. Bonetti, R. Levi Setti, M. Panetti, and G. Tomasini, Nuovo Cimento 10, 345 (1953).
29. R. Marshak and H. Bethe, Phys. Rev. 72, 506 (1947).
30. M. Gell-Mann, Phys. Rev. 92, 833 (1953).
31. K. Nishijima, Prog. Theoret. Phys. (Kyoto) 12, 107 (1954).
32. L. W. Alvarez and S. Goldhaber, Nuovo Cimento 2, 344 (1955).
33. T. O. Lee, Les Prix Nobel en 1957.
34. C. N. Yang, Les Prix Nobel en 1957.
35. L. W. Alvarez, F. S. Crawford, Jr., M. L. Good, and M. L. Stevenson, Phys. Rev. 101, 303 (1956).
36. V. Fitch and R. Motley, Phys. Rev. 101, 496 (1956).

37. S. von Friesen, Ark. Fys. 8, 309 (1954); 10, 460 (1956).
38. R. W. Birge, D. H. Perkins, J. R. Peterson, D. H. Stork, and M. N. Whitehead, Nuovo Cimento 4, 834 (1956).
39. D. Glaser, Les Prix Nobel en 1960.
40. R. H. Hildebrand and D. E. Nagle, Phys. Rev. 92, 517 (1953).
41. J. G. Wood, Phys. Rev. 94, 731 (1954).
42. D. P. Parmentier and A. J. Schwemin, Rev. Sci. Instr. 26, 958 (1955)
43. D. C. Gates, R. W. Kenney, and W. P. Swanson, Phys. Rev. 125, 1310 (1962).
44. L. W. Alvarez, F. S. Crawford, Jr., and M. L. Stevenson, Phys. Rev. 112, 1267 (1958).
45. L. W. Alvarez, P. Davey, R. Hulsizer, J. Snyder, A. J. Schwemin, and R. Zane, UCRL-10109, 1962 (unpublished); P. G. Davey, R. I. Hulsizer, W. E. Humphrey, J. H. Munson, R. R. Ross, and A. J. Schwemin, Rev. Sci. Instr. 35, 1134 (1964).
46. L. W. Alvarez, in Proceedings of the 1966 International Conference on Instrumentation for High Energy Physics, Stanford, California, p. 271.
47. W. K. H. Panofsky, L. Aamodt, and H. F. York, Phys. Rev. 78, 825 (1950).
48. L. W. Alvarez, H. Bradner, P. Falk-Vairant, J. D. Gow, A. H. Rosenfeld, F. T. Solmitz, and R. D. Tripp, Nuovo Cimento 5, 1026 (1957).
49. L. W. Alvarez, H. Bradner, F. S. Crawford, Jr., J. A. Crawford, P. Falk-Vairant, M. L. Good, J. D. Gow, A. H. Rosenfeld, F. T. Solmitz, M. L. Stevenson, H. K. Ticho, and R. D. Tripp, Phys. Rev. 105, 1127 (1957).
50. F. C. Frank, Nature 160, 525 (1947).
51. Ya. B. Zel'dovitch, Dokl. Akad. Nauk SSSR 95, 493 (1954).
52. C. S. Wu, E. Ambler, R. W. Hayward, D. D. Hoppes, and R. P. Hudson, Phys. Rev. 105, 1413 (1957).
53. T. D. Lee and C. N. Yang, Phys. Rev. 104, 254, 822 (1956).
54. F. S. Crawford, Jr., M. Cresti, M. L. Good, K. Gottstein, E. M. Lyman, F. T. Solmitz, M. L. Stevenson, and H. K. Ticho, Phys. Rev. 108, 1102 (1957).
55. P. Eberhard, M. L. Good, and H. Ticho, UCRL-8878, Aug. 1959 (unpublished); J. J. Murray, UCRL-3492, May 1957 (unpublished); J. J. Murray, UCRL-9506, Sept. 1960 (unpublished).
56. R. H. Dalitz, Phil. Mag. 44, 1068 (1953).
57. W. F. Cartwright, C. Richman, M. N. Whitehead and H. A. Wilcox, Phys. Rev. 78, 823 (1950); D. L. Clark, A. Roberts, and R. Wilson, Phys. Rev. 83, 649 (1951); R. Durbin, H. Loar, and J. Steinberger, Phys. Rev. 83, 646 (1951).
58. M. E. L. Oliphant and E. Rutherford, Proc. Roy. Soc. (London) 141, 259 (1933); M. E. L. Oliphant, A. E. Kempton and E. Rutherford, Proc. Roy. Soc. (London) 150, 241 (1935).
59. R. L. Cool, L. Madansky, and O. Piccioni, Phys. Rev. 93, 637 (1954); see also references in R. F. Peierls, Phys. Rev. 118, 325 (1959).
60. M. Alston, L. W. Alvarez, P. Eberhard, M. L. Good, W. Graziano, H. K. Ticho, and S. G. Wojcicki, Phys. Rev. Letters 5, 520 (1960).
61. M. H. Alston, L. W. Alvarez, P. Eberhard, M. L. Good, W. Graziano, H. K. Ticho, and S. G. Wojcicki, Phys. Rev. Letters 6, 300 (1961).
62. M. H. Alston, L. W. Alvarez, P. Eberhard, M. L. Good, W. Graziano, H. K. Ticho, and S. G. Wojcicki, Phys. Rev. Letters 6, 698 (1961).
63. R. Hofstadter, Rev. Mod. Phys. 28, 214 (1956).
64. W. Holladay, Phys. Rev. 101, 1198 (1956); Y. Nambu, Phys. Rev. 106, 1366 (1957); C. F. Chew, Phys. Rev. Letters 4, 142 (1960); W. R. Frazer and J. R. Fulco, Phys. Rev. 117, 1609 (1960); F. J. Bowcock, W. N. Cottingham, and D. Lurie, Phys. Rev. Letters 5, 386 (1960).

65. B. C. Maglić, L. W. Alvarez, A. H. Rosenfeld, and M. L. Stevenson, Phys. Rev. Letters 7, 178 (1961).

66. J. A. Anderson, V. X. Bang, P. G. Burke, D. D. Carmony, and N. Schmitz, Phys. Rev. Letters 6, 365 (1961).

67. A. R. Erwin, R. March, W. D. Walker, and E. West, Phys. Rev. Letters 6, 628 (1961)

68. C. Perrier and E. Segrè, Accad. Naz. Lincei, Rendi. Classe Sci. Fis. Mat. e Nat. 25, 723 (1937).

69. A. Pevsner, R. Kraemer, M. Nussbaum, C. Richardson, P. Schlein, R. Strand, T. Toohig, M. Block, A. Engler, R. Gessaroli, and C. Meltzer, Phys. Rev. Letters 7, 421 (1961).

70. P. L. Bastien, J. P. Berge, O. I. Dahl, M. Ferro-Luzzi, D. H. Miller, J. J. Murray, A. H. Rosenfeld, and M. B. Watson, Phys. Rev. Letters 8, 114 (1962).

71. M. Ferro-Luzzi, R. D. Tripp, and M. B. Watson, Phys. Rev. Letters 8, 28 (1962).

72. G. M. Pjerrou, D. J. Prowse, P. Schlein, W. E. Slater, D. H. Stork, and H. K. Ticho, Phys. Rev. Letters 9, 114 (1962).

73. L. Bertanza, V. Brisson, P. L. Connolly, E. L. Hart, I. S. Mittra, G. C. Moneti, R. R. Rau, N. P. Samios, I. O. Skillicorn, S. S. Yamamoto, M. Goldberg, L. Gray, J. Leitner, S. Lichtman, and J. Westgard, Phys. Rev. Letters 9, 180 (1962).

74. M. Gell-Mann, California Institute of Technology Synchrotron Laboratory Report CTSL-20, 1961 (unpublished).

75. Y. Ne'eman, Nucl. Phys. 26, 222 (1961).

76. V. E. Barnes, P. L. Connolly, D. J. Crennell, B. B. Culwick, W. C. Delaney, W. B. Fowler, P. E. Hagerty, E. L. Hart, N. Horwitz, P. V. C. Hough, J. E. Jensen, J. K. Kopp, K. W. Lai, J. Leitner, J. L. Lloyd, G. W. London, T. W. Morris, Y. Oren, R. B. Palmer, A. G. Prodell, D. Radojičić, D. C. Rahm, C. R. Richardson, N. P. Samios, J. R. Stanford, R. P. Shutt, J. R. Smith, D. L. Stonehill, R. C. Strand, A. M. Thorndike, M. S. Webster, W. J. Willis, and S. S. Yamamoto, Phys. Rev. Letters 12, 204 (1964).

77. A. H. Rosenfeld, A. Barbaro-Galtieri, W. J. Podolsky, L. R. Price, P. Söding, C. G. Wohl, M. Roos, and W. J. Willis, Rev. Mod. Phys. 39, 1 (1967).

13

Muon Catalysis of Fusion

J. David Jackson

Discoveries in experimental science are of several kinds. Some are the results of well-focused, systematic, quantitative studies of phenomena whose qualitative, empirical aspects were at least partially known. Coulomb's and Ampère's discoveries are in this class, as are Rutherford's and Soddy's elucidations of the nature of radioactive transformations. The numerous achievements of the Alvarez bubble chamber group, for which Luis W. Alvarez received the Nobel Prize in Physics, are also of this type, even though the initial motivation for a program of bubble chamber work was the investigation of strange particle decays. As the utility of the technique for the study of hadronic resonant states was being established, a powerful array of detectors and analyzing tools was being developed, making the group preeminent in high-energy physics in the late 1950s and 1960s.

Another class of discoveries occurs when experiments intended (at least in part) to resolve an anomaly of earlier work lead to unexpected results. Two examples are the discovery of CP violation and the discovery of the ψ-particle in $e + e -$ collisions. Another class results from hunches that in retrospect were based on an erroneous interpretation of other relatively new discoveries. Becquerel's incorrect hunch about the true source of Roentgen's x-rays led to the discovery of radioactivity. Still another class stems from the testing of theoretical concepts. Rutherford's desire to test J. J. Thomson's "currant bun" model of the atom led to the discovery of the atomic nucleus (and the destruction of Thomson's model). The discoveries of the Ω^- particle at Brookhaven in 1964 and of the $W^\pm$ and Z^0 particles at CERN in 1983 are similar examples, this time with theoretical expectations triumphantly confirmed. Perhaps the most dramatic in this class is the discovery of nonconservation of parity in weak interactions. Experimenters were instructed in detail on where to look and what to expect to find.

A final class of discoveries is those that, because of the state of our understanding of the laws of nature at the time, are better called "observations." Thus, by the 1940s, our understanding of quantum mechanics and quantum electrodynamics was such that few would doubt that essentially any pair of oppositely charged particles could form hydrogen-like atoms. "Exotic atoms" only awaited the discovery of new charged par-

ticles (the muon and the positron were already at hand) and development of the requisite experimental techniques and interest. The discoveries of positronium, mesic atoms, and muonium are clearly different in character from the discovery of, say, radioactivity. But to call them "observations" in no way diminishes the achievement of discovery, often a technical *tour de force,* or the importance of their consequences.

The observation of the catalysis of nuclear fusion by muons in a liquid hydrogen bubble chamber by Alvarez and coworkers in late 1956, which is described in the accompanying paper, is a classic discovery of the last type but differs from those of mesic atoms or muonium in being a discovery that was not sought. It was entirely accidental, totally unexpected, and peripheral to the group's main study of hadronic interactions. From the point of view of fundamental physics, the phenomenon is completely "understood," yet it was viewed by its discoverer and others (like me) as bizarre. It led to a flurry of speculation on energy production by "cold fusion," as seen in Figure 1 and my 1957 paper.[1] Then, apart from an initial few experiments and a steady but low level of interest in the problem by theorists, twenty years passed before there was a resurgence of interest, as shown in Figure 2.

This commentary presents a brief and spotty survey of the subject of muon catalysis of fusion of hydrogen isotopes from its beginnings to 1984. It begins with some personal recollections. Then follows a very rapid skimming of the theoretical and experimental research, leading to a brief discussion of the reasons for the present level of interest in the subject and the prospects for useful energy production.

During the academic year 1956–57, I was visiting the Physics Department at Princeton University on a John Simon Guggenheim Memorial Fellowship. During the fall I thought about a few problems in nuclear physics but had not settled into any serious research. By early December there were rumors about parity nonconservation in the experiments of Wu and coworkers and also rumors of a μ' meson at Berkeley, but nothing definite, and by the time of the long Christmas vacation, Palmer Laboratory was virtually deserted. Despite Christmas visitors, I seemed to keep

---

1. J. D. Jackson, Phys. Rev. **106**, 330 (1957).

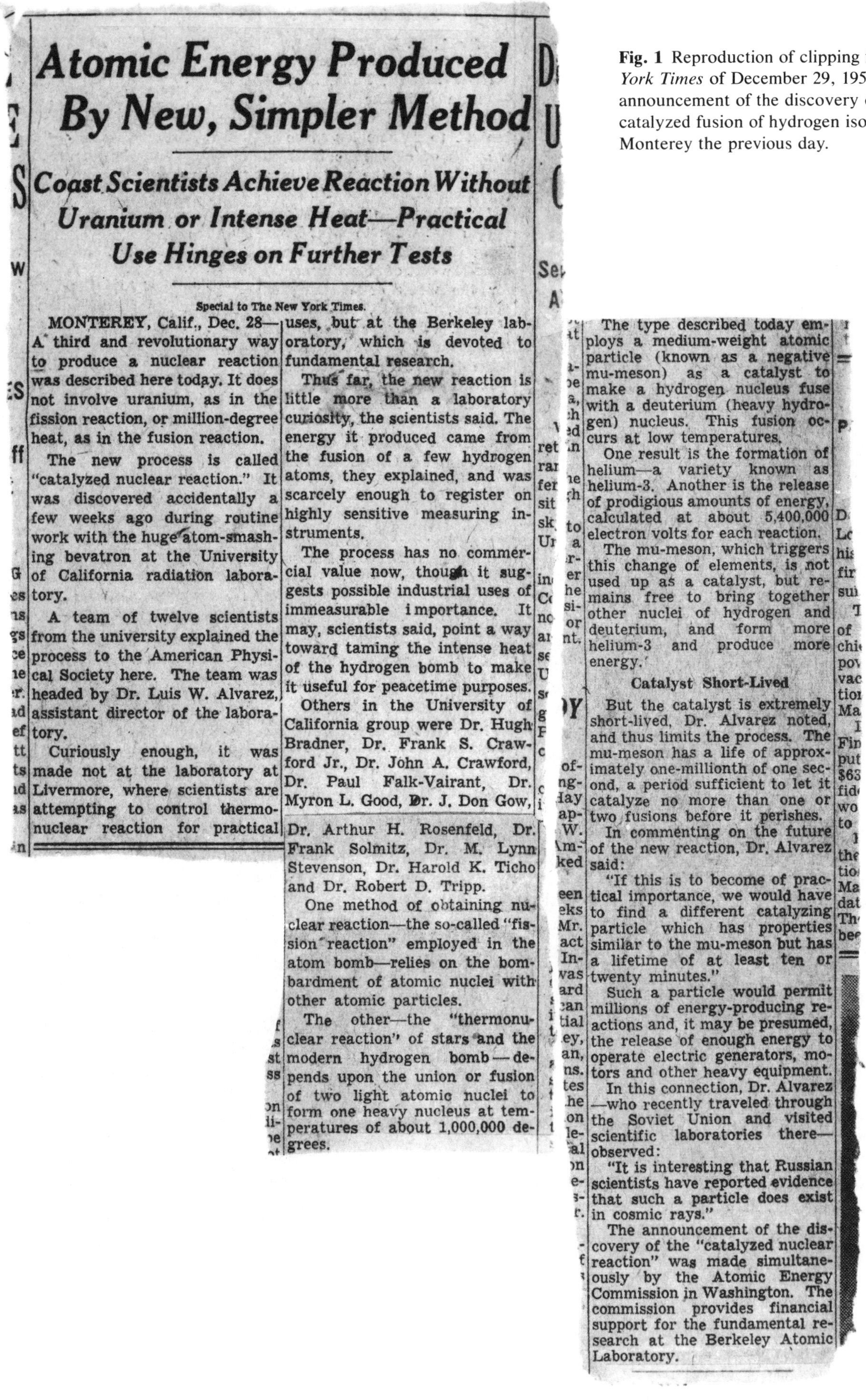

Fig. 1 Reproduction of clipping from the *New York Times* of December 29, 1956, reporting the announcement of the discovery of muon-catalyzed fusion of hydrogen isotopes at Monterey the previous day.

Atomic Energy Produced By New, Simpler Method

Coast Scientists Achieve Reaction Without Uranium or Intense Heat—Practical Use Hinges on Further Tests

Special to The New York Times.

MONTEREY, Calif., Dec. 28—A third and revolutionary way to produce a nuclear reaction was described here today. It does not involve uranium, as in the fission reaction, or million-degree heat, as in the fusion reaction.

The new process is called "catalyzed nuclear reaction." It was discovered accidentally a few weeks ago during routine work with the huge atom-smashing bevatron at the University of California radiation laboratory.

A team of twelve scientists from the university explained the process to the American Physical Society here. The team was headed by Dr. Luis W. Alvarez, assistant director of the laboratory.

Curiously enough, it was made not at the laboratory at Livermore, where scientists are attempting to control thermonuclear reaction for practical uses, but at the Berkeley laboratory, which is devoted to fundamental research.

Thus far, the new reaction is little more than a laboratory curiosity, the scientists said. The energy it produced came from the fusion of a few hydrogen atoms, they explained, and was scarcely enough to register on highly sensitive measuring instruments.

The process has no commercial value now, though it suggests possible industrial uses of immeasurable importance. It may, scientists said, point a way toward taming the intense heat of the hydrogen bomb to make it useful for peacetime purposes.

Others in the University of California group were Dr. Hugh Bradner, Dr. Frank S. Crawford Jr., Dr. John A. Crawford, Dr. Paul Falk-Vairant, Dr. Myron L. Good, Dr. J. Don Gow, Dr. Arthur H. Rosenfeld, Dr. Frank Solmitz, Dr. M. Lynn Stevenson, Dr. Harold K. Ticho and Dr. Robert D. Tripp.

One method of obtaining nuclear reaction—the so-called "fission reaction" employed in the atom bomb—relies on the bombardment of atomic nuclei with other atomic particles.

The other—the "thermonuclear reaction" of stars and the modern hydrogen bomb—depends upon the union or fusion of two light atomic nuclei to form one heavy nucleus at temperatures of about 1,000,000 degrees.

The type described today employs a medium-weight atomic particle (known as a negative mu-meson) as a catalyst to make a hydrogen nucleus fuse with a deuterium (heavy hydrogen) nucleus. This fusion occurs at low temperatures.

One result is the formation of helium—a variety known as helium-3. Another is the release of prodigious amounts of energy, calculated at about 5,400,000 electron volts for each reaction.

The mu-meson, which triggers this change of elements, is not used up as a catalyst, but remains free to bring together other nuclei of hydrogen and deuterium, and form more helium-3 and produce more energy.

Catalyst Short-Lived

But the catalyst is extremely short-lived, Dr. Alvarez noted, and thus limits the process. The mu-meson has a life of approximately one-millionth of one second, a period sufficient to let it catalyze no more than one or two fusions before it perishes.

In commenting on the future of the new reaction, Dr. Alvarez said:

"If this is to become of practical importance, we would have to find a different catalyzing particle which has properties similar to the mu-meson but has a lifetime of at least ten or twenty minutes."

Such a particle would permit millions of energy-producing reactions and, it may be presumed, the release of enough energy to operate electric generators, motors and other heavy equipment.

In this connection, Dr. Alvarez—who recently traveled through the Soviet Union and visited scientific laboratories there—observed:

"It is interesting that Russian scientists have reported evidence that such a particle does exist in cosmic rays."

The announcement of the discovery of the "catalyzed nuclear reaction" was made simultaneously by the Atomic Energy Commission in Washington. The commission provides financial support for the fundamental research at the Berkeley Atomic Laboratory.

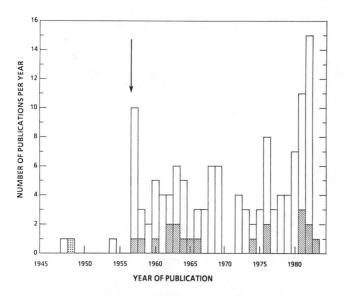

Fig. 2 Histogram of the number of physics journal publications per year on muon-catalyzed fusion or closely related topics from 1945 to 1982. The arrow marks the time of the experimental discovery. Shading indicates experimental papers. The dotted entry for 1948 represents Sakharov's unpublished report. (Source, *Physics Abstracts*.)

going in to work. Now, one of the virtues of being within 100 miles of New York City in those days was the delivery to one's doorstep each morning of the *New York Times*. On the morning of December 29, 1956, I read in the *New York Times* a report of a paper presented the previous day at an American Physical Society meeting in Monterey, California. My yellowed clipping from the *Times* is displayed in Figure 1.

My imagination was stirred by the newspaper article, and I began to work on the nuclear fusion aspects of muonic diatomic molecular ions, the capture of the muon by the moving helium fragments after fusion, and the possibility of its liberation during the slowing down of the fragment, as well as speculating about energy production. The most important conclusions of my work were that in the energetically interesting case of the fusion of deuterons with tritons, the nuclear reaction rate once the molecular ion is formed is extremely fast ($\Gamma \geq 10^{12}$ s^{-1}); and that, whatever the rates of molecular processes, there was an upper limit of the order of 10^2 on the number of possible fusions caused by the muon because of capture by the produced alpha particle, independent of the muon's lifetime or mass. The last conclusion vitiated the remarks attributed to Alvarez at the end of the news story about the efficacy of a possible longer lived lepton.

I must have worked feverishly, for my files show that my paper, "Catalysis of Nuclear Reactions between Hydrogen Isotopes by μ^- Mesons," (cited above) was sent to the *Physical Review* on January 9, 1957.

They also contain a carbon copy of a letter to Alvarez, dated January 5, 1957, that said in part:

> Your μ' meson and its explanation were featured in the newspapers of this area around the New Year. Having nothing better to do, I began playing with the problem. The enclosed rough draft is the result (excuse the typing in the paper—it's my own). It is more than likely that you and your group have done similar and better calculations already. I found it entertaining, anyway, to see what could be done from first principles in the way of estimating the various rates. The speculations on power production are, of course, very wild and probably wrong.

Alvarez replied on January 8. (Note the efficiency of transcontinental postal service in 1957!) I quote the handwritten letter in its entirety. It is necessary to point out here that in the manuscript the first reference read, "'1L. W. Alvarez et al., *New York Times* CVI No. 36, B4,1 (December 29, 1956).''

> Dear Dr. Jackson:
>
> It was good of you to send us your very interesting report. My theoretical friends are fighting over it at the moment. We are trying to estimate the number of μ's which get stuck on the recoil He^3, after a D + D→He^3 + n reaction. If the μ isn't stripped in the recoil, it can't get away. If the stripping cross-section is high enough, we'll look for a big burst of neutrons from a Dewar flask of D$_2$ above the large scintillator tank of the Reines-Cowan type, which is here in Berkeley. We had thought of such experiments before, but never could see how the chains would be long enough to make it interesting.
>
> We have a large group of molecular experts now working hard on the whole problem. Teller and McMillan have been thinking hard since the earliest days of the effect and they have been joined by a bunch of younger people. Much of their work parallels yours and I will let them have the fun of communicating their results directly to you, rather than trying to interpret what they have done.
>
> I am enclosing a preprint with a few minor changes.
>
> I have only one request and that is that you change the reference from the *New York Times*, to the APS meeting—Monterey, Dec. 28, 1956 and UCRL 3620. The lab is a bit sensitive on this point, since in the case of the antiproton, we received lots of letters from people who thought we held the thing too long before publishing and from others who said, "Don't you guys publish anywhere except the *New York Times*?" So in the μ-catalysis, we were careful to follow the book—we presented the thing first at the APS meeting, after sending out preprints. All the news stories came out after the APS talk.—But still, we find a reference to the *N.Y. Times*!!
>
> With many thanks again for a most interesting paper.
>
> Sincerely,
> Luis W. Alvarez

This was my first contact with Luis Alvarez and soon it led to contacts with members of his group, contacts that have flourished over the years. A not totally frivolous speculation: On what university faculty might I be now if I had not read the *New York Times* that December day?

The delightful story of the discovery of the muon catalysis events and their explanation is told in Alvarez's 1968 Nobel Lecture, also reproduced in this volume. In those early days, there were no professional bubble chamber film scanners. The physicists did it themselves. As Luie has noted, the physicists' involvement at the scanning level was crucial for the discovery. Casual instructions to scanners about π-μ decays would not have triggered the selection of the catalysis events for further study, if only because no one dreamt of their existence. In scanning film from the 10-inch hydrogen bubble chamber for strange particle events, the physicists noticed a few anomalously energetic muons. The hypothesis of pion decay in flight (just before stopping) had to be rejected when it was found that the muons all had a kinetic energy of 5.4 MeV. The particle physicists needed an astrophysicist colleague to identify the likely source of 5.4 MeV muons, namely, the fusion reaction, $p + d \rightarrow He^3 + \gamma$, with the muon being responsible for the union of the proton and deuteron in a molecular ion and for sometimes carrying off the energy release, instead of the photon. An apparent problem, the paucity of deuterium (1 part in 5000) in the chamber relative to the frequency of fusions, was solved for the Alvarez group by Edward Teller, who, in a bravura performance upon hearing about the results, pointed out that the reduced mass effect made the μd atom ~ 135 eV more tightly bound than the μp atom, with the consequent high probability of an exothermic charge transfer of the muon from a proton to a deuteron. The neutral recoiling μd atom may travel some distance before forming a $p\mu d$ molecular ion (thereby causing the gaps sometimes seen), after which fusion can occur.

Neither the physicists in Berkeley nor I in Princeton were aware of considerably earlier theoretical work on the subject until after submission of our papers for publication.

As is well known by now, the first published discussion of muon-catalyzed fusion was given by F. C. Frank,[2] more than nine years before its experimental observation. A solid-state physicist and colleague of C. F. Powell's at Bristol, Frank examined and rejected a large number of possible alternative explanations of the π-μ decay events discovered by Lattes, Occhialini, and Powell. One of the alternatives was the formation of muonic molecular ions, among them $p\mu d$, with its fusion energy release of 5.5 MeV. Frank rejected the

mechanism because of insufficient deuterium in the emulsion and the incorrect Q-value but discussed briefly the formation of μp atoms and barrier penetration within the molecule.

The next research on the subject was apparently made in 1948 by Andrei Sakharov. It was the basis of a now legendary unpublished report.[3] In Sakharov's own words, "Having become acquainted with a paper of Frisch [sic] in which he discussed a possible alternative interpretation of the experiment of Powell, Lattes and Occhialini (discovery of the π meson) by means of a μ-catalysis reaction, I wrote a report considering the possibility of realizing μ catalysis of a D + D reaction on a macroscopic scale with a positive energy balance, and I made some calculations."[4]

Some years later, but still before the experimental discovery, Zel'dovich considered some aspects of the muon catalysis process.[5] As well as discussing reactions by μp or μd atoms in flight, he treated the vibrational states of the molecular ion, estimating for the $d\mu d$ ion the lowest and first excited vibrational energies as -330 eV and -30 eV. He then said (in translation), "The presence of the oscillation level, real or virtual, with an energy very near to zero, can greatly increase the amplitude of the wave function in the hole and at the entrance into the barrier. . . . Both the probabilities of the reaction in flight and of the reaction in formation of a molecule will be increased." Zel'dovich acknowledges helpful discussions with Kompaneetz, Landau, and Sakharov. It seems clear that theorists in the Soviet Union were fully cognizant of the idea of muon-induced catalysis of fusion reactions and had done some serious thinking about it by the end of 1953.

In Figure 2 are displayed the number of papers *published* per year on the general subject of muon catalysis of fusion or closely related topics, as listed in *Physics Abstracts*. With the exception of Sakharov's unpublished 1948 report, the data are restricted to journal publications. Clearly, theoretical papers dominate over the years. The experimental papers in the period 1957–1959 were concerned with elaborating on the basic discovery, with experiments in the 1960s mostly having the catalysis secondary to the study of the basic weak interaction process of muon capture in hydrogen. The Dubna group of Dzhelepov studied muon catalysis in its own right. Reviews of the subject was published in

3. A. D. Sakharov, P. N. Lebedev Physics Institute Report, 1948 (unpublished). Recently, a physicist visiting the Soviet Union and interested in seeing Sakharov's report asked a Soviet theorist in the field about it. The reply was that copies were not available, that he had not seen it himself, but that his professor had told him that *he* had indeed seen the report and read it!

4. From Sakharov's introductory commentary in *A. D. Sakharov Collected Scientific Works*, edited by D. ter Haar, D. V. Chudnovsky, and G. V. Chudnovsky (Marcel Dekker, New York, 1982), p. 3.

5. Ya. B. Zel'dovich, Dokl. Akad. Nauk SSSR **95**, 493 (1954).

2. F. C. Frank, Nature (London) **160**, 525 (1947).

1960 by Zel'dovich and Gerstein[6] and in 1975 by Gerstein and Ponomarev.[7]

For almost twenty years the level of interest in the subject remained rather slight. A few theorists made calculations, and an occasional experiment was done. Then, in the mid 1970s, the situation began to change. The Soviet theorists, who had maintained over the years a far greater interest than those elsewhere, began to make highly accurate calculations of the energy states of the various molecular ions consisting of a muon and two hydrogen isotopes. By 1977, Ponomarev and his colleagues had shown[8] that in both $d\mu d$ and $d\mu t$ molecules there were bound excited states with binding energies of less than 2 eV (specifically $J = 1$, $v = 1$ states at -0.64 eV for $d\mu t$ and -1.91 eV for $d\mu d$, compared with -320 eV for $J = 0$, $v = 0$). The presence of states with binding energies in the range of *electronic* molecular energies means that there can be resonant formation of the muonic molecule whereby the thermal neutral $t\mu$ atom, say, enters easily into a D_2 electronic molecule and combines with one of the deuterons. The small energy release of the transition into the $J = 1$, $v = 1$ state is transferred into excitation of the rotational and vibrational degrees of freedom of the electronic molecule.[9] As Zel'dovich stated in 1954, such circumstances lead to very large cross sections for muonic molecule formation. The calculated rate was on the order of 10^8 s^{-1} (normalized to liquid hydrogen density). In a year or two it was verified experimentally at Dubna.

The certainty that molecular formation rates in some circumstances could be 100 or more times faster than the muon's decay rate is the cause of the recent dramatic increase of interest, reflected in the precipitous rise since 1980 shown in Figure 2. (For 1983 and 1984, the numbers are surely far off the scale, but 1982 is the last year of complete data from *Physics Abstracts* at this writing.) Numerous experimental groups began major programs of study of the many fascinating subprocesses involved (Dubna, Gatchina, Los Alamos, Vancouver, and Zurich). The various experimental results now available (temperature dependences, hyperfine effects, etc.) form such a vast and intricate complex that I could not begin to do them justice here. Apart

from a few further comments, I must refer the reader to a variety of conference papers and reviews.[10]

The papers tabulated in Figure 2 do not include those discussing possible practical uses of muon-catalyzed fusion for energy production. These, too, have shown a resurgence in the past five years. Apart from the original speculations, this aspect lay dormant until the mid 1970s[11] and only caught fire with Petrov's idea of an indirect energy source through use of the 14 MeV neutrons from the d-t reaction to cause fast fission and breed fissile material.[12] By 1983, it was a major topic at the Third International Conference on Emerging Nuclear Energy Systems, Helsinki, Finland.[13]

A workshop on muon-catalyzed fusion was held in Jackson Hole, Wyoming, in June 1984. Its purpose was to discuss pressing physics issues as well as practical applications. A backdrop to the workshop was the impressive series of experiments in $D_2 + T_2$ mixtures at high pressures and temperatures performed by an Idaho–Los Alamos group.[14] They reported as many as 90 ± 10 fusions on the average per muon. This number is impressively close to my early estimate of an upper bound of about 100 d-t fusions per muon (confirmed by later, more elaborate calculations by others). With such experimental results, it is not easy to prevent speculation that the theorists may be wrong and that many hundreds of fusions per muon might be possible. If so, direct energy production with "cold fusion" might be economical!

The Jackson Hole workshop showed that muon catalysis of fusion is alive and well some thirty years after its experimental discovery. On the physics side, there is a rich spectrum of exotic atomic and molecular processes, worthy of study in their own right and with impressive experiments meeting the challenges. Theory is hard pressed to explain all the observed effects. For example, preliminary results from the Idaho–Los Alamos group suggest a marked decrease in the "α-sticking fraction" (the fraction of muons per fusion that end up bound to a thermalized alpha particle) with increased D_2-T_2 density or T_2 concentration. Taken at face value, the data are very difficult to understand (at least by this theorist), although there are reasons for expecting values somewhat less than the 0.9% of the most careful estimates so far. On another front, results

6. Ya. B. Zel'dovich and S. S. Gerstein, Usp. Fiz. Nauk **71**, 581 (1960) [transl., Sov. Phys. Usp. **3**, 593 (1961)].

7. S. S. Gerstein and L. I. Ponomarev, in *Muon Physics,* edited by V. W. Hughes and C. S. Wu (Academic, New York, 1975), Vol. 3, p. 141.

8. S. I. Vinitsky *et al.,* Zh. Eksp. Teor. Fiz. **74**, 849 (1978) [trans., Sov. Phys. JETP **47**, 444 (1979)]. See also S. S. Gerstein and L. I. Ponomarev, Phys. Lett. B**72**, 90 (1977).

9. In 1967, E. A. Vesman, Pis'ma Zh. Eksp. Teor. Fiz. **5**, 113 (1967) [transl., Sov. Phys. JETP Lett. **5**, 91 (1967)] had suggested the efficacy of the transfer of energy into vibrations of the electronic molecule for small energy releases but did not have reliable enough muonic molecular calculations to make quantitative statements.

10. L. I. Ponomarev, in *Atomic Physics 6,* edited by R. Damburg and O. Kukaine (Plenum, New York, 1979), pp. 182ff.; J. Rafelski, in *Exotic Atoms '79,* edited by K. M. Crowe, J. Duclos, G. Fiorentini, and G. Torelli (Plenum, New York, 1980), pp. 177ff.; W. H. Breunlich, Nucl. Phys. A**353**, 201 (1981); G. Fiorentini, Nucl. Phys. A**374**, 607c (1982); and L. Bracci and G. Fiorentini, Phys. Rep. **86**, 170 (1982). See also L. I. Ponomarev, Atomkernerg. Kerntech. **43**, 175 (1983).

11. W. P. S. Tan, Nature (London) **263**, 656 (1976).

12. Yu. V. Petrov, Nature (London) **285**, 466 (1980).

13. See the last citation in n. 10, pp. 175–210.

14. S. E. Jones *et al.,* Phys. Rev. Lett. **51**, 1757 (1983).

from Zurich showed a large, temperature-independent rate of formation of the $d\mu t$ molecule by the μt atom in the triplet hyperfine state, while theory predicts a temperature-dependent rate a factor of 4 or more smaller. Clearly, there is work for experimenters and theorists on the basic physics for quite a few years to come.

The practical side looks much less promising, despite the hints of hundreds of fusions per muon. *If* one could attain 500 or 1000 fusions per muon in D_2-T_2 (it requires a cyclic rate in excess of 2×10^8 per second and an α-sticking fraction of less than 0.2%), direct energy production would become feasible. There would still be, however, a limitation on the size of plant one could have, given foreseeable accelerator technology: 20–50 Megawatts is an upper limit often cited. (I find that I stated 10^4 kW in 1957.) The Petrov scheme (with, say, less than 200 fusions per muon) does not seem

terribly attractive. It is similar to the spallation breeder scheme, where a high-intensity proton beam of 1 GeV or more bombards a target of some heavy element, causing spallation reactions with the nuclear evaporation of neutrons, which are then absorbed by a suitable breeding material. In fact, Petrov's idea was for a combined system in which spallation neutrons are produced at the same time as the pions (source of the muons). Modeling of such a scheme indicates that the muon-catalyzed fusion part of the facility would only increase modestly (by less than a factor of 2) the number of neutrons available for breeding. That is not a sufficient advantage. Another negative aspect of the Petrov idea is the breeding itself. Muon-catalyzed fusion as a direct, clean, cold source of energy is one thing; as the source of neutrons for making fuel to burn in fission reactors, it is quite another. Finally, breeding of any sort is not sensible while there are cheap sup-

Fig. 3 A diagram used at colloquia in 1957 to illustrate the catalytic cycle in a D_2-T_2 mixture.

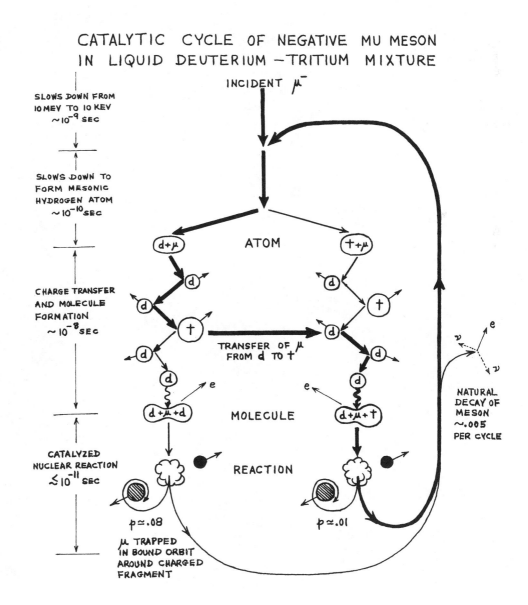

plies of uranium ore adequate for all likely demands. All in all, the prospects of energy production from "cold fusion" seem quite remote.

In the 37 years since its first contemplation and the 28 years since its discovery in a hydrogen bubble chamber, muon catalysis of fusion has had a long and interesting history. The toy of a few theorists at first, it became an experimental reality in late 1956 and sparked a flurry of speculation, published and unpublished, on its potential as a new, exciting source of fusion energy. Some studies were made of the process for its own sake, but then it became a complication that needed to be understood in order to get at fundamental processes like $\mu^- p \rightarrow n\nu$. Theorists, most of them in the Soviet Union, continued an interest, and some experiments were done at Dubna. A dramatic awakening of renewed interest occurred in the late 1970s, when the theorists predicted large, resonant cross sections for formation of $d\mu d$ and $d\mu t$ molecular ions, and they were soon confirmed experimentally. The dreams of a muon-catalyzed fusion power plant were resurrected; numerous experiments to study the rather complicated chains of atomic and molecular interactions and formations were mounted around the world and continue today. Diagrams far more complex than the conceit from 1957 shown in Figure 3 now abound in the discussion of muonic atoms and molecules and the fusion process.

Luie and the colleagues listed in the *New York Times* story (Fig. 1) founded (with their little fingers, one might say) a unique and still flourishing field of experimental science. It is an Alvarez hallmark.

Since the preparation of this commentary in the summer of 1984, muon-catalysis research, both theoretical and experimental, has continued in several laboratories. The vital α-sticking fraction has received considerable attention. The best theoretical estimates for the total capture probability (including dynamic effects beyond the Born-Oppenheimer approximation, crudely estimated by Cahn and myself at the Jackson workshop) are ~0.85%, with the loss of the muon as the system slows down causing the α-sticking fraction to be ~0.6%. The experiments imply a complicated situation, with many-body effects at extreme conditions of density and temperature. Alpha sticking fractions on the order of 0.3% and 150 neutrons per muon have been reported. A brief review of the recent work at Los Alamos has been presented by Jones.[15] Clearly, the field of muon catalysis continues to fascinate and surprise.

15. S. E. Jones, Nature (London) **321**, 127 (1986).

Reprinted from THE PHYSICAL REVIEW, Vol. 105, No. 3, 1127–1128, February 1, 1957
Printed in U. S. A.

Catalysis of Nuclear Reactions
by μ Mesons*

L. W. ALVAREZ, H. BRADNER, F. S. CRAWFORD, JR., J. A.
CRAWFORD,† P. FALK-VAIRANT, M. L. GOOD, J. D. GOW,
A. H. ROSENFELD, F. SOLMITZ, M. L. STEVENSON,
H. K. TICHO, AND R. D. TRIPP

Radiation Laboratory, University of California, Berkeley, California

(Received December 17, 1956)

IN the course of a recent experiment involving the stopping of negative K mesons in a 10-inch liquid hydrogen bubble chamber,[1] an interesting new reaction was observed to take place. The chamber is traversed by many more negative μ mesons than K mesons, so that in the last 75 000 photographs, approximately 2500 μ^- decays at rest have been observed. In the same pictures, several hundred π^- mesons have been observed to disappear at rest, presumably by one of the "Panofsky reactions."[2] For tracks longer than 10 cm, it is possible to distinguish a stopping μ meson from a stopping π meson by comparing its curved path (in a field of 11 000 gauss) with that of a calculated template. In addition to the normal π^- and μ^- stoppings, we have observed 15 cases in which what appears (from curvature measurement) to be a μ^- meson coming to rest in the hydrogen, and then giving rise to a secondary negative particle of 1.7-cm range, which in turn decays by emitting an electron. (A 4.1-Mev μ meson from $\pi-\mu$ decay has a range of 1.0 cm.) The energy spectrum of the electrons from these 15 secondary particles looks remarkably like that of the μ meson: there are four electrons in the energy range 50 to 55 Mev, and none higher; the other electrons have energies varying from 50 Mev to 13 Mev. The most convincing proof of the fact that the primary particle actually comes to rest, and does not—for example—have a large resonant cross section for scattering at a residual range of 1.7 cm, is the following: in five of the fifteen special events, there is a large gap between the last bubble of the primary track and the first bubble of the secondary track. This gap is a real effect, and not merely a statistical fluctuation in the spacing of the bubbles, since in some cases the tracks form a letter X (see Fig. 1), and in another case the secondary track is parallel to the primary, but displaced transversely by about 1 mm at the end of the primary. These real gaps appear also (although perhaps less frequently) between some otherwise normal-looking μ^- endings and the subsequent decay electron; they are thought to be the distance traveled by the small neutral mesonic atom.[3]

One may quickly dispose of the most obvious suggestion that the events are $\pi^--\mu^--e^-$ decays. If, by

FIG. 1. Example of H-D reaction catalyzed by μ^- meson. The incident meson comes to rest, drifts as a neutral mesonic atom, is ejected with 5.4 Mev by the H-D reaction, comes to rest again after 1.7 cm, and decays.

some unknown process, negative π mesons could decay at rest in hydrogen, their secondary μ's would have a range of 1.0 cm, rather than the observed unique range of 1.7 cm. But, most importantly, the curvature of the stopping particles definitely precludes any possibility that they are π's. Therefore, if one is to explain the new observations in terms of known particles, he must say that the primary is a μ meson (as determined by curvature and range), and the secondary is also a μ meson (as determined by its decay-electron spectrum). The problem presented is then to find the source of the energy that "rejuvenates" the μ meson after it has come to rest. The energy that must be supplied to the μ meson is 5.4 Mev, as determined from the range-energy relationship in hydrogen. (We explored the possibility that one of the particles was an ordinary μ meson, while the other was either heavier or lighter by about 6 Mev. In this case, the heavier could not decay into the lighter in free space, as a π decays into a μ, because this process requires more of a mass difference between the two particles than was allowed by the measurements. One could just stay within the experimental limits by assuming that the decay took place in the field of a proton, and that the lighter particle then decayed in the usual μ-meson manner.)

The following explanation seems satisfactory.[4] If the $\mu - p$ mesonic atom referred to above finds a deuteron, and the deuteron becomes bound in the mesonic equivalent of an H-D molecular ion, then the mean H-D spacing is about 1/200 as large as that in the ordinary H-D molecular ion. The meson, in effect, confines the two nuclei in a small box. Rough estimates of the barrier penetration factor (approximately 10^{-5}) and the vibration frequency (approximately 10^{17} per second) indicate that the time required for a nuclear reaction between H and D should be small compared with the life of the μ meson. In some yet unknown fraction of the cases, the reaction energy is taken up by the μ meson, which appears in the bubble chamber with a kinetic energy of 5.4 Mev, i.e., nearly the mass difference between H+D and He^3. (The recoil He^3 should not be visible in any case.)

If, as we believe, the explanation outlined above is correct, several apparent discrepancies must be resolved. For example, early suggestions that deuterium might have something to do with the observations were discarded because the ratio of 1.7-cm μ's to decay electrons is about 1/200 whereas the deuteron contamination in the bubble chamber is only about 1/5000. It seems possible to overcome this difficulty if a deuteron is able to rob the meson from a proton. The μ mesons will be bound more tightly by deuterons than by protons, because of the 5% larger reduced mass. This amounts to 135 ev for the ground state. This effect, and several others of a similar nature, are being investigated experimentally by increasing the concentration of deuterium in the bubble chamber.

It may also be that the surprisingly long gaps at the end of some of the stopping μ's can be understood by invoking the 135 ev available for recoil when the deuteron robs the μ^- from a $\mu - H$ mesonic atom during a collision.

It is interesting to speculate on the practical importance of this process if a sufficiently heavy, negatively charged, weakly interacting particle more long-lived than the μ is ever found. The particle observed by Alikanian et al.[5] in the cosmic rays has a mass of about 500 m_e, and was observed to come to rest in a cloud chamber without interacting or ejecting a decay fragment. A bubble chamber filled with liquid deuterium should be an excellent detector for such particles. One might expect to see large "stars" at the end of the heavy meson track, due to a sequence of catalyzed reactions that would continue until the meson disappeared by decay.

We wish to express our thanks to the bubble chamber crews, under the direction of R. Watt and G. Eckman, and to our scanners. We are also indebted to the three new members of our group, M. Cresti, L. Goldzahl, and K. Gottstein; and to E. Teller for an interesting discussion.

Note added in proof.—We have obtained preliminary results on the effect of increasing the deuterium concentration. The following numbers come from spot-checking and fast scanning only:

Deuterium Concentration:	Natural	0.3%	4.3%
$\mu^- \rightarrow e^-$	2541	2959	1269
H+D→He^3+μ^-	15	57	32
He^3+μ^- per μ^- ending	0.6%	2%	2.5%

Preliminary analysis indicates that the frequency of visible gaps in $\mu - e$ decays at first increases with increasing deuterium concentration; however, at the 4.3% concentration, gaps are no longer seen. We have seen one case where the same μ^- catalyzes the He^3+μ^- reaction twice. We have seen a few events which we interpret as the reaction D+D→H^3+H^1.

* This work was done under the auspices of the U. S. Atomic Energy Commission.

† Astronomy Department, University of California, Berkeley, California.

[1] Alvarez, Bradner, Falk-Vairant, Gow, Rosenfeld, Solmitz, and Tripp, University of California Radiation Laboratory Report UCRL-3583, 1956 (unpublished).

[2] Panofsky, Aamodt, and Hadley, Phys. Rev. **81**, 565 (1951).

[3] We have telephoned to inquire if other groups observe these gaps. R. H. Hildebrand has noticed occasional 1-mm gaps of $\mu - e$ decays in the Chicago hydrogen bubble chamber. Leon Lederman reports that no surprising gaps have been noticed by the Columbia diffusion chamber group [C. P. Sargent, thesis, Columbia University, 1951 (unpublished)].

[4] F. C. Frank, Nature **160**, 525 (1947); Ya. B. Zel'dovitch, Doklady Akad. Nauk U.S.S.R. **95**, 493 (1954).

[5] A. I. Alikanian et al., 1956 Moscow Conference on High-Energy Physics (unpublished).

14

My First Days in the Alvarez Group

Stanley G. Wojcicki

This is a personal story of several early years in the Alvarez group. Nominally the topic of this chapter is the discovery of a $\Lambda\pi$ resonance, commonly known today as the Y^* (1385). However, as with most of the bubble chamber discoveries, the eventual publication was the culmination of many years of work by many people: physicists, engineers, technicians, scanners, and programmers. But above all, it was the result of the foresight, scientific leadership, and organizational skill of Luis W. Alvarez. Thus, the full story of Y^* (1385) can only be told in the context of this historical background.

I joined the Alvarez group as a first-year graduate student in May 1958. My first job was to help in the data analysis in the customary last-minute frantic efforts to get results ready for a high-energy conference at CERN, in Geneva, Switzerland. This last-minute rush to obtain the "final" results for a conference still takes place before all major meetings, but in those days the data analysis was quite different. To give some flavor of the computational technology of that era, I need only mention that the geometrical reconstruction was performed on an IBM 650 computer: a decimal machine with 2000 10-digit words of memory on a rotating magnetized drum. The sole input was IBM cards, and the output was either on cards or on paper. The geometry reconstruction was relatively advanced compared to the kinematic fitting, which was done by graduate students and young Ph.D.'s assisted by such artifacts as the "Wolf charts," kinematics tables, and many kinematical graphs generated on the IBM 650.

But these modest beginnings should not obscure Luie's vision about the future of bubble chamber physics. Immediately after the announcement of the success of Don Glaser's first bubble chamber, Luie saw the bubble chamber as the ideal detector for exploiting the new particle accelerators then being planned or constructed. He saw the power of hydrogen for bubble chambers, the need to build them very large, and the simultaneous need to develop a large-scale organization to run them and process the massive amounts of film that were to come. From the very beginning Luie dreamed of the 72-inch hydrogen bubble chamber, an instrument that became a reality in 1959 and proved during the next decade to be the most productive high-energy physics detector in the world. Luie's ability to

extrapolate into the future and not be afraid of great leaps is illustrated by a now well-known anecdote that I heard from Luie. Clearly, one of the major problems in constructing the 72-inch bubble chamber was the large window of high optical quality cooled to liquid hydrogen temperature. As solutions to that problem were being pondered, somebody saw a technical abstract entitled something like "A large optical window at cryogenic temperatures." Excitedly, Luie and his coworkers went to the library, only to discover that the window in question was only five centimeters long! Luie loved new challenges and had that particular disdain of bureaucracy that characterizes many scientists. For example, the 72-inch bubble chamber was built and operational before the building to house it was ready. When I once asked Luie about this, he said something to the effect of "It's not surprising. Nobody ever built a big hydrogen bubble chamber before, so there are no rules to follow. But people have built buildings for centuries, so there are hundreds of volumes of regulations that have to be satisfied."

Those years of the late 1950s were exciting ones in Alvarez's group. The beginnings of the great effort to systematically use hydrogen bubble chambers to study particle physics were already there: Don Gow headed the bubble chamber operations group, and Hugh Bradner the scanners group. Frank Solmitz and Art Rosenfeld, two relatively young Ph.D.'s who had recently joined the group, started thinking about the mathematics and computing of data reconstruction. The excitement and the novel way things were done were known throughout the world; many young European Ph.D.'s spent a year in Berkeley during that epoch. I particularly remember Paul Falk-Vairant (France), Ferdnand Grard (Belgium), Philippe Eberhard (Switzerland), Phil Burke (England), and Marcello Cresti (Italy). But maybe the most exciting aspect was that the organization was still small and very personal, and thus one could feel very much a part of it. For example, I remember my first scanners' meeting, an organization that eventually grew to 100 people. We all sat around a table in Luie's office and for refreshments had home-baked cookies brought by Janet Landis, the future Mrs. Alvarez, who then worked as a data analyst.

In the late 1950s, the Gell-Mann–Nishijima elementary particle formalism appeared to be reasonably successful, but a key question remained about the nature of the so-called cascade (Ξ) particles. Some examples of Ξ^- particles had been observed in the cloud chambers, but under such conditions that a precise understanding of their production and decay kinematics was impossible. The latter was absolutely necessary for testing the theory. It was appreciated that the best way to test this scheme was to observe the production and decay of the then unobserved, but postulated, neutral Ξ^0 hyperon in a hydrogen bubble chamber.

Luie believed that the best way to produce Ξ particles was with beams of K^- mesons rather than the much more copious π^- mesons, since the K^--Ξ strangeness difference is only one unit, as opposed to two units in the case of π^--Ξ production. A few years earlier, Luie had suggested to Harold Ticho of UCLA that he investigate the possibility of tagging K^- particles by their slower time of flight in a predominantly π^- beam. Harold tried this in an exposure of the 10-inch bubble chamber to a π^- beam, but with inconclusive results.

However, a much better way to accomplish these goals soon became apparent. The first parallel-plate electrostatic separators had just been built and tried at Berkeley by Bruce Cork and Bill Wenzel, which could, in principle, be used to achieve a separated beam of K^- particles. The optimal way to observe Ξ particles in a bubble chamber would be to expose it to a separated beam of K^- mesons. To achieve such a beam, however, quite a bit more was required. The particle trajectories entering the separators had to be made parallel and sharply focused at the exit, so that the pion and kaon images would appear separated in space. The latter could be accomplished by devices known as magnetic quadrupoles, first introduced in 1952 by Courant, Livingston, and Snyder. Today, it is widely known that such magnets can be made to act as lenses for charged particles, and indeed they are routinely used in that fashion. In the late fifties, however, their use was limited mainly to increasing particle fluxes, and their application as optical elements in particle beams was still far from being standard experimental procedure.

Thus the problems connected with the design and construction of the needed high-energy K^- separated beam were indeed formidable.[1] First, a large fraction of the Bevatron proton beam had to be spilled rapidly on the target, since bubble chambers need a short pulse (about a millisecond), and because of the low K^- production rate and its relatively short decay path. Second, the separation itself required high electric field in the separators, about 200 kilovolts over a five centimeter distance. Third, the optics in the beam had to

be of high quality, since the spatial separation between the kaons and pions produced by the separators was only $\sim$ 1 cm. Finally, all this equipment had to function reliably over a long period of time, since the production cross section was expected to be rather low. I should also add that this was to be only the second physics experiment with the recently completed 15-inch hydrogen bubble chamber, a chamber built as a prototype for the 72-inch in which a lot of the ideas planned for the 72-inch were tried for the first time.

The two operating principals in the experiment were Harold Ticho, who was spending that summer at the Rad Lab, and Bud Good, a physicist in Luie's group. Harold's primary responsibility was the beam, and it was mainly he who conceptualized the K^- beam and put together the diverse technologies into its design. Bud wrote the first memo suggesting the feasibility of the separated beam and subsequently made several key contributions during the design stages of the experiment. Later on, he assumed responsibility for the myriad problems associated with the electrostatic separators. I was only told later that the original idea for the experiment came from Luie, who appreciated at an early stage the potential power of the hydrogen bubble chambers and saw such a chamber as the ideal instrument to use for this experiment. Many others were quite skeptical at the idea that by this technique you could measure the mass of an unknown, uncharged particle that decayed into two other uncharged particles, since neutral particles would not leave any visible tracks in the bubble chamber.

Some time during the summer of 1958, Philippe Eberhard and my fellow graduate student, Bill Graziano, joined the experiment. I became involved in the fall. Luie was a senior statesman, uninvolved in the day-to-day problems but interested and frequently giving invaluable advice. The active members were three Ph.D. physicists and two very green graduate students. The smallness of the group meant long hours: thirty or forty hour stints at the laboratory were not uncommon, but it also meant great excitement and a feeling of being right in the middle of it. Most of our time during the first few months was spent on floating-wire measurements, shimming magnets, and fighting the electrostatic separators. Many times their sparking was so frequent while we were trying to "bake them in" that we had to pull the plates out and clean them thoroughly with acetone. At that time, high-voltage technology in this application was rather poorly understood, and many of our procedures were highly empirical.

My close interaction with Bud and Harold during those exciting months allowed me to learn more experimental physics than at any other period in my life. Even though my interaction with Luie was close, I was impressed by his keen interest and his ability to penetrate to the heart of problems. For example, when we

1. P. Eberhard, M. L. Good, and H. K. Ticho, Rev. Sci. Instrum. **31**, 1054 (1960).

were putting finishing touches on tuning the beam, Luie became very upset when he heard we had never tested the quality of the optics with the spectrometers on. "How do you know that the separators will maintain the quality of the beam?" he asked. Another time, when we tried to check the purity of the beam with counters and electronics, Luie asked, "Why don't you turn the bubble chamber on and see what's really there? After all, it is the best detector known to man."

One evening when I was on shift Luie came in to see how things were going. When our conversation turned to the power supplies that were driving our magnets, Luie asked whether these were motor generator sets or the new solid-state supplies. I confessed I did not know and, frankly, that I was not even aware that the Lab had large-scale solid-state supplies. My ignorance set Luie off on a long lecture about what it takes to become a good physicist and the shortcomings of my generation. Luie was emphatic about the need for curiosity, for always trying to find out what other physicists are doing ("snooping" was his word), and for constantly striving to understand new techniques. He contrasted trying to become a good physicist with trying to become the president of General Motors: the latter was a political task requiring great care not to offend key people, while the most important ingredient in becoming a good physicist was curiosity.

Several days later, Luie followed up on our conversation by taking me and Bill Graziano on a tour of some of his old haunts. We visited various nooks and crannies at the Bevatron as well as the old Radiation Laboratory on campus. Luie pointed out where he did some nuclear physics experiments in the 1930s, commenting that if only he had but his apparatus a bit closer to the target, he might have discovered nuclear fission. All in all, it was a fascinating and nostalgic tour.

There were other follow-ups to our conversation that evening. Both Bill and I felt that the University of California Physics Department could do a better job in graduate student training, a universally familiar sentiment of graduate students. Luie listened to our comments and encouraged us to write up some suggestions for improvement, which he forwarded to Francis Jenkins, the department vice-chairman. Several days later we were invited by Francis and Luie to discuss with them some of our suggestions. Francis was generally sympathetic, but I am not sure that anything concrete was ever done. One of the points we brought up, however, was the lack of opportunities for a variety of hands-on experience with experimental hardware. As a result, the next summer there was at the Rad Lab a series of lectures for graduate students about counter techniques, discussing photo-tubes, discriminators, transmission lines, amplifiers, coincidence circuits, and the like. The lectures were given mainly by the Rad Lab employees, and some equipment was made available for students to play with.

Luie's day-to-day involvement in the experiment became more intense when the data began to come in. We all excitedly looked at film, scanning for two V's, a potential signature of the reaction $K^-p \to \Xi^0 K^0$, with a subsequent decay $\Xi^0 \to \Lambda\pi^0$. No other K^- induced reaction could give two simultaneous V's, but the background π^- in the beam could simulate it via the associated production process $\pi^-p \to \Lambda(\text{or } \Sigma^0) K^0$. Thus the ultimate identification of the Ξ^0 could be made only after kinematical fitting, even though the biggest single "footprint" of the Ξ^0 production was that a Λ did not point to the production vertex. This fact could sometimes be identified already on the scanning table.

The first double-V event was found by Bud Good after a few days of data-taking. It appeared to pass very detailed analysis and had a mass agreeing within experimental errors with the known Ξ^- mass. We were all excited and eagerly scanned for more events. The next candidate (I believe) was found by Luie; it had one V that clearly appeared not to point to the vertex. It was felt by the senior people that two events, if they stood up to further tests, would be enough to establish the existence of the Ξ^0. This called for celebration, and Luie took Bud and Harold out to dinner; I stayed on shift at the Bevatron taking more data and began a detailed analysis of the second event (graduate students have always stayed in the trenches). Before the evening was over, it was clear to me that the second event was not a Ξ^0 but merely an example of $K^-p \to \Lambda\pi^+\pi^-\pi^0$ where the last bubbles of the K^- were missing, making the charged pions look like the decay prongs of a detached V. The momentary excitement subsided, more film was taken, and more scanning hours were logged. Finally, after several months of running, the experiment came to a premature end when the bubble chamber broke down.

A major decision was then made by the laboratory director while Luie was on his honeymoon somewhere in the Caribbean. Two options presented themselves. One was to take a delay, fix the chamber, and continue data-taking under the same conditions. The second was to substitute the 30-inch propane chamber of Wilson Powell's group as the primary detector. The physics arguments on both sides were indecisive. It was already clear from the hydrogen data taken that the Ξ production cross-section at this energy was much lower than initially anticipated. Thus even a long hydrogen run could not produce a large number of Ξ's. On the other hand, the larger and denser propane chamber could yield many more Ξ events in the same amount of running, but it was doubtful that Ξ^0 events could be successfully identified in propane. Thus the experiment would be basically concentrating on Ξ^- events and giving up on Ξ^0's. I was not a party to these deliberations; thus I cannot relate the substance of arguments in detail. But in the end the propane chamber was moved in. Luie was very unhappy with the decision and subsequently would frequently argue that this was the wrong thing to have done. Parenthetically, I should mention that even though the Ξ^0 search in propane was abandoned after a short time because of inherent difficulties, several clean Ξ^- events were found. In addition, about a year after we published our discovery of the Y^* (1385), their larger number of the $K^- p \to \Lambda \pi^+ \pi^-$ events enabled the Powell-Birge group to determine the Y^* (1385) spin.

Meanwhile, the search for double V's in the hydrogen film continued. I believe that five more examples were found, only to be identified as interactions of the π^- contamination in the beam. The original event remained. It was remeasured many times, kinematical analysis was refined, and we all searched for a possible short proton recoil that could spoil the Λ coplanarity. The event remained as good as ever after all this careful scrutiny, and no other explanation appeared possible. The senior members of the group then decided that the event should be published as the first observation of the neutral cascade hyperon.

This was an important physics discovery, as it gave very strong support to the Gell-Mann–Nishijima scheme. A press conference followed its publication,[2] and several photographs of the six members of our experimental team were taken. Luie, as always, was concerned that proper credit be given where due and that everybody be treated fairly. Most of the photographs were taken in groups of three, with Luie, Philippe, and me in one and Bud, Harold, and Bill in the other. Most of the local press published, of course, the picture with Luie, but some papers published the other. Luie was happy because everybody got his picture in the paper.

Thus ended the first phase of the experiment. As it was designed to look for the cascade particles, it was known officially in the group as the Cascade Experiment. But all it had to show was one example, albeit the first, of a neutral Ξ-particle. The natural question was, What else could we get out of the film? Or, more pragmatically, What could we two graduate students do for our theses? The problem of extracting more useful physics out of the film was frequently discussed with resident and visiting theorists. To my knowledge, the possibility of looking at invariant masses and searching for bumps was never brought up in those conversations. There was a strong feeling, expressed for example by A. Pais, that extraction of topological cross sections could be quite interesting.

Before discussing the next developments, I should relate some of the inevitable personnel changes that

took place in the experiment at this time. Soon after the run was over, Harold Ticho returned to UCLA from his nine-month "summer" at Berkeley. Philippe Eberhard joined the antiproton experiment which first used the new 72-inch chamber, and a year later returned to Europe. Finally, late in 1959, Bud Good joined the faculty at the University of Wisconsin. We were joined by Margaret Alston, a physicist from England who became a permanent member of the Alvarez group. Even though both Bud and Harold retained an active interest and involvement in the experiment after their departure and Luie periodically inquired about the status of things, practically all the analysis details (supervision of scanning and measuring, and data analysis) fell to Margaret and us two graduate students.

Two very important developments took place then. The first was the success of the extensive geometrical reconstruction and kinematical fitting program development, directed mainly by Frank Solmitz and Art Rosenfeld, which allowed us to start thinking about analyzing all the K^- interactions in the film, including those with three or more final state particles. Most of the physicists and graduate students in the group participated in this effort at some level. For example, Bill and I fitted all the Λ decays in our film both by hand and through the new programs and spent many hours understanding the discrepancies and thus help root out program bugs.

The second was the initiation of Luie's Monday night seminars. This event, subsequently copied at many other places but to my knowledge never with equal success, probably contributed more to the general physics education of graduate students and Ph.D. physicists than any other activity in the Alvarez group. The ground rules and their informal atmosphere contributed to the seminars' success, but the most important ingredient was undoubtedly Luie's presence and the high standards that he required of the speakers. One of the group's physicists was the organizer for a year or so, responsible for choosing the speakers. There was a seminar every Monday night unless Christmas fell on Monday, and only the organizer and the speaker knew who would speak.

When one of those ground rules was broken, it was always to Luie's great displeasure. We did not miss a single Monday during the year I served as an organizer, just before my departure from the group. On one occasion, when there was a high probability that the speaker's wife might go into labor that Monday evening, the speaker and I agreed that this was not a sufficient reason to cancel the meeting.

The meeting format was simple. The seminars were held in Luie's home, in his living room-dining room. If Luie and his wife were out of town, the organizer had the key to his house. Luie provided beer and pretzels, and the talks lasted about an hour and a half. No holds were barred in the questioning, and if the speaker

2. L. W. Alvarez, P. Eberhard, M. L. Good, W. Graziano, H. K. Ticho, and S. G. Wojcicki, Phys. Rev. Lett. **2**, 215 (1959).

did not really understand what he was talking about, that fact became apparent after a few minutes. Luie sat in his chair in the front row and was never shy about asking even trivial questions if things were said that were not clear to him. The majority of the talks dealt with the work being done in the group, and it was considered standard that any future public talk by a group member be first rehearsed before this audience. But other topics were also presented: reviews of conferences, talks on new high-energy physics technology, talks by visitors, discussions of interesting developments in science outside of physics, and talks by theorists. It was the responsibility of the organizer to present a varied and interesting program. The attendance was almost always high; some thirty or forty people, filling Luie's living room pretty much to capacity.

Shortly before his departure for Wisconsin, Bud Good talked at the first Monday night seminar. In our small "cascade" subgroup we had been thinking about what we should do with our data, and Bud presented some of those ideas. One of the things that he discussed was the analysis of the $\Lambda\pi\pi$ final state to see if there might be a resonance there in analogy with the (3,3) resonance in the pion nucleon system. The idea that symmetry transformations can be thought of as linking all the known baryons had not yet become firmly established. There was, however, a rather qualitative concept of "global symmetry" that was thought to possibly provide the explanation for the multiplicity of baryons observed.[3] In that framework, it was natural to expect $T=1$ and $T=2$ $\Sigma\pi$ resonances, a $T=1$ $\Lambda\pi$ resonance, and a $T=3/2$ $\Xi\pi$ resonance. These resonances would presumably exhibit themselves as bumps in the hyperon-pion mass spectra. To my knowledge, Bud's presentation was the first public discussion within the group of this possibility and the first public speculation that this might be a fruitful investigation to pursue.

During 1959 and the first half of 1960, I was involved in a lot of drudgery and routine work that frequently tended to discourage me. Scanning instructions had to be written, scanners and measurers had to be supervised, failing measurements had to be understood and most often remeasured, and above all a working system of programs had to be obtained so that the data analysis could proceed. Margaret, Bill, and I tried to get some physics results for the International High Energy Conference in Kiev in the summer of 1959. Those results, presented by Luie as the rapporteur for strange particle interactions, consisted mainly of topological cross sections based primarily on the results of scanning.[4] But by that time the geometry program (PANG) worked

quite well and could be used to analyze crudely some of the events. Thus we were able to look at all the V's to see if any of them could be noncoplanar with the vertex and thus represent a possible candidate for Ξ^0 production and decay. None were found. In addition, the geometrical information could be used to separate K^0's from Λ's on a statistical basis and thus allowed us in some cases to attribute topological configuration to specific reaction channels.

During this period my interaction with Luie was mainly at the Monday night seminars. He complained once or twice about my spending too much time in the computer room and not enough "snooping" around the Lab. I found those comments frustrating, since like all graduate students I was anxious to finish my thesis and proceed to other things. I can now appreciate Luie's concern about the quality and completeness of my graduate education. Luie was seriously concerned that graduate students in the bubble chamber groups at Berkeley could easily become mainly "computer jocks." This point of view was frequently expressed by many of the senior people at Berkeley and became an ingredient in the intense intergroup rivalries. Luie said that it was not important whether students are forced to do certain tasks as part of their education as long as they are curious about what goes on around them and take the trouble to find out what others are doing and how they are doing it. I sensed a strong internal conflict in Luie. On one hand, he intellectually realized that if high-energy physics was to progress it must be done within the framework of large organizations. On the other hand, he worried that such organizations might stifle originality and provide less than optimal training for students. Down deep, he always remained a gadgeteer and an individualist and very far from an "organization man."

The next annual "Rochester" high energy physics conference was actually to be held in Rochester in the summer of 1960. We concentrated our analysis efforts on two topologies, namely the three-prong K-decays (τ-decays) and the interactions involving a V and two accompanying charged particles in the final state. The first channel was being analyzed mainly to understand our measurement errors and eliminate the remaining bugs in the analysis program. There was also some physics interest in this decay, since it could test some of the symmetry properties in the K-decay.

The other channel, hopefully, would yield some interesting physics results. Our primary goal was to isolate the well-constrained reaction $K^-p \to \Lambda\pi^+\pi^-$ and look for any strong final-state interactions. These would exhibit themselves as bumps in the center-of-mass kinetic energies of the pions (or in other words, invariant masses of the $\Lambda\pi$ systems) which could be the "global symmetry" analogs of the (3,3) pion-nucleon resonance.

There were about 250 examples of the V two-prong topology in our film sample. I realize that this number

3. M. Gell-Mann, Phys. Rev. **106**, 1296 (1957).

4. L. W. Alvarez, in *Proceedings of the 1959 International Conference on High Energy Physics* (Moscow, 1960), p. 471.

appears insignificant when compared with the multi-million event samples that are common in today's high-energy experiments, but since each event had to be treated pretty much individually through the measurement-analysis chain, the whole data analysis task represented a formidable task. About a month before the Conference our data sample was almost completely analyzed, and about 150 examples of $\Lambda\pi\pi$ events were found. That particular data set was quite pure because of the four constraints of momentum and energy conservation, with possibly a very small contamination due to the $\Sigma^0\pi^+\pi^-$ events, which were kinematically quite similar.

We worried for a while about this contamination and tried to estimate it in a quantitative way. We decided that the only way to do this was to generate random $\Sigma^0\pi^+\pi^-$ events and see how often their kinematics would look like $\Lambda\pi^+\pi^-$. And thus Bill Graziano and I embarked on what was probably the first Monte Carlo simulation program in particle physics. I am not sure whether there existed at that time a random number generator for the IBM 704 (we were already using this magnetic core, albeit vacuum tube, machine). At any rate, we used a published book of random numbers and had one of the scanners laboriously keypunch a whole series of them on IBM cards, subsequently to be used as input to our Monte Carlo. After two or three weeks of intensive work, we produced a working version of the code, but it became obvious to us that the program had to be made much more detailed if it was to serve as a means of obtaining precise quantitative answers. Our work further convinced us that the original estimates, based on back-of-the-envelope calculations, that the background was small were indeed correct. A few months later, Gerry Lynch embarked on creating a much more sophisticated and general program along those lines, called FAKE. It had general utility and eventually became a standard tool of bubble chamber physicists, but only after several man-years of programming effort.

Returning now to the data analysis, our final computer output for the accepted events consisted of the kinematical quantities resulting from the fit and some variables calculated in the K^-p center-of-mass system. The quantities of primary interest were the center-of-mass momenta of the two pions. We had no computer histogramming routines, so when the raw computer output was generated, we would usually have one of our student scanning technicians plot the variables of interest. When the final π^+ and π^- momentum plots were generated, each displayed a prominent peak around 400 MeVc^{-1}, corresponding to a $\Lambda\pi$ mass of 1385 MeV. In addition, the π^- momentum plot showed a broader second bump around 200 MeVc^{-1}. The much smaller sample of $\Sigma^0\pi\pi$ events showed no interesting structure in these variables.

The next day we went to talk to Luie about the results. Luie showed great interest and strongly encouraged us to come up with a simple, coherent explanation of the observations. I believe that Harold Ticho was visiting Berkeley the next day, and we spent some time together discussing the data. I remember Harold saying that this may be more interesting than the observation of the neutral cascade event. We talked to Bud Good about the results on the phone, but I do not recall that Bud had a chance to come to Berkeley before the Rochester meeting to discuss the data in person.

After more reflection and discussion it became clear that the data could be understood in terms of the production of an excited state, to be called Y^*, which then decayed rapidly via strong interactions into a Λ and a pion. The lower lying bump in the π^- spectrum could be understood as a kinematical reflection of the decay $Y^* \to \Lambda\pi$. The fact that the positively charged Y^* appeared to be produced less copiously than its oppositely charged partner accounted for the absence of a similar reflection in the π^+ momentum spectrum. The absence of peaks in the $\Sigma^0\pi^+\pi^-$ channel could be due to a lower branching ratio of the Y^* into $\Sigma\pi$ than into $\Lambda\pi$, a fact which could at least be partly understood on the basis of the phase space considerations. This explanation, though certainly ad hoc, appeared reasonable and accounted for all features of the data. Luie said this reminded him of the situation in the interaction of a proton with boron-12 going to three α-particles. Most of the α's are distributed smoothly in energy, but some come out in an "energy bump." This indicates that sometimes pairs of α-particles stick together long enough to give the third α a relatively unique energy. The nuclear physicists had measured the lifetimes of the excited beryllium-8 state from the width of the "bump"; similarly, later on, we spent a lot of time measuring the lifetimes of our resonances by measuring their energy widths.

In a conversation with a couple of resident theorists about our data, one of them told us that an explanation in terms of a new particle was very ugly, and most likely what we were seeing was some complicated effect in a crossed channel. I recall that conversation when tempted to place too much confidence in theorists' foresight and in the validity of theoretical models. Physics is, after all, an experimental science.

Most of these exciting developments occurred in the last few days before the Conference. One or two days before Luie's departure for Rochester, we stayed up late at night writing a paper about our observations. Luie was to take the paper with him to the Conference, but he felt that Bud Good should make the presentation. Incidentally, Bud did not know that he was to give the talk until he arrived at Rochester.

There was another channel among the V two-prong events that was also strongly constrained and thus rather free of background, $K^-p \rightarrow \bar{K}^0 p \pi^-$. There were fewer of these events, about 50, so we did not look at them in detail at first. But after we crudely digested the implications of the $\Lambda\pi\pi$ data and started looking at the $\bar{K}^0 p \pi^-$ events, a prominent resonant structure became obvious, this time in the proton momentum spectrum, corresponding to a well-defined mass of the $\bar{K}^0 \pi^-$ system. If true, this was even more surprising, since no $K\pi$ resonance was postulated by any theory at that time. Since the $K\pi$ mass bump occurred very near the edge of phase space, we were rather leery of any simple interpretation. So we agreed to emphasize only the two new "particles," Y^{*+} and Y^{*-}.

The data presented by Bud generated a great deal of interest. I understand from those present that the excitement in the informal hallway conversations was even stronger than one might gather from the formal questions.[5] The talk was essentially about the $\Lambda\pi\pi$ channel, with only a brief mention at the end of the $\bar{K}^0 p \pi^-$ data and the τ-decay results.

After the Conference, we concentrated on understanding the data in more detail with the goal of preparing a paper for publication. There were several questions that had to be attacked. First, What is the best way to display the data so that the kinematical reflections would be most readily apparent? The late Frank Solmitz made the very valuable suggestion that we follow Dalitz's idea from the θ-τ puzzle days on how one should display the $K^+ \rightarrow \pi^+\pi^+\pi^-$ data. By plotting each event as a point, the two axes labeled by the two center-of-mass kinetic energies, the dynamical effects and the kinematical reflections became much more apparent. This Dalitz plot technique became the standard one in the "bump-hunting" business during the next two decades. Harold and Bud's recollections are, however, that the Dalitz plot idea was actually brought up first by Frank in Rochester during a private discussion between Harold, Luie, Bud, and Frank.

The second very important question was the spin of the Y^*. There was a feeling in the group that the Y^* was very likely a "relative" of the (3,3) resonance, this point of view being strengthened by the similarity of the energies released by, and the intrinsic widths of, the two phenomena. For this to be true, the spin of the Y^* had to be 3/2, and thus the determination of that quantum number became a very important experimental task. A well-known technique to measure spins of the particles was the so-called Adair analysis. This very elegant idea, however, was limited to particles produced very close to the forward direction. Thus, it was not surprising that with our small number of events,

we could not discriminate between spin 1/2 and 3/2. On the other hand, a spin 1/2 Y^* had to decay isotropically along any direction; thus, *anisotropy* along *any* direction could be used to rule out spin 1/2 provided that one could be convinced that this anisotropy was not generated by some complicated interference effect due to the fact that the Y^* lives such a short time. We did indeed observe some anisotropy, but the statistical significance was marginal, so that even though the data favored a higher spin, no firm conclusion was possible.

Finally, after several iterations, many discussions, and some more calculations, the accompanying paper was written and ready to be submitted for publication.[6] I remember one incident during this process that illustrates Luie's sense of fairness and integrity. In one of the final iterations of the paper, a reference to relevant theoretical work was inadvertently omitted, even though we had decided previously that it should be included. Luie was upset about that omission and made us search through the recent literature to make sure that nobody else's work that should be cited was left out. Along similar lines I might recall another story, from a time that we were writing our third paper,[7] that one on $\Sigma\pi$ resonances announcing the evidence for Y^{*0} (1405) and an upper limit on the branching ratio of Y^* (1385) into $\Sigma\pi$. The first two papers had the authors listed alphabetically, with Margaret's name appearing first and mine last. Luie felt that this was unfair to me and suggested reversing the author order, so that the paper would be known as Wojcicki *et al.* We decided not to do this because the majority of us (myself included) felt that it would be better and less confusing to outsiders to maintain the same byline. Nevertheless, I felt that Luie's suggestion was characteristic of his concern that the credit be properly distributed.

Those two papers on the $\bar{K}^0 \pi^-$ and $\Sigma^\pm \pi^\mp$ resonances followed within the next few months and initiated a whole series of other discoveries in this general field.[8] They also stimulated a great deal of theoretical work that led to the formulation of SU3, a group theory classification of all particles, by Gell-Mann and independently by Neeman and subsequently to enunciation of the quark model by Gell-Mann and independently by Zweig.

Insofar as these discoveries were not predicted, or even suggested, by the theorists, they were accidental. What was, however, certainly not accidental was Luie's

5. M. Alston, L. W. Alvarez, P. Eberhard, M. L. Good, W. Graziano, H. K. Ticho, and S. G. Wojcicki, in *Proceedings of the 1960 Annual International Conference on High Energy Physics at Rochester,* edited by E. G. C. Sudarshan, J. H. Tinlot, and A. C. Melissinos (Interscience, New York, 1960), p. 445.

6. M. Alston, L. W. Alvarez, P. Eberhard, M. L. Good, W. Graziano, H. K. Ticho, and S. G. Wojcicki, Phys. Rev. Lett. **5**, 520 (1960).

7. M. Alston, L. W. Alvarez, P. Eberhard, M. L. Good, W. Graziano, H. K. Ticho, and S. G. Wojcicki, Phys. Rev. Lett. **6**, 698 (1961).

8. M. Alston, L. W. Alvarez, P. Eberhard, M. L. Good, W. Graziano, H. K. Ticho, and S. G. Wojcicki, Phys. Rev. Lett. **6**, 300 (1961).

vision of large-scale bubble chamber organization as the ideal way to do particle physics in the 1950s and 1960s. None of the discoveries that the group so prolifically produced would have been possible without it. Luie also provided the essential ingredients of overall intellectual leadership, very high scientific standards, excellent physics taste, and a strong sense of fairness.

To me he remains a first-rate role model of a superb scientist whose extensive interests in a variety of problems are only matched by his ability to use his ingenuity to attack and find solutions to those problems. But above all, because of his great impact on my career, I most remember him as a superb teacher and a good friend.

RESONANCE IN THE Λπ SYSTEM[*]

Margaret Alston, Luis W. Alvarez, Philippe Eberhard,[†] Myron L. Good,[‡]
William Graziano, Harold K. Ticho,[‖] and Stanley G. Wojcicki
Lawrence Radiation Laboratory and Department of Physics, University of California, Berkeley, California
(Received October 31, 1960)

We report a study of the reaction

$$K^- + p \rightarrow \Lambda^0 + \pi^+ + \pi^- \qquad (1)$$

produced by 1.15-Bev/c K^- mesons and observed in the Lawrence Radiation Laboratory's 15-in. hydrogen bubble chamber. A preliminary report of these results was presented at the 1960 Rochester Conference.[1] The beam was purified by two velocity spectrometers.[2] A Ξ^0 hyperon observed during the run[3] and the preliminary cross sections[4] for various K^- reactions at 1.15 Bev/c have been reported previously. Reaction (1) was the first one selected for detailed study, because it appeared to take place with relatively large probability and because the event, a 2-prong interaction accompanied by a V, was easily identified. In a volume of the chamber sufficiently restricted so that the scanning efficiency was near 100%, 255 such events were found. These events were measured, and the track data supplied to a computer which tested each event for goodness of fit to various kinematic hypotheses. The possible reactions, the distribution of events, and the corresponding cross sections are given in Table I. An event was placed in a given category of Table I if

the χ^2 probability for the other hypotheses was < 1%. It appears likely that the majority of the events in group (e) are also reactions of type (1). This belief is based on the following arguments:

1. Since the kinematics of a $\Lambda\pi\pi$ fit (four constraints) are more overdetermined than those of a $\Sigma^0\pi\pi$ fit (two constraints), it is relatively easy for a $\Lambda\pi\pi$ reaction to fit a $\Sigma^0\pi\pi$ reaction, but only

Table I. Distribution of events among different reactions.

Reaction	No. of events	Cross section (mb)
(a) $K^- + p \rightarrow \overline{K}^0 + p + \pi^-$	48	2.0 ±0.3
(b) $K^- + p \rightarrow (\Lambda$ or $\Sigma^0) + \pi^+ + \pi^- + \pi^0$	39	1.1 ±0.2
(c) $K^- + p \rightarrow \Sigma^0 + \pi^+ + \pi^-$	27	4.1 ±0.4
(d) $K^- + p \rightarrow \Lambda + \pi^+ + \pi^-$	49	
(e) $K^- + p \rightarrow (\Lambda$ or $\Sigma^0) + \pi^+ + \pi^-$	92	
Total	255	7.2 ±0.5

very few Σ^0 configurations can fit the $\Lambda\pi\pi$ reactions.

2. The events of group (e) when treated as $\Sigma^0\pi\pi$ reactions give a χ^2 distribution which is much worse than that obtained when they are treated as $\Lambda\pi\pi$ reactions.

In what follows, the 141 events of groups (d) and (e) are treated as examples of reaction (1). We estimate that 10 to 15% are actually Σ^0 events.

The energy distribution of the two pions in the K^--p barycentric system is shown in Fig. 1. If the cross section were dominated by phase space alone, the distribution of the points on the two-dimensional plot of Fig. 1 should be uniform. This is clearly not the case. On the contrary, both the π^+ and the π^- distributions have peaks near 285 Mev, such as would be expected from a quasi-two-body reaction of the type

$$K^- + p \to Y^{*\pm} + \pi^{\mp}, \qquad (2)$$

the Y^* having a mass spectrum peaking at ~1380 Mev. If the Y^* of mass 1380 Mev breaks up according to

$$Y^{*\pm} \to \Lambda^0 + \pi^{\pm}, \qquad (3)$$

the pions from this breakup are expected to have energies ranging from 58 to 175 Mev in the K^-p

rest system. Those pions from (3) are well separated from the pions arising from reaction (2) in the energy histograms.

The isotopic spin of this excited hyperon must be one, since it breaks up into a Λ and a π. Since the Y^* is produced with a pion, also of isotopic spin one, the reaction could proceed either in the $I = 0$ or the $I = 1$ state. Therefore the ratio of Y^{*+} to Y^{*-} will depend on the relative magnitude and phase of the two isotopic-spin amplitudes and thus could differ from unity. We observed 59 Y^{*+} events and 82 Y^{*-} events, using the criterion for separation that the high-momentum π meson is the pion from reaction (2).

Figure 2 shows the distribution in mass of the Y^* state (both Y^{*+} and Y^{*-}) including all 141 events, again using the higher energy pion in each event to calculate the Y^* mass. The experimental uncertainty in the mass for each event is small compared to the observed width of the curve. The curves of Fig. 2 are discussed later.

Figure 3 shows production angular distributions for Y^{*+} and Y^{*-} in the K^-p rest system. Partial waves with $l > 0$ appear to be present, as would be expected since $\hbar k/m_\pi c$ approximately equals 3. The difference between the Y^{*+} and Y^{*-} angular distributions may reflect the different superposi-

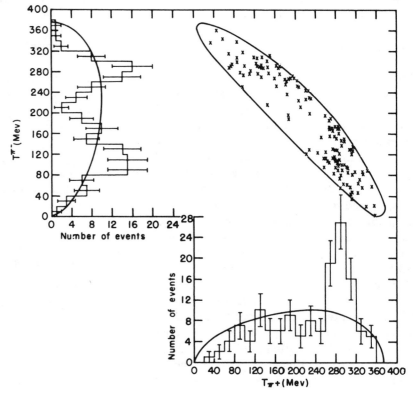

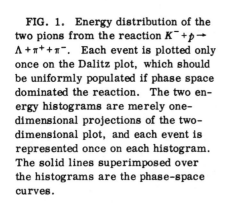

FIG. 1. Energy distribution of the two pions from the reaction $K^- + p \to \Lambda + \pi^+ + \pi^-$. Each event is plotted only once on the Dalitz plot, which should be uniformly populated if phase space dominated the reaction. The two energy histograms are merely one-dimensional projections of the two-dimensional plot, and each event is represented once on each histogram. The solid lines superimposed over the histograms are the phase-space curves.

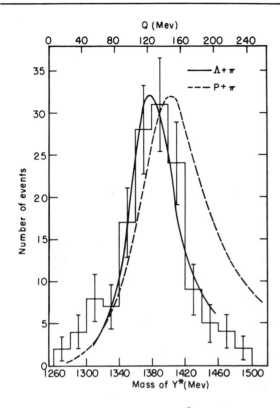

FIG. 2. Mass distribution for Y^* and fitted curves for $\pi\Lambda$ and πp resonances. The lower scale refers only to the $\pi\Lambda$ resonance. Q is the kinetic energy released when either isobar dissociates. The curve for the $\pi\Lambda$ resonances is fitted to the center eight histogram intervals of our data. The πp curve is the fit obtained by Gell-Mann and Watson,[7] to πp scattering data. Both fits are to the formula $\sigma \propto \lambdabar^2 \Gamma^2/[(E - E_0)^2 + \frac{1}{4}\Gamma^2]$, where $\Gamma = 2b(a/\lambdabar)^3/[1 + (a/\lambdabar)^2]$.

tions of the isotopic-spin zero and one amplitudes for the two cases.

The following two methods were used in an effort to determine the spin of Y^*.

(a) The angular momentum of Y^* was investigated by means of an Adair analysis.[5] We first restricted ourselves to production angles with $|\cos\theta| \geqslant 0.8$. For this angular range the Adair analysis should be valid if only S and P waves are present in the production process. We then computed η for each event, where

$$\eta = \vec{P}_{K^-} \cdot \vec{P}_\Lambda/(|\vec{P}_{K^-}||\vec{P}_\Lambda|).$$

Of the 29 events with $|\cos\theta| \geqslant 0.8$, the fraction 0.62 ± 0.09 has $|\eta| \geqslant 0.5$. If the above-mentioned restriction on the angular interval is sufficient to insure the validity of the Adair analysis, this ratio is expected to be 0.50 for $j = 1/2$ and 0.73 for $j = 3/2$. The experimental result is thus ~ 1.3 standard deviations from both possibilities, and no conclusion may be drawn from the data. Similar results were obtained for several larger values of the cutoff angle. Presence of D waves, however, cannot be excluded by the production angular distributions (Fig. 3). If they are present, indeed, then none of these choices of angle would be sufficiently restrictive to guarantee the success of the Adair analysis.

(b) Since Y^* may be polarized perpendicular to its plane of production, correlations can exist between the decay angle of the Y^* and the polarization of the resulting Λ. Also, a net Λ polarization

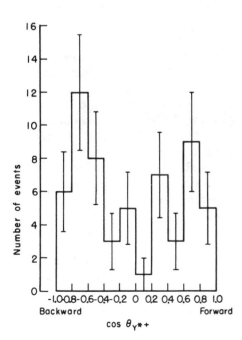

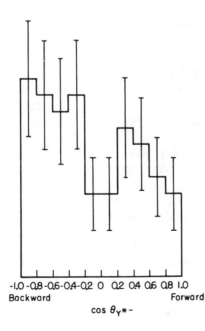

FIG. 3. Angular distribution of $Y^{*\pm}$ in the K^-p barycentric system for the reactions $K^- + p \rightarrow Y^{*\pm} + \pi^\mp$.

can result. With our limited data, we see no statistically significant Λ polarization or angular correlations.

However, one can also look for anisotropy, i.e., a polar-to-equatorial ratio, in the decay angle of Y^* with respect to the normal to the plane of production. For spin 3/2, the distribution must be of the form $A + B\xi^2$ by the Sachs-Eisler theorem,[6] independent of the Y^* parity, where we have

$$\xi = (\vec{P}_{K^-} \times \vec{P}_{Y^*}) \cdot \vec{P}_\Lambda / (|\vec{P}_{K^-} \times \vec{P}_{Y^*}||\vec{P}_\Lambda|),$$

and $\vec{P}_a$ is the momentum of the particle a in the K^-p barycentric system.

Since the coefficient B is a function of the production angle, we want to restrict ourselves to that range of the solid angle where the polar-to-equatorial anisotropy is probably greatest along the normal to the production plane. For production angles near 0 deg and 180 deg (Adair-analysis region), one expects the polar-to-equatorial ratio to be most different from unity in another direction (namely along the direction of the beam). Thus the equatorial region of production angles is more likely to show a large anisotropy along the direction in question. Therefore the production-angle range $\sin\theta \gtrsim 0.866$ was selected for study. We find the ratio of events with $|\xi| > 0.5$ to all events is 0.355. If the distribution is isotropic, as is required for spin 1/2, we expect 0.500 ± 0.063 for our 62 events. The result is thus 2.3 standard deviations from isotropy. The 45-to-1 odds against isotropy overstate the case for higher spin because this is the fourth anisotropy looked for.

Since Y^* may be regarded as a hyperon isobar, which decays into a π and a Λ, it evidently corresponds to a resonance in pion-hyperon scattering. The mass distribution of Fig. 2 then invites a comparison to the cross section for pion-nucleon scattering in the 3/2 - 3/2 state. For this purpose a p-wave resonance formula employed by Gell-Mann and Watson[7] for pion-nucleon scattering was fitted to our $\pi\Lambda$ data by using the eight central histogram intervals of Fig. 2. In fitting the curve, it was found that the interaction radius (a) could be varied over a wide range without changing the goodness of fit appreciably, provided that the reduced width (b) was also changed appropriately. The radius parameter was therefore fixed arbitrarily at $\hbar/m_\pi c$. Table II summarizes our results for Y^*, along with those of Gell-Mann and Watson for the 3-3 resonance.

Even if Y^* does turn out to be a p-wave reso-

Table II. Parameters for $\pi-\Lambda$ and $\pi-p$ resonance fitted to $\sigma \propto \lambdabar^2 \Gamma^2 / [(E - E_0)^2 + \frac{1}{4}\Gamma^2]$, where $\Gamma = 2b(a/\lambdabar)^3 / [1 + (a/\lambdabar)^2]$.

Parameter	$\pi-p$	$\pi-\Lambda$
Interaction radius a in units of $\hbar/m_\pi c$	0.88	1
Reduced width b (Mev)	58	33.4
Resonance energy E_0 (Mev)	159	129.3
Full width at half maximum (Mev)	100	64
Lifetime (sec)		$\sim 10^{-23}$

nance, there are still many reasons why the $\pi-\Lambda$ resonance parameters must not be taken too literally: (a) There is a small contamination of $\Sigma^0\pi\pi$ events in our data. (b) A nonresonant background may be present. (c) The production matrix element for reaction (2) might well depend on the outgoing momentum, and hence distort the mass distribution of Y^*. (d) Two thresholds for other possible decay modes of Y^* appear within the mass interval covered by the resonance curve; i.e., the $\Sigma\pi$ mode threshold around 1330 Mev and the $\bar{K}N$ threshold around 1435 Mev. This must have some effect on the shape of the mass spectrum as observed via the $\Lambda\pi$ decay mode. (e) Final-state pion-pion interaction could disturb the spectrum. (f) Even when the two resonances, Y^{*+} and Y^{*-}, are well resolved in terms of intensity— as in our experiment—there can still be an appreciable interference between the amplitude in which the π^+ arises from reaction (2) and the π^- from reaction (3) and the amplitude in which the roles of the two pions are reversed.

If we bear all these uncertainties in mind, the resemblance to the 3-3 resonance is certainly remarkable (Fig. 2). The resonance energies when expressed in terms of barycentric kinetic energies differ by only 30 Mev, which is much less than the width of either resonance. Furthermore, the widths are at least comparable.

These results are strongly reminiscent of the concept of global symmetry which predicts two spin 3/2 pion-hyperon resonances, one with $T = 1$, the other with $T = 2$.[8] These are the hyperon counterparts of the $J = T = 3/2$ resonance of the pion-nucleon system. On the other hand, the possibility that Y^* is a $J = 1/2$ resonance cannot be excluded on the basis of our data. The concept of pion-hyperon resonance in either $J = 1/2$ or $3/2$ state has been discussed recently by several authors.[9]

A study of $\Sigma^{\mp} \pi^{\pm} \pi^0$ events in our experiment is under way at present. The results, however, are too incomplete for us to be able to draw any definite conclusions.

The authors are greatly indebted to the bubble chamber crew under the direction of James D. Gow for their fine job in operating the chamber, especially Robert D. Watt and Glen J. Eckman for their invaluable help with the velocity spectrometers. We also gratefully acknowledge the cooperation of Dr. Edward J. Lofgren and the Bevatron crew, as well as the skilled work and cooperation of our scanning and measuring staff. Special thanks are due the many colleagues in our group who developed the PANG and KICK computer programs—especially Dr. Arthur H. Rosenfeld, and to Dr. Frank Solmitz for many helpful discussions.

One of us (P.E.) is grateful to the Philippe's Foundation Inc. and to the Commisariat à l'Energie Atomique for a fellowship.

*This work was done under the auspices of the U. S. Atomic Energy Commission.

†Presently at Laboratoire de Physique Atomique, College de France, Paris, France.

‡ Presently at University of Wisconsin, Madison, Wisconsin.

‖ Presently at University of California at Los Angeles, Los Angeles, California.

[1] Margaret Alston, L. W. Alvarez, P. Eberhard, M. L. Good, W. Graziano, H. K. Ticho, and S. Wojcicki, paper presented at the Tenth Annual Rochester Conference on High-Energy Nuclear Physics, 1960 (to be published).

[2] P. Eberhard, M. L. Good, and H. K. Ticho, Lawrence Radiation Laboratory Report UCRL-8878 Rev, December, 1959 (unpublished); also Rev. Sci. Instr. (to be published).

[3] L. W. Alvarez, P. Eberhard, M. L. Good, W. Graziano, H. K. Ticho, and S. Wojcicki, Phys. Rev. Letters $\underline{2}$, 215 (1959).

[4] L. W. Alvarez, in Proceedings of the 1959 International Conference on High-Energy Physics at Kiev (unpublished); also Lawrence Radiation Laboratory Report UCRL-9354, August, 1960 (unpublished).

[5] R. K. Adair, Phys. Rev. $\underline{100}$, 1540 (1955).

[6] E. Eisler and R. G. Sachs, Phys. Rev. $\underline{72}$, 680 (1947).

[7] M. Gell-Mann and K. Watson, Annual Review of Nuclear Science (Annual Reviews, Inc., Palo Alto, California, 1954), Vol. 4.

[8] M. Gell-Mann, Phys. Rev. $\underline{106}$, 1297 (1957).

[9] R. H. Capps, Phys. Rev. $\underline{119}$, 1753 (1960); R. H. Capps and M. Nauenberg, Phys. Rev. $\underline{118}$, 593 (1960); R. H. Dalitz and S. F. Tuan, Ann. Phys. $\underline{10}$, 307 (1960); M. Nauenberg, Phys. Rev. Letters $\underline{2}$, 351 (1959); A. Komatsuzawa, R. Sugano, and Y. Nogami, Progr. Theoret. Phys. (Kyoto) $\underline{21}$, 151 (1959); Y. Nogami, Progr. Theoret. Phys. (Kyoto) $\underline{22}$, 25 (1959); D. Amati, A. Stanghellini, and B. Vitale, Nuovo cimento $\underline{13}$, 1143 (1959); L. F. Landovitz and B. Margolis, Phys. Rev. Letters $\underline{2}$, 318 (1959); M. H. Ross and C. L. Shaw, Ann. Phys. (to be published).

15

Magnetic Monopoles

Philippe H. Eberhard

At the same time the Alvarez Group was "farming" the 72-inch bubble chamber and collecting an abundant "crop" of particles, it was also supporting a small "pioneering" experiment. The participants were Luie, Pete Schwemin, Robbie Smits, and Bob Watt. The goal was a search for magnetic monopoles using a completely novel approach.

Previous searches for single monopoles looked for particles with magnetic charge in motion with respect to their environment, similar to the way we detect other elementary particles, but with the appropriate signature.[1] Such searches would have been inadequate if, as was conceivable, monopoles were rare particles that would only end up firmly trapped in big chunks of material.

The idea that there should be better detecting techniques was Luie's. The magnetic flux quantum of superconducting loops had just been measured, and, as predicted by theory, it was half the minimum value of the flux expected from a monopole. Luie concluded that technology was then ready to measure the flux of even one single monopole trapped in samples as large as the superconducting loops used in the flux quantum experiment. If such a technique could be developed, monopoles could be found, even if they could not be extracted from matter. Furthermore, the method of detection would be very safe, since the measurement would not perturb the sample and could be repeated again and again till the last uncertainty was resolved. There was no risk, then, of making false claims of discovery, as happened only too often in the case of monopoles.

The above-mentioned team went to work, with everyone making significant contributions. The direct current induction effect of a moving magnetic charge was soon recognized by Robbie and Pete as the key phenomenon to be used.[2] A detector was built using a rapidly spinning ring carrying gram samples. As it went round and round, it always traversed the coil in the same direction. If the sample contained even a single monopole of minimum charge, a measurable direct current voltage would have been detected across the coil. Many samples of all sorts of material were investigated, including pieces of meteorites that were supposed to be good places to look for monopoles. Meteorites were exposed to cosmic rays for a very long time, during which cosmic-ray monopoles could have been stopped or a pair of monopoles could have been produced by very energetic protons.

From the beginning, I found that experiment interesting; it was important in any case. If a monopole was found, it would have an impact on theory and possible technology. If none were found, it could demonstrate magnetic neutrality of matter in all sorts of conditions, thus severely constraining the theories that predicted monopoles. Either way, it supplied an important piece of information. However, to join in, it seemed better to be able to make a significant contribution too. The opportunity came because Luie had seen the possibility of expanding the search for monopoles to even more valuable material than meteorites. The Moon is older than most meteorites, and its surface has been exposed to cosmic rays for a longer time. The astronauts of the Apollo program were to bring rocks from the Moon very soon, and a search for monopoles there had a chance to be authorized if one could be absolutely sure that it would not disturb the lunar material in any way. The direct current induction technique for detecting monopoles seemed promising in this respect; however, the spinning ring had to be slowed down so as not to submit the sample to excessive acceleration, and the coil inner diameter had to be increased to accept larger samples. Thus the whole detection system had to be made more sensitive.

For everyone, it was obvious that the sensitivity would be improved if the coil were to be cooled down and its resistance reduced. How much could be gained this way was not so clear. Only after study did the behavior of the apparatus with a coil resistance equal or nearly equal to zero became clear.[3] Order-of-magnitude improvements in sensitivity could be obtained by making the coil superconducting and by short-circuiting it. At each pass of a sample containing a

1. E. Goto, H. H. Kolm, and K. W. Ford, Phys. Rev. 132, 387 (1963); E. Amaldi, G. Baroni, A. Manfredini, and H. Bradner, Nuovo Cimento 28, 773 (1963); and E. M. Purcell, G. B. Collins, T. Fujii, J. Hornbostel, and F. Turkot, Phys. Rev. 129, 2326 (1963).

2. L. W. Alvarez, Lawrence Berkeley Laboratory Phys. Note 470 (1963), unpublished.

3. P. H. Eberhard, Lawrence Berkeley Laboratory Phys. Note 506 (1964), unpublished.

monopole, the current in the closed circuit would steadily increase by a finite amount proportional to the magnetic charge. Therefore it would be only a matter of time for the current to reach a detectable value, whatever the speed of the sample, the size of the coil, or any other parameter of the apparatus. All the other advantages of the original apparatus were preserved. This detection technique seemed appropriate for the experiment with the lunar sample.

As soon as Luie learned about this new approach, he was enthusiastic. A proposal was submitted to NASA and was accepted. Luie took care of all the contacts with NASA at the highest levels and was very successful. John Taylor, then Ron Ross and many others, joined us. They got involved with me to organize the building of the equipment and the running of the experiment. The collaboration worked out very well, and many problems were solved. During the summer of 1969 the first monopole experiment with 8 kilograms of lunar sample was run in the Lunar Receiving Lab-

oratory in Houston.[4] Later other searches in the lunar sample and elsewhere[5] were conducted with a further improved apparatus.[6]

We never found any monopoles, but we made sure that each sample we looked at was magnetically neutral. In many respects, our experiment was the most reliable monopole search made. Our direct current induction technique using closed superconducting circuits were used by others later on, but before then, our experiment had already set strict upper limits to monopole density in many important areas. It gives constraints to many theories that predict monopoles. It is an experiment that had to be done. We are all proud our names are associated with it.

4. L. Alvarez, P. Eberhard, R. Ross, and R. Watt, Science **167**, 701 (1970); Geochem. Cosmochim. Acta **3**, 1953 (1970), Suppl. 1; and Phys. Rev. **D4**, 3260 (1971).

5. R. R. Ross, P. H. Eberhard, L. W. Alvarez, and R. D. Watt, Phys. Rev. **D8**, 698 (1973) and P. H. Eberhard, R. R. Ross, J. C. Taylor, L. W. Alvarez, and H. Oberlack, Phys. Rev. **D11**, 3099 (1975).

6. P. H. Eberhard, R. R. Ross, and J. D. Taylor, Rev. Sci. Instrum. **46**, 362 (1975).

Reprinted from
30 January 1970, Volume 167, pp. 701-703

Search for Magnetic Monopoles in the Lunar Sample

Luis W. Alvarez, Philippe H. Eberhard, Ronald R. Ross and Robert D. Watt

Abstract. *An electromagnetic search for magnetic monopoles of the minimum size predicted by Dirac, or of any larger magnitude, has been performed on 8.37 kilograms of lunar surface material. No monopole was found. This experiment sets new limits on the production cross section for monopoles and on their occurrence in cosmic radiation.*

For several years now, the hunt has been on for particles that would interact with the magnetic field just as electric charges interact with the electric field, acting as a source for the field and being accelerated by it. These particles, called monopoles, would be stable. They would have a magnetic charge measured by an integer v, the Dirac charge 3×10^{-8} emu being used as a unit (1). Their existence would give credence to the only known explanation for the extraordinarily accurate phenomenon of charge quantization (2). According to a recent theory (3), they would be the most fundamental particles, the building blocks of the universe. However, no such particle or combination with a net nonzero magnetic charge has ever been found (4–6).

In view of the negative results of these experiments (4–6), the lunar surface is considered to be the most likely hiding place for monopoles, whether they belonged to the primary cosmic rays or were produced in the collision of a high-energy cosmic ray particle with a nucleon of the lunar surface. In either case, the lunar material would slow the monopole down and trap it. The reasoning that favors the lunar sample involves its great age, 3 to 4×10^9 years, and the small depth to which the surface has been churned during the long period of time. These two factors give the lunar surface the longest known exposure to cosmic rays. Furthermore, the absence of both an atmosphere and a magnetic field on the moon allows the fate of a monopole after it has been slowed down to be assessed with more certainty than it could be on the earth.

Our detection technique relies on the electromotive force induced in a coil by a moving monopole. As in previous work (7), the sample was transported along a continuous path threading the windings of a coil. In this experiment the coil was made of superconducting material and was short-circuited by a superconducting switch. A small current was stored in the superconducting loop before a sample was run. If a sample containing a monopole had been run, the induced electromotive force would have modified this current. After each sample had been circulated 400 times, the superconducting switch was opened and a signal proportional to the current in the loop was transferred electrically out of the cryostat, amplified, and finally recorded on an oscilloscope. A real magnetic charge would have been detected as a difference between the signal obtained when the switch was opened and the one normally observed when the opening of the switch interrupted the "standard current" that had been introduced as an overall check on the apparatus. A zero magnetic charge therefore corresponded to a nonzero standard signal. This technique assured us that the equipment was working at all times.

An overall calibration was obtained from a long solenoid in which a known change of current simulated the "missing term" in Maxwell's equations—the one describing the contribution of a "magnetic current density." A statistical study of our signals shows that the measurement of the magnetic charge was affected by a 1 standard deviation error of about 1/8 of a Dirac unit, when a ride of 400 passes was given the sample. Therefore, the smallest monopole compatible with Dirac theory was expected to produce an 8 standard deviation signal. There are reasons to believe the smallest actual charge would have twice the Dirac value (8), and this would correspond to twice as big a signal.

The lunar surface material analyzed in this experiment consisted of 28 individual samples. One sample was composed of three rocks (NASA 10022-1, 10023-1, and 10024-3) weighing all together 213 g. The remaining 27 samples were all fines from the bulk sample (NASA 10002). The individual samples of fines ranged in weight from 261 to 356 g, and weighed all together 8.13 kg.

The measured magnetic charge of each sample was consistent with zero, and statistically incompatible with the hypothesis that the absolute value of the magnetic charge was as large as, or larger than, a single Dirac unit of magnetic charge. We can therefore set upper limits on the number of monopoles present in the primary cosmic rays *and* on the number of monopoles produced by high energy cosmic ray particles interacting with nucleons of the lunar surface material. We quote our results at the 95 percent confidence level, including a correction of 10 percent to the monopole density to take into account the possibility that any individual sample may have contained paired monopoles of opposite charge.

The actual values of both upper limits depend upon unknown properties of the hypothetical magnetic monopole—namely, its charge, its mass, and all the parameters that determine its range inside the lunar material before it comes to rest. Therefore we express our results as a function of n, a parameter which relates the approximate range R, in grams per square centimeter, to the kinetic energy E, in Gev, by $R = 0.1\ E/n^2$. For low velocities, when the monopole loses energy by ionization only, n in this formula is the magnetic charge ν measured in Dirac units. At higher velocities, the effective value of n is expected to increase with E due to bremsstrahlung.

The values of both upper limits depend also upon the assumption we make regarding the depth, D, to which the lunar surface has been churned. We have represented our results in Figs. 1 and 2 for an assumed exposure time of 3×10^9 years and for two mixing depths, (i) 5 cm (solid curve), which represents effectively no mixing depth, and (ii) 100 cm (dashed curve).

In Fig. 1, we plot our upper limit for the flux of monopoles per square centimeter per second, per steradian, in the cosmic rays. The curves are displayed as a function of the kinetic energy of the monopole with n and D as parameters. Curves A and B represent upper limits known from the most extensive previous searches for monopoles in cosmic rays. Curve A results from examination of deep ocean deposits (5), and curve B results from an analysis of tracks in obsidian and mica (6). The results of the most extensive search carried out in the earth's atmosphere are given by curve C (9).

The production of monopoles by proton-nucleon interactions depends upon the monopole pair-production cross section, σ. In Fig. 2 we have plotted the upper limit for σ, as it results from our experiment, as a function of the monopole mass for different values of D and n. The flux of primary cosmic rays above the energy E, in Gev, was assumed to be $1.4 \times E^{-1.67}$ cm^{-2} sec^{-1} sr^{-1} (4).

The incident proton was assumed to lose 40 percent of its energy at each proton-nucleon interaction (9). The monopole pair-production cross section was assumed to be constant above the threshold energy for monopole pair production. Curves A and C represent the limits for σ as known from previous work (5 and 9, respectively). Curve D comes from a search for monopoles in a meteorite (10) as interpreted in reference (4). Curve E corresponds to the most extensive accelerator study made to date (11).

The search for monopoles in the lunar sample of Apollo 11 resulted in the finding that there was neither an unpaired north or south monopole in any of the 28 samples studied. This result sets upper limits on the presence of monopoles both in the primary cosmic rays and in the proton-nucleon interactions, without any assumption concerning the migration of the monopoles through matter under the influence of the magnetic field. If the lunar mixing depth is less than 10 meters over a

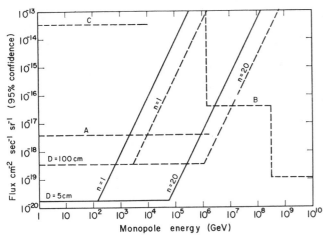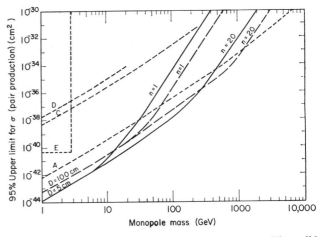

Fig. 1 (left). Ninety-five percent confidence level upper limit on the flux of monopoles as a function of monopole energy. The solid and dashed curves for D=5 and 100 cm are from this work. The parameters n and D are defined in the text. Data for curves: A, reference (5); B, reference (6); C, reference (9). Fig. 2 (right). Ninety-five percent confidence level upper limit on the monopole pair-production cross section in proton-nucleon collisions. The solid and dashed curves for D=5 and 100 cm are from this work. The parameters n and D are defined in the text. Data for curves: A, reference (5); C, reference (9), D, reference (10); E, reference (11).

period of 3×10^9 years, our upper limits are lower than any previous values except in high ranges of mass and energy, as shown in Figs. 1 and 2.

Luis W. Alvarez
Philippe H. Eberhard
Ronald R. Ross

Lawrence Radiation Laboratory and Space Science Laboratory, University of California, Berkeley 94720

Robert D. Watt

Stanford Linear Accelerator Center, Stanford, California

References and Notes

1. P. A. M. Dirac, *Phys. Rev.* **74**, 817 (1948); *Proc. Roy. Soc.* **A133**, 60 (1931).
2. L. J. Fraser, E. R. Carlson, V. W. Hughes, *Bull. Amer. Phys. Soc.* **13**, 636 (1968) (paper E 17); J. G. King, *Phys. Rev. Lett.* **5**, 562 (1960).
3. J. Schwinger, *Phys. Rev.* **173**, 1536 (1968); *Science* **165**, 757 (1969).
4. For a systematic review of experimental searches for Dirac monopoles up to 1968, see E. Amaldi "On the Dirac magnetic poles," in *Old and New Problems in Elementary Particles*, G. Puppi, Ed. (Academic Press, New York, 1968), p. 20. Two more recent searches are reported in references (*5* and *6*).
5. R. L. Fleischer, H. R. Hart, I. S. Jacobs, P. B. Price, W. M. Schwarz, F. Aumento, *Phys. Rev.* **184**, 1393 (1969).
6. R. L. Fleischer, P. B. Price, R. T. Woods, *ibid.* **184**, 1398 (1969).
7. L. W. Alvarez, A. J. Schwemin, R. G. Smits, R. D. Watt, Semi-annual Report, UCRL 11466, October 1964, p. 6 (unpublished).
8. L. W. Alvarez, Physics Notes 479, Lawrence Radiation Laboratory, 1963 (unpublished); J. Schwinger, *Phys. Rev.* **144**, 1087 (1965).
9. W. C. Carithers, R. Stefanski, R. K. Adair, *Phys. Rev.* **149**, 1070 (1966).
10. V. A. Petukhov and M. N. Yakimenko, *Nucl. Phys.* **49**, 87 (1963).
11. E. Amaldi, G. Baroni, A. Manfredini, H. Bradner, L. Hoffman, G. Vanderhaeghe, *Nuovo Cimento* **28**, 773 (1963).
12. We are indebted to Roscoe Byrns for the design of our cryogenic equipment, to John Taylor and Leo Foley for the design and test of many parts of the apparatus, to Maurilio Antuna, Robert Gilmer, and Hans Stellrecht for the construction of the electronic equipment, to Egon Hoyer for the design of the sample containers, and to all the members of the team that worked on this experiment in Berkeley and Houston. We acknowledge the help we have received from the staff of the Lunar Receiving Laboratory and the Lunar Sample Analysis Planning Team. Work done under NASA contract NAS 9-8806.

5 January 1970

16

Using Cosmic Rays in the Search for Hidden Chambers in the Pyramids

Luis W. Alvarez

When I first saw the pyramids in the summer of 1962, on the way to a high energy physics conference in Geneva, I tried to imagine how they were built. Not until I found myself on Antarctica for ten days in early 1964, however, did I find time to think more deeply about the pyramids. I then became so excited that when I returned to Berkeley, I took a large stack of books from the library and poured over them in the next weeks.

The elevation views of the interiors of the three largest pyramids show Chephren's Pyramid has no chambers, while those of his father (Cheops) and his grandfather (Sneferu) both have large chambers in the main pyramid mass. When I learned that Cheop's chambers were discovered quite accidentally in about AD 900—3,500 years after the pyramid was built—it occurred to me that there might still be undiscovered chambers in the Chephren Pyramid. My "original idea," I later learned, had been anticipated by the early archeologists, some of whom had looked for undiscovered chambers with gunpower. That Chephren would have built no chambers seemed contrary to everything I know about human behavior. Since Chephren's grandfather had put two chambers in his pyramid and Chephren's father had three, I reasoned that Chephren would more logically have put four chambers in his pyramid! So the Pyramid Project, as it was to be known, was based on my simple guess concerning the way people behave.

I decided to use cosmic rays to probe the body of the Chephren Pyramid for undiscovered chambers. The idea of using sound waves or radio waves also occurred to me, but I rejected both of them because the regular pyramid structure would scatter waves short enough to detect a chamber, while waves long enough to penetrate would be unable to pinpoint a chamber.

In the summer of 1964, while Director of a summer school on Lake Como, I drove over the Alps to talk with friends at CERN. When I explained my ideas to Bogdan Maglich, he introduced me to an Egyptian member of his research group, Fikhry Hassan. Hassan was interested in the proposed experiment and contacted Professor El Bedewi, Head of the Physics Department at the Ein Shams University in Cairo, who

proposed cooperating with my group in carrying out the project. El Bedewi and I then started a voluminous correspondence, which led to the establishment of the Joint UAR-USA Pyramid Project. El Bedewi talked with archeologists and government leaders in his country to enlist their support and I did the same in this country. A very important event that helped greatly to establish the program was the coincidental meeting of El Bedewi with my old friend Glenn Seaborg, chairman of the Atomic Energy Commission, at the International Atomic Energy Conference in Tokyo in September 1965. El Bedewi cleverly included mention of the Pyramid Project as a peaceful international collaboration between California and Ein Shams Universities in the Egyptian statement delivered in the plenary session. El Bedewi also arranged for Seaborg to visit Cairo on his way back to the States.

The project administration was entrusted to a committee of distinguished Egyptian scholars, while the day-to-day operations were the responsibility of a three-man executive committee composed of El Bedewi, Egyptologist Ahmed Fakhry, and myself. In September 1966, my young Berkeley colleague Jerry Anderson and I, with our wives, visited Cairo where we met the Egyptians who were to play an important part in the joint program. Anderson brought mock-ups of all the equipment to make sure that it would fit through the pyramid passages.

As soon as the Atomic Energy Commission funded the project in 1966, the work of designing and building the equipment started at Berkeley under Anderson's leadership. Four physicists from the Ein Shams University came to Berkeley for a three-month stay, during which they participated in the assembly and testing of the equipment.

There is no simple word or phrase to describe the process we used to probe the pyramid interior. "X-raying" comes closest, because our measurements resulted in photographic transparencies showing chambers and passages. In fact, our computer programs produced x-ray-like photographs which would show a chamber in the same way an abdominal x-ray would show a gas bubble in the intestines.

To map a small model of the pyramid onto an x-ray film, an industrial radiologist would have placed a radioactive x-ray source in a chamber under the pyramid and balanced the x-ray film on the pyramid's peak. We did not have a point source of the very penetrating radiation, as the cosmic radiation comes with equal brightness from all directions in the sky. We could use this isotropic radiation by measuring the direction of each cosmic ray muon recorded in the subterranean chamber. The projection operation is then mathematically equivalent to that of the industrial radiologist, but the directions of all rays are reversed. The cosmic ray detector, located in a chamber near the center and just under the ground-level section of the Chephren Pyramid, was capable of recording the azimuth and elevation angles of each recorded cosmic ray.

Muons, the penetrating component of the cosmic ray, are born high in the earth's atmosphere and result from the collisions between still more energetic primary cosmic rays (mostly protons) and the nuclei of air atoms (mostly nitrogen). These muons lose energy by ionizing the matter through which they pass, similarly to the way a rifle bullet loses energy by friction as it bores its way through a fence post. The muon loses energy of roughly 200 million electron volts (MeV) for every meter of water through which it passes. If a particular muon is stopped upon passing through rock of density 3, its range would have been three times greater in water. Because of the simplicity of the energy-range law for muons, it is customary to express the muon range in meters of water equivalent (m.w.e.). Thus, a single range-energy curve tells us all that is known about the penetration of muons in any material, if the range is expressed in m.w.e. The Cheops Pyramid has a present height of 137 meters, the density of limestone is about 2.5, the height of the pyramid is 342 m.w.e., so all muons with at least 70 GeV of energy passing straight down through the peak will make it to the bottom of the pyramid.

If we make our measurements at sea level with only air above the detector, we observe the same counting rate in all directions above the horizon. This isotropy is true for muons of interest to a pyramid prober. For low energy muons, the intensity decreases from the zenith. In contrast, higher energy muons come more frequently from angles closer to the horizontal. It is further true that the counting rates from the vertical and to the horizontal direction will be equal when seen through the same thickness of rock. It is this fortunate aspect of cosmic radiation that makes it possible to measure the thickness of rock from the subterranean chamber to the surface of the pyramid.

However, for this method to be successful the counting rate in apparatus in the subterranean chamber had to be high enough to yield a "grainless" x-ray picture in less than a year of operation. Statistical fluctuations in the counting rate could be treated in standard ways,

and the conclusion was reached that such fluctuations would not be large enough to indicate the presence of non-existent chambers. Also, random deflections of the cosmic ray muons as they pass through the rock (multiple scattering) would be small enough to keep the images of the chambers from being washed out. So we were fortunate not only in the penetration nature of our muons, but also in their intensity, which allowed us to make useful measurements in less than a year.

To reconstruct an "x-ray picture," we must measure the two angles that specify the muon's arrival direction. The then recently invented spark chamber could do this job beautifully. The simplest spark chamber consists of two flat metal plates spaced about a centimeter apart. When a muon penetrates the two plates, an electronic circuit suddenly applies a high voltage between the two plates and a spark jumps between the plates, very nearly along the muon trajectory. The spark, easily seen with the naked eye, was originally recorded by two cameras looking into the gap, their optical axes at right angles to each other so that the spark position could be located by stereoscopic reprojection. We fortunately started to use spark chambers just after the need to photograph the sparks had been bypassed. The coordinates of our sparks were recorded electronically and stored on magnetic tape.

There were many ways to display our data. The most meaningful would have been to print out the number of counts obtained in each elementary unit of solid angle, (Pixel) much like the pressure map printed by the weather bureau. We showed that 1600 counts for each 8.5×8.5 square foot area on the pyramic surface would be adequate to locate a "standard" burial chamber. The last digit in this count is of no significance, but the tens digit is of some significance as we expected an extra 160 counts if we were looking through a standard chamber. Since the standard deviation of a measured number of counts is the square root of the number, the expected number of counts per element is $1,600 \pm 40$. So if we found an additional 160 counts, then that number is $1,760 \pm 42$. By the laws of statistics, such a 4-standard deviation effect will happen by chance only once in every 15,000 trials. This is the argument we used to say that our x-ray picture was not grainy. If we saw a 10% increase in the number of counts per element, we could be virtually certain that the detected muons had passed through a chamber. And by counting a longer time and still seeing the 10% effect we could be *absolutely* sure that we had found a previously unknown chamber.

Through the courtesy of the Department of Antiquities, and the friendly interest of the Chief Inspector of the Pyramids, Ali Hassan, the Ein Shams-Berkeley team of scientists and technicians installed our cosmic ray detector in the Belzoni Chamber near the center of the base of the Chephren Pyramid. The detector was then connected by cable to a magnetic tape-recorder in our laboratory building about a quarter of a mile from the pyramid. There, complete electronics and mechanical shops were located.

Certainly things in this project didn't always go smoothly, but they went at least as smoothly as on the floor of any U.S. accelerator laboratory. The different ways of thinking, Arab and Western, had the potential to create problems, but these were offset because English was a "second language" for the Egyptians. If an American said something offensive, an Egyptian could simply ignore it, giving the impression that he did not understand, and allowing time for de-escalation. The Egyptians could vent their frustration by cursing in Arabic, which was similarly ignored by the Americans. General Eisenhower was once reported to have sent an American officer home from England during the planning of the invasion of Europe for calling his counterpart "you British S.O.B." The General is reported to have said that it was proper in his unified command for one officer to call another an S.O.B., but unacceptable for him to be singled out by nationality. That same spirit pervaded the Joint UAR-USA Pyramid Project.

In the spring of 1967, we shipped the equipment to Cairo, and Jerry Anderson and four colleagues, with their wives, settled down for what they thought would be a year or more in Cairo. The Ein Shams-Berkeley team had the equipment in place by the first of June, but the day immediately following our first full operation, the Six-Day War broke out. At the direction of the U.S. Embassy, all the wives had already left Cairo for Athens. Shortly after the outbreak of war, all Americans in Egypt were taken into "protective custody" and sent by train to Alexandria to be transported by chartered ship to Athens. In spite of that, communications between El Bedewi and me never stopped.

By September 1967, the Ein Shams group had done a fine job of reactivating the Project. Lauren Yazolino returned to Cairo at the end of February 1968 and, with his Ein Shams colleagues, collected a large amount of data. In August 1968 he and Amr Goneid began analysing the data on the newly installed IBM computer. A month later, I received a cablegram from Yazolino indicating that they had a strong indication that there was a grand gallery under the eastern face. Within about an hour, I was on my way to the airport and two days later I was looking at the data in the Ein Shams University Computing Center. However, the data indicated a cavity so large that the pyramid should collapse!

Within a few hours, it was apparent what had happened: "double binning." The raw data had been binned in angular intervals, and then replotted as the number of muons passing through an area of the pyramid's surface. In this process some data had been counted twice. I am particularly glad, however, that my young colleagues didn't spot this as a difficulty immediately, because if they had, I would have missed two of the most exciting days in my life.

One of the most rewarding things about research is that in addition to the real things one finds, one "discovers" many things that turn out on a more careful examination to be false. If human beings were "linear systems," the disappointment on learning that a discovery was false would just cancel out the elation built up when one believed the discovery to be real. But fortunately, we are not linear systems, so the overall time-integral of excitement and pleasure is strongly positive—the enormous area under the elation vs. time curve is hardly diminished by the small negative area under the let down phase. For me this is a very powerful effect.

I reported to the American Physical Society in the spring of 1969 that there were no chambers in the 19% of the pyramid we had probed—a vertically oriented cone above our spark chamber, with a half angle of 35 degrees. These results were published in the accompanying article. In September 1970, El Bedewi, Fakhry, and I concluded that we should reactivate the program and explore the remaining 81% of the pyramid volume. As the first step, the Ein Shams University group redesigned the mechanical mount so that our cosmic ray telescope could be tilted at 45 degrees from the zenith and turned to any desired azimuth. I talked with a large number of potential patrons without success, in marked contrast to the ease with which we had initially raised money. The National Science Foundation eventually provided us with Egyptian pounds to be spent there.

In May 1971, visits by Buck Buckingham, Bob Graven, and Pete Schwemin put our equipment in working order. These men, however, could not stay in Cairo for the many months we expected these new measurements to last. But Nick Chakakis, on a year's leave from Lawrence Livermore Laboratory, took the job of keeping the operation running smoothly and providing the liaison between the Berkeley and Ein Shams groups. The data tapes were taken to the Ein Shams Computer Center for preliminary analysis, twenty-five miles on the opposite side of Cairo, with its normally crowded streets. Then, after the data had been "compacted," a new tape was shipped to Berkeley for final analysis under the watchful eye of Gerry Lynch.

The angular coordinates of each cosmic ray muon were calculated from the recorded x and y spark positions in the two chambers at Ein Shams. Gerry Lynch then analyzed them and made the density plots, which we watched as the statistics accumulated. When the statistical fluctuations were small enough to rule out any chambers in a field of view, Chakakis and our Ein Shams colleagues rotated the system to a new angle.

Once for several weeks we thought that we had indications of a chamber in the southwestern direction at a low elevation angle. The Belzoni Chamber is well to the east of the pyramid centerline, so our cosmic ray measurements seemed to indicate a chamber directly under the pyramid apex. We were quite excited and decided to point the cosmic ray telescope directly at it, to see if it showed up more clearly. After a very long period of taking data at the new angle, we concluded that the earlier indication had been a statistical fluctuation, and the pyramid was quite solid in that direction.

A visual inspection of our density plots showed no evidence of a chamber anywhere in the pyramid volume. As doctors looking at x-ray pictures of a patient's bowels, we could have said, "There are no gas bubbles," and that would have concluded the analysis. But because the pyramid has such a regular and accurately known shape, we could carry our analysis much farther; we could fit the known cosmic ray "range spectrum" to the known geometry of the pyramid, and see if the observed and calculated counting rates were equal.

Lynch, after devoting a great deal of time to this analysis, concluded that the pyramid contained no standard chambers. He did, however, discover a large chamber, using the cosmic ray data! That chamber was the one in which our spark chambers were mounted close to the south and east walls. In his simulation of the pyramid structures, he had fed in the position of the apparatus, and the coordinates of the four triangular pyramid faces, including those of its distinctive limestone cap. All was well until at the lower elevation angles, he looked west and couldn't get agreement between the measured and calculated intensities. There were too many cosmic rays coming in. Lynch soon found that he hadn't put the Belzoni chamber geometry into the computer simulation of the pyramid. These cosmic rays had a free ride through about 30 feet of chamber air that made the number of counts observed greater than those calculated. We didn't see this effect since the eye is not sensitive to slow variations in intensity with direction. So it is quite correct to say that Lynch discovered a chamber using the cosmic rays. That Belzoni discovered that same chamber 154 years earlier does not diminish our demonstration that with cosmic rays one can find a chamber if there is one to be discovered!

A second difficulty involved another mundane problem, but also showed that the cosmic ray method had great analytical power. The pyramids are universally believed to be made of limestone with a density between 2.68 and 2.76. So Lynch used 2.72 in his calculations, but could not get satisfactory fits to the counting rates at various azimuth and elevation angles unless he used what seemed to be an unrealistically low value for the density of pyramid rock. The resolution of this disturbing situation came unexpectedly. Lambert Dolphin of SRI International, who had been attempting to probe the pyramids using radio and sonic techniques, had brought home a small piece of the

Chephren Pyramid to measure its radio frequency loss properties in his laboratory. We asked him to measure its density and were surprised when he reported that it was only 1.8—unexpectedly low for limestone. When Lynch recalculated the counting rates letting the density of the rock vary, he got excellent agreement between the measured and theoretical values, with an average density of rock of 1.9 ± 0.2 grams per cm^3. This is surely the hard way to measure density!

Lynch finished his analysis just in time for me to report to the Washington meeting of the American Physical Society, on April 22, 1974, that the Chephren Pyramid contained no "standard" chambers as seen in other pyramids. The ancient Egyptians had no testing machines to measure stress-strain curves for the ultimate strength of their building materials. What they knew of structural engineering, they had learned in the school of hard knocks. They kept building ever larger pyramids until one of them collapsed by plastic flow, according to the distinguished British physicist Kurt Mendelssohn in his book *The Riddle of the Pyramids* (New York: Prager, 1974). What we know about how those architects incorporated chambers into the body of the large pyramids is from the work of Colonel Howard-Vyse, a retired British Army officer who spent two years (1837–1838) and much of his wealth in a brutal attack on the Giza pyramids. He believed that the true burial chambers of Cheops was above the King's Chamber, and he used gunpowder to blast holes in the rock ceiling. He discovered four of the five so-called relieving chambers incorporated to relieve the pressure on the ceiling at the King's Chamber. The invention of the arch by the Romans was still two thousand years away.

The ceilings of smaller passageways were kept from collapsing by using the technique exhibited in the original entrance to the Cheops Pyramid. Here a 3 foot wide passageway was supported by an enormous structure which extends 10 meters above its floor, although there is almost no ordinary pyramid rock above. The top of the King's Chamber relieving structure extends 24 meters above the floor. Between the roof of the smaller Queen's Chamber and the floor of the Grand Gallery, there is room for a relieving chamber almost as tall as that over the King's Chamber, though it has not yet been discovered. In searching the Chephren Pyramid, we assumed there would be relieving structures of this same size. I believed that the architects of the later Chephren Pyramid would use this same successful non-collapsed design. With these best guesses, we then examined the Chephren Pyramid to see if the architects could have located a King's Chamber that would have escaped our notice.

First, we excluded the pyramid volume that lies within 24 meters of the square edge of the pyramid base, since a King's Chamber with its relieving chambers and gabled stone cap would be within 2 meters of the present irregular surface. The casing stones were removed in the Middle Ages to build the Cairo mosques. So an undetectable King's Chamber would have to be located somewhere above the centered square 168 meters on a side. We showed in our first phase search that there were no chambers in a vertical cone with a half angle of 35 degrees. Since we saw the details of the four diagonal ridges and the lower edge of the limestone cap, we had more than enough sensitivity to see a "standard" chamber. (Geologists would say that we had tested our method by "ground truth," a method used to check the findings from space-based sensors, such as cameras that photograph geographical and agricultural features of the earth's surface.)

If a King's Chamber complex was located under one of the four diagonal ridges, and as close to the corners as the height of the complex would allow (24 meters), then we might have missed such a chamber as its solid angle could have been as small as 4 degrees × 4 degrees. It would seem quite unlikely, however, that architects who were worried about grave robbers equipped with digging tools would have made it easier for such diggers by locating their chamber at a low elevation near the outer surface of the pyramid, just where cosmic ray detectors would have the greatest difficulty in finding it.

We concluded our program by pointing the cosmic ray telescope to the northeast where muons coming in at low angles had to pass through the Cheops Pyramid, before traversing 130 meters of Chephren Pyramid rock.

We saw a blurred picture with sloping sides on the Cheops Pyramid because we didn't do the subtraction of computer predicted counts from actual counts. I am disappointed that we didn't do that proper subtraction that would have made this historic photograph sharp. It will probably be a long time before any other object is "photographed" through 130 meters of rock.

A surprising feature of this picture is that cosmic rays appear to come from below the base of the Great Pyramid, which is obviously impossible. Those muons actually penetrated the spark chambers from the back side, in a direction nearly opposite to that of the Cheops Pyramid. If it had been important, we could have eliminated these muons by demanding that the upper spark chamber be struck before the lower one, using a time-of-flight circuit.

We attempted to write a second article for *Science* detailing the work performed in the second phase, but all of us found more pressing uses of our time. Those who would have authored this second paper, a draft of which was entitled "The Structure of the Second Pyramid at Giza, as Determined by Cosmic Ray Absorption," were Luis W. Alvarez, Nick J. Chakakis, and Gerald Lynch, Lawrence Berkeley Laboratory, University of California, Berkeley, and F. El Bedewi, Amr Goneid, Abdel Wahab, and Mohamed Tolba, Faculty of Science, Ein Shams University, Cairo, Arab Republic of Egypt. We all hope to release that publication in the future, as interest in probing the interior by other techniques has recently revived.

Search for Hidden Chambers in the Pyramids

The structure of the Second Pyramid of Giza
is determined by cosmic-ray absorption.

Luis W. Alvarez, Jared A. Anderson, F. El Bedwei,
James Burkhard, Ahmed Fakhry, Adib Girgis, Amr Goneid,
Fikhry Hassan, Dennis Iverson, Gerald Lynch, Zenab Miligy,
Ali Hilmy Moussa, Mohammed-Sharkawi, Lauren Yazolino

The three pyramids of Giza are situated a few miles southwest of Cairo, Egypt. The two largest pyramids stand within a few hundred meters of each other. They were originally of almost exactly the same height (145 meters), but the Great Pyramid of Cheops has a slightly larger square base (230 meters on a side) than the Second Pyramid of Chephren (215.5 meters on a side). A photograph of the pyramids at Giza is shown as Fig. 1. Figure 2 shows the elevation cross sections of the two pyramids and indicates the contrast in architectural design. The simplicity of Chephren's pyramid, compared with the elaborate structure of his father's Great Pyramid, is explained by archeologists in terms of a "period of experimentation," ending with the construction of Cheops's pyramid (1). (The complexity of the internal architecture of the pyramids increased during the Fourth Dynasty until the time of Cheops and then gave way to quite simple designs after his time.)

An alternative explanation for the sudden decrease in internal complexity from the Great Pyramid to the Second Pyramid suggested itself to us: perhaps Chephren's architects had been more successful in hiding their upper chambers than were Cheops's. The interior of the Great Pyramid was reached by the tunneling laborers of Caliph Ma-

mun in the 9th century A.D., almost 3400 years after its construction. Of our group only Ahmed Fakhry (author of *The Pyramids*, professor emeritus of archeology, University of Cairo, and member of the Supreme Council of Archeology, Cairo) was trained in archeology. As laymen, we thought it not unlikely that unknown chambers might still be present in the limestone above the "Belzoni Chamber," which is near the center of the base of Chephren's Second Pyramid, and that these chambers had survived undetected for 4500 years. [We learned later that such ideas had occurred to early 19th-century investigators (2), who blasted holes in the pyramids with gunpowder in attempts to locate new chambers.]

In 1965 a proposal to probe the Second Pyramid with cosmic rays (3) was sent to a representative group of cosmic-ray physicists and archeologists with a request for comments concerning its technical feasibility and archeological interest. The principal novelty of the proposed cosmic-ray detectors involved their ability to measure the angles of arrival of penetrating cosmic-ray muons with great precision, over a large sensitive area. The properties of the penetrating cosmic rays have been sufficiently well known for 30 years to suggest their use in a pyramid-probing experiment, but it was not until the invention of spark chambers with digital read-out features (4) that such a use could be considered as a real possibility. [Cosmic-ray detectors with low angular resolution had been used in 1955 to give an independent measure

of the thickness of rock overlying an underground powerhouse in Australia's Snowy Mountains Scheme (5)].

The favorable response to the proposal led to the establishment by the United Arab Republic and the United States of America of the Joint U.A.R.–U.S.A. Pyramid Project on 14 June 1966. Cosmic-ray detectors were installed in the Belzoni Chamber of the Second Pyramid at Giza in the spring of 1967 by physicists from the Ein Shams University and the University of California, in cooperation with archeologists from the U.A.R. Department of Antiquities. Initial operation had been scheduled for the middle of June 1967, but for reasons beyond our control the schedule was delayed for several months. In early 1968 cosmic-ray data began to be recorded on magnetic tape in our laboratory building, a few hundred meters from the two largest pyramids. Since that time we have accumulated accurate angular measurements on more than a million cosmic-ray muons that have penetrated an average of about 100 meters of limestone on their way to the detectors in the Belzoni Chamber.

Proof of the Method

Before any new technique is used in an exploratory mode, it is essential that the capabilities of the technique be demonstrated on a known system. We gave serious consideration to a proposal that the cosmic-ray detectors be tested first in the Queen's Chamber of the Great Pyramid, to demonstrate that the King's Chamber and the Grand Gallery could be detected. But this suggestion was abandoned because the King's Chamber is so close to the Queen's Chamber and because it subtends such a large solid angle that earlier (low resolution) cosmic-ray experiments had already shown that the upper chamber would give a large signal. It was apparent that the only untested feature of the new technique involved the magnitude of the scattering of high energy muons in solid matter. (An anomalously large scattering would nullify the high angular resolution that had been built into the detectors, in the same way that frosted glass destroys our ability to see distant objects.) We had no reason to doubt the calculated scattering, but we were anxious to be able to demonstrate to our colleagues in the U.A.R. Depart-

The authors are affiliated with the Joint Pyramid Project of the United Arab Republic and the United States of America. They reside either in Cairo, United Arab Republic, or in Berkeley, California. The article is adapted from an address presented by Luis W. Alvarez at the Washington Meeting of the American Physical Society, 30 April 1969.

ment of Antiquities in a convincing manner that the technique really worked as we had calculated. For this purpose we required as our test objects not large features that were nearby but, instead, small features separated from the detectors by the greatest possible thickness of limestone. Fortunately, such features are available in the Second Pyramid; the four diagonal ridges that mark the intersections of neighboring plane faces were farther from the detectors than any other points on the individual faces. (From now on, we will refer to these ridges as the "corners.")

From the known geometry of the Second Pyramid, the trajectories of cosmic-ray muons that pass through a point on a face 10 meters from a corner and then down to the detectors can be shown to traverse 2.3 fewer meters of limestone than do muons that strike the corner. They should therefore arrive with 5 percent greater intensity than the muons from the corner. Such an increase in intensity, corresponding to such a decrease in path through the limestone, is about half of what would be expected to result from the presence of a chamber of "typical size" (5 meters high) in the pyramid. Since such a chamber would necessarily be closer to the detectors, it would for these two reasons be a much "easier object to see" than the corner.

The detection equipment was therefore installed in the southeast corner of the Belzoni Chamber, with the expectation that it would first show the corners in a convincing manner, so that the presence or absence of unknown chambers could later be demonstrated to the satisfaction of all concerned. In September 1968 the IBM-1130 computer at the Ein Shams University Computing Center produced the data

Fig. 2 (bottom right). Cross sections of (a) the Great Pyramid of Cheops and (b) the Pyramid of Chephren, showing the known chambers: (*A*) Smooth limestone cap, (*B*) the Belzoni Chamber, (*C*) Belzoni's entrance, (*D*) Howard-Vyse's entrance, (*E*) descending passageway, (*F*) ascending passageway, (*G*) underground chamber, (*H*) Grand Gallery, (*I*) King's Chamber, (*J*) Queen's Chamber, (*K*) center line of the pyramid.

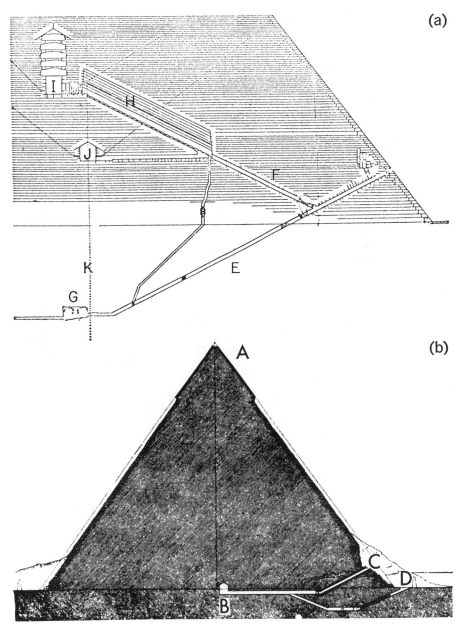

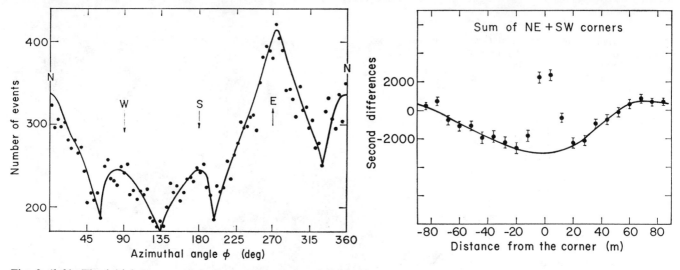

Fig. 3 (left). The initial measurement (with zenith angle of counts from 20 to 40 degrees) of the variation of cosmic-ray intensity with azimuthal angle, as observed from the Belzoni Chamber underneath the Second Pyramid of Chephren.   Fig. 4 (right). Detection of the northeast and southwest corners of the pyramid obtained by plotting the second differences of the counting rate on the planes tangent to the corners as a function of distance from the corners.

for Fig. 3, which shows the variation of cosmic-ray intensity with azimuthal angle (compass direction). The expected rapid changes in cosmic-ray intensity in the vicinity of the corners were clearly shown, and the capability of the method could no longer be doubted. An analysis of more data was later made on the Lawrence Radiation Laboratory's CDC-6600 computer and is shown in Fig. 4. Here the "second differences" of the counting rate with distance from each corner are plotted on planes that are located symmetrically with respect to adjacent faces and that are tangent to the corner. Mathematically, we would expect to see a sharp spike at the corner of a sharply defined pyramid in the plot of the second derivative of counting rate with respect to distance. The second derivative becomes a second difference curve when we use bins of a finite size. The sharpness of the peaks in the second difference curves shows that the effect of the scattering of muons in limestone is somewhat smaller than the conservative estimate made in the original proposal.

We were at first surprised by the large variations in maximum counting rate through the four faces of the pyramid. We knew that the Belzoni Chamber was not at the exact center of the base of the pyramid, but we had not appreciated what large changes in counting rate would be occasioned by the actual displacement of the detector from the center of the base; the equipment is 15.5 meters east and 4 meters north of the center. There are two independent ways to use cosmic-ray data

to determine the location of the detector with respect to the exterior features of the pyramid.

1) The difference in the maximum counting rate through the east and west faces gives the displacement of the detector toward the east, and similar measurements in the north-south directions give the displacement to the north.

2) The azimuthal angles of the dips corresponding to the corners give a second, quite independent, and more sensitive measure of the displacements. We can report that from cosmic-ray observations alone, "looking through" 100 meters of limestone, we can locate the position of our detectors to within 1 meter. To the best of our knowledge, no such measurement has ever been made before. Our cosmic-ray-derived position agrees to within less than 1 meter in the north-south direction with a recently surveyed position obtained by the U.A.R. Surveying Department, but it differs by 2 meters (that is, it indicates 13.5 rather than 15.6 meters) in the east-west direction.

Simulated X-ray Photographs

We have presented the cosmic-ray data in two different ways, one photographic and the other numerical. Both these methods involve the projection of each recorded muon back along its trajectory to its intersection with either a horizontal plane or a sphere that touches the peak of the pyramid. Figure 5a is a diagram representing the Second Pyramid with the horizontal "film

plane" touching the peak of the pyramid and with a dashed line (representing the path of a cosmic ray) passing from the detector through a hypothetical chamber to the image of the chamber on the "film plane." (The mapping of the pyramid structure by this technique is identical to what we would obtain by x-raying a small model of the pyramid, with an x-ray source in the Belzoni Chamber and with an x-ray film touching the peak of the model pyramid.) Figure 5b represents the spherical shell onto which cosmic rays were projected for numerical analysis.

Figure 6 is a view of all the equipment, which occupied most of the southeastern part of the Belzoni Chamber. Figure 7 is a closer view of the detector. The two spark chambers, each 6 feet (1.8 meters) square, are separated vertically by a distance of 1 foot (0.3 meter). Above and below the spark chambers and just above the floor level were scintillation counters, which triggered the spark chambers when all three counters signaled the passage of a penetrating muon. The 4 feet (1.2 meters) of iron between the bottom two scintillators was installed to minimize the effects of muon-scattering in the limestone.

The simulated x-ray photograph of the pyramid shown in Fig. 13a is an uncorrected (raw data) scatter plot of 700,000 recorded cosmic-ray muons as they passed through the "film plane." The four corners of the pyramid are very clearly indicated. If a Grand Gallery and a King's Chamber were located in the Second Pyramid as they are in the Great Pyramid, the Grand Gallery

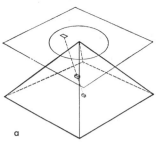
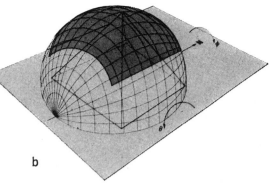
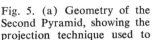

Fig. 5. (a) Geometry of the Second Pyramid, showing the projection technique used to produce a simulated x-ray photograph. The plane on the top of the pyramid can be thought of as the "film plane." (b) The spherical surface on which the events were projected for the numerical analysis of the data.

would have shown up clearly but the King's Chamber would probably have required some computer assistance to be made visible. There is one unexpected feature in Fig. 13a: on the north face, there appears to be a narrow north-south-oriented region that has a lower cosmic-ray intensity than is found in surrounding areas. We were at first hopeful that the north-south streak indicated the presence of a Grand Gallery above and north of the Belzoni Chamber, just as the Grand Gallery is above and north of the Queen's Chamber in the Great Pyramid. But we later found a satisfactory explanation of this feature in the picture that did not involve any interior structure in the pyramid. The region of lower cosmic-ray intensity resulted from the construction of the spark chambers. Since we could not transport square chambers 6 feet (1.8 meters) on a side through the small passage-

ways of the pyramid, each square chamber comprised two chambers 3 by 6 feet (0.9 by 1.8 meters) in area. Also, each of the large scintillation counters was divided into sections. The inactive areas between the two pairs of spark chambers and between the sections of the counters led in a predictable way to the unexpected signal shown in Fig. 13a.

Numerical Analysis

We concluded from our study of the simulated x-ray picture that no unexpected features were discernible. But since we had been looking for an increase in intensity of approximately 10 percent over a region larger than that to which the eye responds easily, we then turned to a more detailed numerical analysis of the data. (The reason for expecting a 10 percent increase in in-

tensity in the direction of a new chamber is simply that the integral range spectrum of the muons is represented by a power law with an exponent equal to -2. Therefore, if the rock thickness is changed by an amount ΔX, out of an original thickness X, the relative change in intensity is $\Delta I/I = -2\Delta X/X$. The four known chambers in the two large pyramids have an average height of about 5 meters. Therefore $\Delta X/X$ should be -5 percent, and the corresponding value of $\Delta I/I$ should be $+10$ percent.)

Since the counting equipment was sensitive out to approximately ± 45 degrees from the vertical, our data were plotted in a matrix with 900 entries, 30×30 bins, each 3 by 3 degrees. Figure 5b illustrates this system of binning on a sphere that encircles the pyramid. We wrote a computer program to simulate the counting rate expected in each of these bins. As the simulation program became more sophisticated with time, it took into account the most detailed features of the measured exterior surface of the pyramid, including the "cap" of original limestone casing blocks near the top, the surveyed position of the detectors in the Belzoni Chamber, the positions of the walls and ceiling of the Belzoni Chamber, and the sizes and positions of each of the four spark chambers and the fourteen scintillation counters.

An important control on the quality of the experimental data being compared with the simulated data came from scatter plots showing the exact x and y coordinates of each recorded

Fig. 6 (left). The equipment in place in the Belzoni Chamber under the pyramid.
Fig. 7 (right). The detection apparatus containing the spark chambers.

muon as it passed through each of the five planes containing scintillators or spark chambers. Unsatisfactory operation of the spark chambers showed up as small blank areas in the scatter plots of muons passing through the chambers. Such unsatisfactory operation was found to be correlated with contaminated neon in the spark chambers; the log books show that whenever the chambers were flushed with fresh neon they recovered their substantially uniform sensitivity. By examining the scatter plots on a day-by-day basis, we eliminated from the data base about one-third of the measured muons.

The scatter plots also served as a check on the resolution and accuracy of the angle measurements. The edges of the counters showed up on these plots as sharp lines at positions that agreed well with the direct measurements of the counter locations. Neither the direct measurements of the counter positions nor the inferred positions of these counters as obtained from the data themselves were good enough to permit the program to make sufficiently accurate calculations. In a typical 3- by 3-degree bin there are 1600 events. The statistical uncertainty in this number of events is 2.5 percent. It was necessary to make calculations to at least such an accuracy to make full use of the data. We first varied the assumed positions of the scintillators by small amounts in an effort to fit the expected counts to the measured counts. This approach was unsatisfactory. Calculations of the desired accuracy were obtained only after we eliminated the events that passed near the edge of at least one of the counters. In effect, each counter was defined to be slightly smaller than it actually was, and only the recorded muons that passed through these defined counter positions were accepted. This method eliminated the problems associated with small displacements of the counters during the experiment, with small-angle scattering of muons in the iron, and with decreased sensitivity of the counters near their edges. About 15 percent of the events were eliminated in this way. We believe that the 650,000 muons in the final selected sample are free of important biases resulting from improper functioning of the equipment.

In the course of the computer analysis, about 40 fits were made to minimize the difference between the matrices of actual and simulated counts. Although the matrices contain 30 × 30 bins each, some of the bins at the edges contain so few counts (or none at all) that the effective number of bins is close to 750. If we knew all the physi-

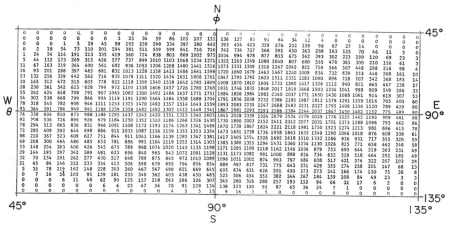

Fig. 8. An array containing the numbers of events (uncorrected) observed during several months of operation in each 3- by 3-degree bin.

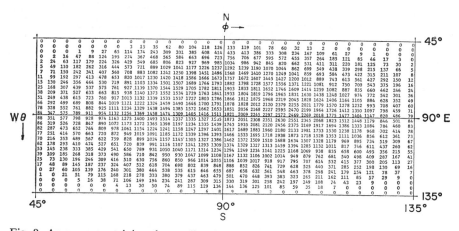

Fig. 9. An array containing the predicted number of events in each 3- by 3-degree bin for the best fit to the data.

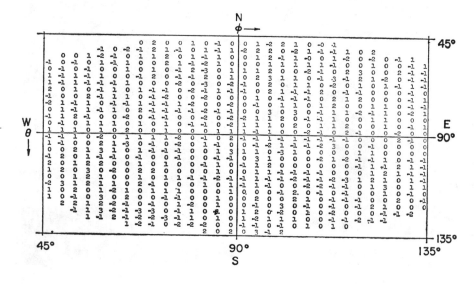

Fig. 10 (bottom left). The differences between the numbers of events measured and predicted expressed in integral numbers of standard deviations for the best fit to the data for which the χ^2 was 905. (The bins for which the predicted number of events was less than 30 were not used in calculating the χ^2.)

cal parameters of our detection equipment, if we were equally sure of the equations describing the cosmic-ray spectrum, and if we were, in addition, sure that the pyramid was made of solid limestone, then we would expect the χ^2 of the fit between the actual and the simulated data to be about 750. The earliest fits had χ^2's of close to 3000, but this important parameter dropped to approximately 1400 by the time the stereophotographically determined contours of the pyramid exterior were made available to us through the courtesy of the U.A.R. Surveying Department.

Figure 8 is a matrix showing the total number of real counts recorded in each of the 900 3- by 3-degree bins. Figure 9 is one of the final simulation runs, and Fig. 10 is the difference between Figs. 8 and 9 expressed as the closest integral number of standard deviations. [For a bin in which the number of counts was 2500, an entry for +2 standard deviations means that the actual count exceeded the expected count by $2(2500)^{1/2} = 100$.]

If these deviations are only statistical in nature, one expects about 87 percent of the bins to have contents of -1, 0, or $+1$, about 12 percent of the bins to have ± 2, and 1 percent of the bins to have ± 3. There is one chance in three of finding one bin having ± 4, one chance in 200 of finding one bin with ± 5, and only one chance in 3×10^4 of finding one bin with ± 6, if the deviations are due only to statistical fluctuations. Thus no single bin has a significant effect unless its contents are at least ± 4. Figure 10 contains no bins showing ± 4.

Detection of the Cap

The most distinctive feature of the Second Pyramid is the cap of original limestone casing blocks near the top. All the casing was removed from the Great Pyramid in the Middle Ages, but the builders of Cairo, who "quarried" the pyramids at that time, stopped before completely stripping the Second Pyramid.

Before the simulation program in the computer took account of the presence of the cap of limestone casing blocks on the pyramid, the difference plots (like the plot given in Fig. 11) always contained a central region with a preponderance of negative entries. When the cap was properly allowed for in the

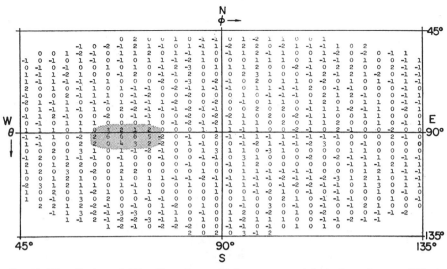

Fig. 11. The display of Fig. 10 as it would have appeared had there been a "King's Chamber" in the pyramid 40 meters above the apparatus. The group of numbers larger than 3 at the center-left (shaded area) indicates the chamber's position.

simulation, the actual counts in this region were no longer systematically lower than predicted by the computer, and the value of χ^2 dropped accordingly.

Although the χ^2 of the fit between the actual and simulated data was lowered when the features of the cap were introduced into the simulation, this drop does not constitute the strongest proof that we were in fact detecting the cap. Figure 12 compares the measured and cosmic-ray-determined variation in thickness of the limestone cap in two 24-degree-wide strips that run over the top of the pyramid in the north-south and east-west directions. In the absence of a cap both on the real pyramid and in the simulation, we

would expect the experimental points to lie along the zero lines of deviation. The smooth curved lines are obtained from the simulation program by utilizing the recently determined contours of the pyramid. The generally good agreement of the data points with the prediction (Fig. 12) shows clearly that we have detected the presence of the cap through more than 100 meters of limestone.

The detection of the cap was much more difficult than detection of the corners; together, these two "proofs of the method" convinced us that we could have seen any previously unknown chamber that might exist in our "field of view."

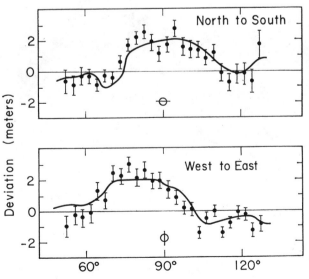

Fig. 12. This graph shows that the cap on the top of Chephren's Pyramid is observed in this experiment. The quantity plotted is the difference between the measured distance from the detector to the surface of the pyramid and the distance calculated under the assumption that the pyramid has no surface irregularities. The data points represent the distances indicated by the cosmic rays and are to be compared with the solid line, which represents the same distance measured by the aerial survey. These distances have been averaged over 24-degree-wide bands centered on the middle of the pyramid, one running north-south, the other east-west.

Search for Cavities

As the analysis proceeded during about 2 months, the value of χ^2 dropped slowly to about 1200 as the computer's simulation program was provided with better geometrical data. [In the course of this work we confirmed that the cosmic-ray intensity in the momentum range of concern (40 to 70 Gev) is isotropic and has an integral-range power law index of -2.1.] For most of this time we were excited by the presence of two "positive regions" on our matrix of differences (experimental minus simulated counts). One of these regions apparently signaled the presence of a "King's Chamber" directly under the apex of the pyramid, about 30 meters above the Belzoni Chamber. Because of the displacement of the Belzoni Chamber to the north and east

of the center of the pyramid's base, the apparent chamber mapped itself onto the southern part of the western face of the pyramid. The relative increase in counting rate was about 10 percent, as expected. The angular size of the anomaly could be related to distance only by assuming a certain size for the floor area of the "chamber." If we assumed that the anomaly came from a room the size of Cheops's King's Chamber, it had to be about 30 meters away, and its plan position turned out to be almost exactly central.

Unfortunately, this large and persistent signal, together with a larger signal over a smaller angular range, disappeared as we learned more exactly all the dimensions of the apparatus and of the pyramid that were important in the simulation program. (We had not anticipated the need for such accurate

data.) The artifacts we observed are mentioned only to show that far from "seeing nothing" throughout the analysis period, we had three very exciting signals that disappeared only after the greatest care had been taken to make the simulation program correspond exactly to the geometry of both the apparatus and the pyramid.

When the simulation program was as complete and as correct as we could make it, the fit between the recorded and the simulated counts was described by a χ^2 of about 1100. The formal rules of statistical analysis say quite unequivocally that such a fit is very unsatisfactory. But a careful look at the matrix of differences showed that the increase in χ^2 (over the expected value of about 750) came primarily from a rather uniform increase in difference values from south to north. If we assumed that the cosmic-ray intensity varied as $1 + d \cos \theta$, where θ is 90 degrees in a vertically oriented east-west plane, and 0 degree for rays approaching horizontally from the north, the χ^2 dropped to 905 when d had the value of 0.15 (Fig. 10). Such a value of d would correspond to a smooth variation in cosmic-ray intensity, from 30 degrees north to 30 degrees south, of ± 7 percent. We do not believe, of course, that the cosmic-ray intensity changes in such a manner, but it is quite reasonable to assume that our spark chamber systems had such a small and systematic change in sensitivity.

Our confidence in such an explanation was increased when we found that the required value of the constant d was different when we analyzed the data in two separate samples, one measured by each pair of 3- by 6-foot (0.9- by 1.8-meter) spark chambers. We know that the spark chambers were not uniformly sensitive over their whole areas, and we discarded all data from runs in which there were gross changes in sensitivity from point to point in the chambers. But we have no technique available to compensate for slow variations in sensitivity with position. (In our next operations, the apparatus will be arranged so that it rotates about a vertical axis; the linear variation in sensitivity that we have just postulated will average out in the improved apparatus.)

We made several attempts to simulate the $1 + d \sin \theta$ behavior of the apparent cosmic-ray intensity by the presence of a chamber above and very close to the apparatus. But the more

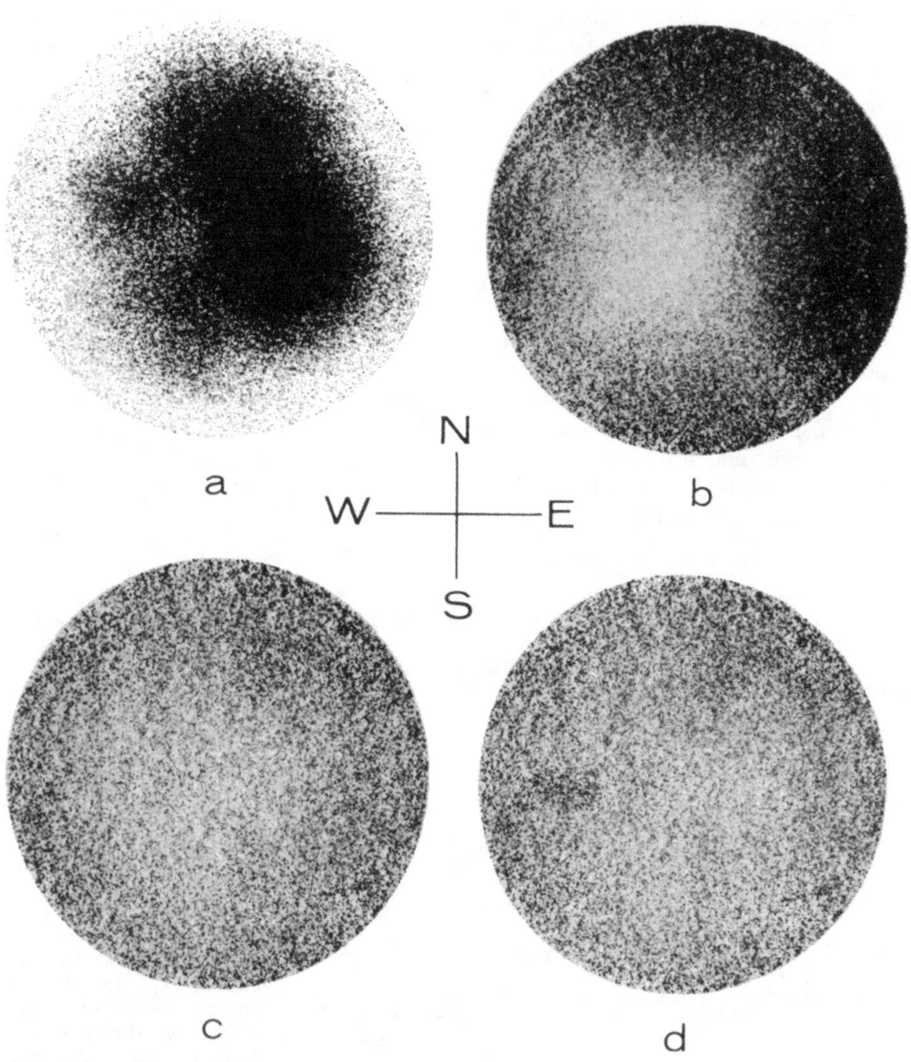

Fig. 13. Scatter plots showing the three stages in the combined analytic and visual analysis of the data and a plot with a simulated chamber. (a) Simulated "x-ray photograph" of uncorrected data. (b) Data corrected for the geometrical acceptance of the apparatus. (c) Data corrected for pyramid structure as well as geometrical acceptance. (d) Same as (c) but with simulated chamber, as in Fig. 12.

we tried, the more evident it became that the slow north-south variation was an instrumental effect, unrelated to any cavities in the pyramid rock. The fit obtained by using this slow north-south variation shows no statistically significant deviations from a solid pyramid.

Conclusions

To be sure that we could have detected a chamber of "average size" above the Belzoni Chamber, we programmed the computer to believe that its simulated pyramid had such a chamber filled with material whose density was twice that of limestone. The program then predicted that fewer muons than usual would come from the direction of the "chamber," so that the difference matrix showed positive numbers, as expected for a hollow chamber in the real pyramid and a pyramid of uniform density in the simulator. Figure 11 shows what we would have seen had there been a King's Chamber 40 meters above the Belzoni Chamber. (The scattering of muons in the rock is simulated by the computer.) If the King's Chamber were moved farther away, the angular width of the region having an excess of counts would drop inversely with distance. We have no doubt that we could detect a King's Chamber anywhere above the Belzoni Chamber within a cone of half-angle 35 degrees from the vertical. If the Second Pyramid architects had placed a Grand Gallery, King's Chamber, and Queen's Chamber in the same location as they did in Cheops's Pyramid, the signals from each of these three cavities would have been enormous. We therefore conclude that no chambers of the size seen in the four large pyramids of the Fourth Dynasty are in our "field of view" above the Belzoni Chamber.

We started by using two methods of analysis, one photographic and the other analytic. By itself the photographic method was unsatisfactory, because effects due to the apparatus itself were so large that they obscured any effects from the pyramid. Although the analytic method could succeed because it was able to take these instrumental effects into account, it was necessary to "invent" a north-south variation in sensitivity to obtain success. By combining the two methods of analysis, it is possible to obtain the sensitivity of

the analytic method in a photographic simulation. The combined method consists of plotting the ratio of the observed number of events to the predicted number, using bins that are about 0.15 by 0.15 degrees. Figure 13 shows three such scatter plots for the data in the cone of 35-degree half-angle centered on the vertical. Plot 13b corrects for instrumental effects but does not take into account the fact that the instrument is covered by a pyramid. The corners of the pyramid are clearly indicated in it. The next scatter plot (Fig. 13c) further corrects for the presence of the pyramid with all its surface irregularities. Figure 13d shows what we would have seen if a "King's Chamber" had been in the pyramid, a chamber the same in size and location as that used to obtain Fig. 11. It is evident that no effect in the data approaches the magnitude of the effect produced by a King's Chamber.

We have explored 19 percent of the volume of the Second Pyramid. We now hope to rearrange the equipment so that we can look through the remaining 81 percent of the volume of the pyramid. This operation will be greatly simplified by our new understanding of the effects of muon scattering on angular resolution. It is now apparent that the iron absorber was not necessary for the success of the experiment, and we will omit it in the rebuilt apparatus. Rotation of the detectors about the vertical will thus be facilitated, and the possibility of "x-raying" some of the other large pyramids will be enhanced.

Summary

Because there are two chambers in the pyramid of Chephren's father (Cheops) and the same number in the pyramid of his grandfather (Sneferu), the absence of any known chambers in the stonework of Chephren's Second Pyramid at Giza suggests that unknown chambers might exist in this apparently solid structure. Cosmic-ray detectors with active areas of 4 square meters and high angular resolution have been installed in the Belzoni Chamber of the Second Pyramid; the chamber is just below the base of the pyramid, near its center. Cosmic-ray measurements extending over several months of operation clearly show the four diagonal

ridges of the pyramid and also outline the shape of the cap of original limestone facing blocks, which gives the pyramid its distinctive appearance. We can say with confidence that no chambers with volumes similar to the four known chambers in Cheops's and Sneferu's pyramids exist in the mass of limestone investigated by cosmic-ray absorption. The volume of the pyramid probed in this manner is defined by a vertically oriented cone, of half-angle 35 degrees, with its point resting in the Belzoni Chamber. The explored volume is 19 percent of the pyramid's volume. We hope that with minor modifications to the apparatus the complete mass of limestone can be searched for chambers.

References and Notes

1. A. Fakhry, *The Pyramids* (Univ. of Chicago Press, Chicago, 1961).
2. R. Howard-Vyse and J. S. Perring, *Operations Carried On at the Pyramids of Gizeh* (London, 1840–42), 3 vols.
3. L. W. Alvarez, *Lawrence Radiation Laboratory Physics Note 544* (1 March 1965).
4. V. Perez-Mendez and J. M. Pfab, *Nucl. Instr. Methods* 33, 141 (1965).
5. E. P. George, *Commonw. Eng.* (July 1955).
6. The Joint U.A.R.–U.S.A. Pyramid Project has been fortunate in the strong support it has received from many individuals and organizations on two continents. The initial and continuing interest of Dr. Glenn T. Seaborg was reflected in the financial support of the project by the U.S. Atomic Energy Commission during the period when the detection equipment was designed and built at the University of California's Lawrence Radiation Laboratory. The U.A.R. Department of Antiquities extended every courtesy to members of the project, and, in addition to inviting us to install our equipment in the Second Pyramid, they put at our disposal an attractive and conveniently located building which served as our laboratory and general headquarters. The Smithsonian Institution was generous in its support of the project, particularly in making available travel funds that permitted project members from our two countries to work together both in Berkeley and in Cairo. The Ein Shams University and the University of California both supported the project in many ways, and we give special thanks to Vice Rector Salah Kotb and Chancellor Roger Heyns. The IBM Corporation, the National Geographic Society, the Hewlett-Packard Company, and Mr. William T. Golden made important contributions to the project for which we will always be grateful. The U.A.R. Surveying Department made accurate measurements of all external and internal features of the Second Pyramid, without which we would have been severely handicapped. The overall guidance of the project was vested by our two governments in a "Committee of the Pyramids," under the chairmanship of Dr. Salah Kotb. We are indebted to the distinguished members of this committee for the warmth and understanding they displayed in the exercise of their responsibilities. All of us take pride in the fact that the friendly spirit in which we started the project survived not only the perils that confront any interdisciplinary or intergovernmental effort but also a break in diplomatic relations between our two countries. In conclusion, we acknowledge important assistance from Raymond Edwards, Sharon Buckingham, Fred Kreiss, William E. Nolan, Kamal Arafa, Sayed Abdel Wahab, Mohamed Tolba, Aly Hassan, August Manza, and Vice Chancellor Loy Sammet.

17

Looking for Antimatter in the Cosmic Rays

Andrew Buffington

The Alvarez balloon project started in 1964 in association with Bill Humphrey when it appeared that funding for the National Accelerator Laboratory, now called Fermilab, might not be forthcoming. Accelerator development in this country would then have been stopped at machines which could produce a few particles with tens of GeV. Luis hoped to extend particle physics interaction studies to significantly higher energies by using the "beam" of high-energy cosmic-ray protons that exists at the top of the atmosphere. This approach was not unique, as a number of mountaintop laboratories were established or upgraded at this time. However, at these sites only a hundredth of the proton flux incident at the top of the atmosphere remains, and worse, it is contaminated by reaction product particles. On the other hand, these laboratories are more easily accessible than either balloons or space flights, and their very long dwell time can compensate for the reduced flux.

The objective of the High Altitude Particle Physics Experiment (HAPPE) was a measurement of the types and energies of the incident particles and their interaction products. To instrument HAPPE, a complete set of superconducting magnets, emulsions, spark chambers, scintillators, and Cerenkov counters suitable for balloon flight was developed, together with an automated measuring stage for the emulsion data. Part of this apparatus was, for that time, the world's largest superconducting magnet. The instrument weighed more than 10,000 pounds and thus required expensive and, as it turned out, unreliable mylar-scrim balloons to lift it to an altitude of 120,000 feet. The excitement was intense as these diverse ideas came together in the hopes of achieving a dramatic new scientific advance.

By early 1968, Luie and HAPPE project members realized that approaching NASA funding reductions, coupled with good progress in building the Fermilab accelerator, made HAPPE's future uncertain. Hope was further diminished when, at the conclusion of the August 1967 test flight, the payload was lost at sea. As a result, the program was redirected toward smaller, lighter instruments optimized for a cosmic-ray "beam survey." The intellectual focus moved accordingly from interaction physics to the astrophysical processes that produce and propagate cosmic rays. The fundamental technical objective, flight of a superconducting magnet

by balloon (and later in space), was unchanged. The magnet design shifted from the double coil, familiar from accelerator experiments, to a single coil. These simpler magnets permitted a substantial savings in weight and complexity, while their more inhomogeneous field could be modeled in computer data analysis codes. The average value of $\int B \cdot dl$ for a typical experiment using one of these coils was 5 kilogauss-meters, about the same as that of the previously planned magnets and incidentally about the same as that of much larger and heavier conventional magnets with iron flux returns. The large integral value is due to the strong magnetic field intensity near the superconducting magnet coils. The geometry of the associated detectors had to be configured to restrict triggering events to those passing suitably near the magnet. To further simplify this, the trajectory-locating emulsions were eliminated, and the particle paths were determined only by optically viewed spark chambers.

A search for cosmic-ray antimatter was one obvious objective for this scaled-down apparatus. A few years earlier, Hannes Alfvén had advanced his idea of a matter-antimatter symmetric universe, in which matter and antimatter are kept from mixing by an intervening electron plasma. Whether by Alfvén's explanation or those of others, regions of bulk antimatter in or near our Galaxy would be revealed by observation of antinuclei such as antihelium, anticarbon, or antiiron.

Heavy antinuclei, if observed, would also indicate that nucleosynthetic processes are taking place in these antimatter regions, since the usual matter-symmetric cosmologies have no other provision for manufacturing antimatter heavier than antilithium. One can think of cosmic rays arriving at Earth with energies of 1–10 GeV as being a sample of material mostly drawn from a region within several hundred parsecs in our Galaxy. Whether some cosmic rays come to us from other galaxies is not known, but this is a possibility. We cannot be certain of the matter or antimatter nature of even nearby stars because photons, the channels through which we normally observe these objects, do not directly carry this information. In principle, one could determine whether a given region is made of antimatter by contacting another civilization in that region (using photons) and asking them to perform and report the results of a CP-violating experiment. However, this question would initially be of relatively minor importance, in the event that such a contact were made! In any case, failing a practical means of doing this, or of traveling beyond the solar system, searching for antimatter in cosmic rays appears to be the only direct means of probing the region around us for bulk antimatter.

In 1968, when I joined the Balloon Project, the group members besides Luie were Jerry Anderson, Ben Clawson, Roy Colombe, Phil Dauber, Jack Lloyd, Larry Smith, and Mike Wahlig, with two to three dozen support personnel. Frank Solmitz helped with the difficult problem of tracing the trajectories through the inhomogenous magnetic field. Later, Terry Mast, Rich Muller, Charles Orth, and George Smoot joined, while Jerry and Jack departed to form a small electronics company. I remember with considerable affection my days in our cramped quarters in both ends of Building 46, at the Lawrence Berkeley Laboratory. Luie kept contact as the equipment for the small superconducting magnet (SSCM) experiment was designed and built. It was then that I became aware of his remarkable abilities to discern the essentials in a technical discussion, to identify the strategic decisions and get them made, and to bring techniques and ideas from his many previous careers to bear on present difficulties.

In the end, our successful cosmic-ray program included not only the antimatter measurement reprinted here, but also measurements of "noncosmological" electrons and antiprotons, isotopes, and the energy spectra of elements from hydrogen through iron.[1] A discovery in this final category raised a major question about cosmic-ray propagation. The energy spectra of lithium, beryllium, and boron, which are created by collisions of heavier nuclei in interstellar space, all drop off more rapidly than those of their parent elements, carbon, nitrogen, and oxygen. This discovery was soon confirmed by a group from the University of Chicago and has since been observed by many groups. Thus, the amount of interstellar material traversed, for cosmic-ray carbon and heavier elements, diminishes slowly with increasing energy above a few GeV per nucleon. Cosmic-ray propagation models have been refined to

1. A. Buffington, S. M. Schindler, and C. R. Pennypacker, Astrophys J. **248**, 1179 (1981); C. D. Orth, A. Buffington, G. F. Smoot, and T. S. Mast, Astrophys J. **226**, 1147 (1978); A. Buffington, C. D. Orth, and T. S. Mast, Astrophys J. **226**, 355 (1978); C. D. Orth and A. Buffington, Astrophys J. **206**, 312 (1976); L. W. Alvarez, Astrophysics Group Internal Memo 328 (1976); A. Buffington, C. D. Orth, and G. F. Smoot, Astrophys J. **199**, 669 (1975); W. L. Pope, G. F. Smoot, L. H. Smith, and C. E. Taylor, Advances in Cryogenic Engineering **20**, 47 (1975); G. F. Smoot, A. Buffington, and C. D. Orth, Phys. Rev. Lett. **35**, 258 (1975); A. Buffington, C. D. Orth, and G. F. Smoot, Nucl. Inst. and Meth. **122**, 575 (1974); C. R. Pennypacker, G. F. Smoot, A. Buffington, R. A. Muller, and L. H. Smith, J. Geophys. Res. **78**, 1515 (1973); and L. H. Smith, A. Buffington, G. F. Smoot, L. W. Alvarez, and M. A. Wahlig, Astrophys J. **180**, 987 (1973).

fit these data, but no consensus as to an explanation has emerged.

One cosmic ray, the "antioxygen event" described on the third page of the accompanying article, has an interesting story behind it. During the summer of 1972, Larry Smith and I attended a cosmic-ray conference in Tasmania. There we reported a null antimatter search based on ~85% of the data from the flights of the first-generation instrument. The rest of the data included events for which the three-constraint fit of trajectories through the magnetic field yielded a poor value of χ^2, the sum of squares of the spark location deviations from a smooth trajectory. A fragmenting event tends to have a large χ^2, as do events having an erroneous spark measurement. After we returned from the conference, Luie said that he would like to see a dossier on each of our antimatter candidates: coordinates, fitting results, and pulse heights. If there were any real antimatter events, they must be in that sample, since our previous criteria were applied generally to the whole data set. He wanted to be sure some excuse wasn't found to throw out a real antimatter event! Since there were only about four dozen such events, compiling the list was an easy task, and Luie was soon poring over it, asking questions about how the numbers were calculated and what they meant. It didn't take him long to discover event No. 26262.

This event caught Luie's eye because it deposited more energy in the bottom scintillator than it had in the two above. The apparent "charge" had increased from oxygen ($Z = 8$) to neon ($Z = 10$). Usually a fragmentation reaction of a particle this heavy leads to a diminished effective charge and, as a result, a smaller energy deposition in succeeding layers. Luie was thinking of the many annihilation pions that might occur in an antioxygen interaction. We hastened to examine the original film on the scanning table and found that an interaction had indeed occurred in the middle of the bottom spark chamber. Extra tracks in the bottom spark chambers showed an interaction star with at least three secondary prongs. More prongs could have been present, especially in the forward direction, since this type of detector can only resolve tracks separated by more than a millimeter or so. Confusion about which sparks to measure in the bottom chamber had been responsible for the bad χ^2 of the first measurement. To get the most precise view of this event, a variety of people measured it a total of fifteen times, and we averaged the results. There was some question about whether we could use the top two sparks of the bottom spark chamber, since the interaction vertex was so close. However, the χ^2 including these was acceptable: The analysis with these sparks yielded a rigidity of 40 GeV/c and a negative charge, while the analysis with the top three chambers and no sparks from the bottom chamber yielded a positive charge, but the trajectory was only a small fraction of a standard deviation from a straight track. The full four-chamber trajectory was about two standard deviations from straight. No claim of cosmic-ray antimatter could be based only on the curvature, since small residual errors in mapping the experiment's optical mirrors could occasionally cause such an event. These residual optical errors were the major contributors to the χ^2 for events that had been measured several times by hand; these optical errors were determined from measuring straight-track trajectories obtained with the magnet turned off. On the other hand, such variations could only seldom move a negative-curvature antimatter event over to the positive-curvature region of the magnetic spectrometer response. Thus an antimatter event having such a rigidity ought to appear in the final sample Luie and the rest of us scrutinized. If the antimatter event were to distinguish itself in some additional way, we might be able to single it out. Luie had the insight that this sample, enriched in potential antimatter events, might contain an event with some unexpected extra character.

Attention now focused on the probability that the common-matter background in the experiment might generate an event in which the energy deposition beyond a fragmentation reaction was greater by the observed amount. The calculation included the spectrum of potential high-energy oxygen events and contributions to energy deposition from both pions created in the collision and an internuclear cascade in the struck nucleus, which we took to be aluminum. Our disappointment grew as this calculation converged toward an explanation for the event, but we felt it was best to discover this privately. Several times, Luie said to me while work was in progress, "Let's be sure to give this event a decent burial," by which he meant we should be certain of our explanation and not just dredge up some excuse to get rid of it! We found this event only an insignificant three times easier to explain as antimatter, as compared with the mundane explanation. Thus, if someone else's experiment had already established the existence of such particles in the cosmic rays at the proper flux, we could have said that this event probably was an antioxygen. On the other hand, the event was sufficiently easy to explain otherwise that we could never use it as a basis for discovery of antioxygen; it probably was caused by the process we had just calculated.

Once started in this business of searching for cosmic-ray antimatter, we kept going. A second-generation instrument pushed the 95% confidence upper limits of Figure 4 in the accompanying article down by another factor of 3. To the present, searches for heavy antinuclei can be summarized by saying that none have been found in $\sim 10^5$ events examined. This second-generation instrument also provided a measurement of cosmic-ray positrons in a heretofore unexplored energy regime. Finally, a rebuilt SSCM gondola, without a magnet, searched for antiprotons with energy of a few hundred MeV, using the annihilation daughters rather than the sign of the charge to provide the identification.[2] This work, completed after I moved to Caltech, was done with Steve Schindler and Carl Pennypacker. We saw between a hundred and a thousand times more antiprotons at these low energies than were predicted by current models of cosmic rays, which assume that all antiprotons are made by high-energy proton collisions with the expected ~ 5 g/cm^2 of interstellar material.

Local observations give us remarkably little information about whether bulk antimatter exists in the universe. We know with high probability, using direct probes, that it does not exist within the solar system. Gamma-ray measurements greatly constrain the amount of mixing and subsequent annihilation between matter and antimatter regions that could take place on larger size scales. Various astronomical observations lead us to believe that the universe was much hotter and more dense in the distant past than it is at present and without most of the heavier elements. As the early universe thinned and cooled, evolution of galaxies and nucleosynthetic processes led to the present configuration. Observation of symmetric rules for the creation of matter and antimatter in accelerator experiments led some cosmologists to wonder whether the total baryon number of the universe might be zero. The conventional interpretation is that it is not, and that the apparent domination of matter in the universe was either built in from the start, or was the result of symmetry-breaking reactions occurring soon after the Big Bang.[3]

On the other hand, for those who believe that the universe is half antimatter, the principal constraint is on mixing with the other half. If our Galaxy is composed of lumps of antimatter and matter kept apart by Alfvén's mechanism, cosmic rays probably could not penetrate this barrier unless they have energies significantly greater than those measured in present-day experiments. The only hope is that a few may leak through somehow and be detected here. Thus, future experiments must either push down the limits on the ratio of antimatter to matter in the cosmic rays, or extend the range of energy or charge being observed. Implicit is the assumption that the antimatter regions have the same general character as the region in which we find ourselves. In other words, these regions must be large enough to support the formation of antistars capable of nucleosynthesis and to accelerate and propagate the resulting antimatter cosmic rays. In particular, their antimatter cosmic-ray composition must be symmetric

2. A. Buffington, S. M. Schindler, and C. R. Pennypacker, Astrophys J. **248**, 1179 (1981).

3. G. Steigman, Ann. Rev. Astron. Astrophys. **14**, 339 (1976).

to what it is here and not be significantly altered in the process of reaching us. It would seem that many potential pitfalls lie in this line of reasoning, most of which have the effect of diminishing the relative abundance of the heavier antimatter cosmic rays. For example, if a conventional composition of cosmic-ray antimatter had to pass through a thick barrier of some kind, few of the heavier constituents would survive without fragmentation reactions. Thus, the only surviving evidence of the existence of antimatter regions protected by such barriers would be enhanced daughter particle fluxes. Of course, if the barrier were thick enough, nothing would get through at all!

Steigman says that bulk antimatter cannot have important consequences in the present universe, or there would be more observed effects.[4] However, a more sensitive test might be provided by a study of rare antimatter cosmic-ray constituents with unit charge, positrons and antiprotons. As observed above, these particles might exit from the antimatter regions more successfully than the heavy antinuclei. On the other hand, interpretation of the fluxes of positrons and antiprotons is difficult, because noncosmological sources exist for these particles. Small fluxes of positrons and antiprotons are expected from high-energy reactions of common-matter cosmic rays with the interstellar material. Antiprotons are produced by the same processes that, when observed in accelerator experiments, disclosed the baryon number conservation law in the first place. Positrons are final decay daughters of mesons, mostly pions, produced in other high-energy collisions. Because of the positron's small mass, its spectrum is significantly altered, after the particles are produced, by interaction with the cosmic background radiation and other photons, unless the antimatter propagation time is less than 10^7 years. The latter seems unlikely, since this is the age of the cosmic rays coming from the nearby region of the Galaxy. Thus, the prospect of using positron measurements as a probe for bulk antimatter is diminished, although they may prove valuable in mapping the cosmic-ray lifetime distribution. Antiprotons, on the other hand, are much less subject to the subsequent interactions. The antiproton flux at 8–12 GeV has been found by Bob Golden and collaborators to be about five times larger than expected,[5] and we have already mentioned the unexpectedly large flux we saw at low energy.

Thus, antiproton fluxes are the only measurements severely out of accord with the conventional view of no bulk antimatter nearby, or with the parent-proton cosmic rays propagating through only ~5 g/cm² of interstellar material. Bob's higher energy antiproton flux measurement, which was performed by a balloon-borne superconducting magnet spectrometer, seems to require ~30 g/cm² of interstellar material. It is possible that alternations in the cosmic-ray model could accommodate that result. However, even more material is needed to fit the low- energy result. Clearly, confirming experiments of these interesting results are required, preferably covering the energy interval between the two experiments and extending results to higher energies.

Unfortunately, there is at present no way of knowing whether these excess antiproton fluxes, if confirmed, indeed demonstrate nearby bulk antimatter regions, or whether instead they could be explained by an altered theory of cosmic rays. Although a definitive calculation has yet to gain acceptance by the cosmic-ray community, it appears that the observed antiproton spectrum is likely to be fitted successfully by a model without bulk antimatter, in which the cosmic-ray protons have propagated through an average interstellar material of ~50 g/cm² thickness. This large amount of material could be supplied by having a substantially different history for the predominant proton component of the cosmic rays, as compared with the history for the heavier nuclei. Thus, the protons could be perhaps ten times older than the heavy nuclei and occupy a proportionally larger volume in the Galaxy. The spectrum shape for the secondary antiprotons would also be significantly altered in their propagation through such a large amount of material, with further interactions transferring antiprotons to lower energies. On the other hand, the protons could have come from a significantly different source than the heavier nuclei and have passed at the source through a thick shell of some kind. In addition, several exotic mechanisms described in the literature may contribute extra antiprotons. These include neutron oscillations, evaporating black holes, and high-energy interactions near accretion disks, as well as regions of bulk antimatter.[6]

Most recent cosmic-ray antimatter experiments have been carried out by people associated with Luie. Bob Golden was a member of the Alvarez group before I joined it and doubtless took with him many of the ideas on which he later capitalized. Our group developed the measurement techniques and objectives through several stages. If Luie had not been around, most of this work would probably not have been done. This is not because he directly contributed the majority of ideas

4. Ibid.

5. R.Golden, et al., Phys. Rev. Lett. **43**, 1196 (1979).

6. L. C. Tan and L. K. Ng, Astrophys J. **269**, 751 (1983); C. Sivaram and V. Krishan, Nature (London) **299**, 427 (1982); D. Eichler, Nature (London) **295**, 391 (1982); R. Protheroe, Astrophys J. **251**, 387 (1981); O. Sawada, J. Arafune, and M. Fukugita, Astrophys J. **248**, 1162 (1981); V. L. Ginzburg and V. S. Ptuskin, Astrophys J. (USSR) **7**, 585 (1981); P. Király, et al., Nature (London) **293**, 120 (1981); F. Stecker, et al., Proc. 17th Int. Cosmic Ray Conf., **9**, 211 (1981); J. Szabelski, et al., ibid., **2**, 206 (1981); R. Cowsik and T. Gaisser, ibid., **2**, 218 (1981); and A. Stephens, Astrophys. Space Sci. **76**, 87 (1981).

necessary for progress, although he obviously has done plenty of that in his career, but because he created the intellectual environment in which the people who did this work could thrive. He did this by taking an interest in our activities and progress and by asking penetrating questions and not suffering our occasional foolishness gladly. Otherwise we would have joined other groups and done other things and perhaps succeeded and perhaps not. It is hard for me to believe that I could have had a more stimulating and exciting time!

Now we speculate on what types of experiments might be used to further the search for cosmic-ray antimatter in the next few decades. Experiments taking the SSCM concept into space for a longer exposure have been proposed for the Space Shuttle and later for a space platform. Luie was the principal investigator on an experiment to fly a magnetic spectrometer on a *High Energy Astronomical Observatory* satellite; extension of the antimatter search was a prime objective of the experiment. Unfortunately, the program had to be scaled down, and near-term prospects are uncertain for another superconducting magnetic spectrometer experiment in space. However, a magnet and cryostat system designed for space use was built under this program. It demonstrated the expected performance, yielding a cryogen lifetime of the requisite one year. After a few years, the opportunities for space flight using this type of experiment may improve and a new attempt be made.

For the shorter term, I know of only one experiment to further the antimatter search. This is being prepared at the Space Sciences Laboratory (SSL) in Berkeley, employing plastic scintillators, track-etch detectors, and Cerenkov counters.[7] The technique makes use of the expected delta-ray structure function for antimatter, which differs from that of common matter. The separation technique appears to be practical for antimatter as heavy as iron.

In the longer term, one is less constrained by practical considerations. It is not inconceivable to have room-sized experiments with magnets, total absorption calorimeters, and Cerenkov counters, in orbits removed from the constraints of Earth's magnetic field, which would carry the present-day philosophy well beyond present-day possibilities. In addition, a new generation of antimatter detectors may evolve. Thus new, low-noise amplifiers could detect the sign of the charge of particles passing through a series of superconducting loops, by analysis of the induced supercurrents in the superconductor. In addition, one can imagine city-block–sized detectors, as Luie first suggested,[8] with low-pressure gaseous xenon and numerous readout wires, as a "dilute emulsion" with electronic readout. Such a detector would be a giant extrapolation of the SSL experiment. It seems that there is no end of adventure for future experimenters, who hopefully will be limited only by the scope of their imaginations.

7. S. P. Ahlen, P. B. Price, M. H. Salamon, and G. Tarlé, Astrophys J. **260**, 20 (1982).

8. L. W. Alvarez, Astrophysics Group Internal Memo 328 (1976).

(*Reprinted from Nature, Vol. 236, No. 5346, pp. 335–338, April 14, 1972*)

Search for Antimatter in Primary Cosmic Rays

A. BUFFINGTON, L. H. SMITH, G. F. SMOOT &

L. W. ALVAREZ

Space Sciences Laboratory, University of California, Berkeley

M. A. WAHLIG

Lawrence Berkeley Laboratory, University of California

A new upper limit on the amount of antimatter in primary cosmic rays has been established. The limits are considerably lower than those for any previous experiment.

THE symmetry between matter and antimatter observed at accelerators has led many astrophysicists[1-3] to suppose that there is a similar balance in the universe. If, however, matter and antimatter were mixed, considerable annihilation reactions would occur. Gamma-ray measurements[4] set a stringent upper limit on any mixing of matter and antimatter within our galaxy. Although Alfvén[5] has advanced a mechanism which might permit regions of matter and of antimatter to coexist within our galaxy without appreciable mixing, it is also possible that matter and antimatter are concentrated in whole galaxies, in which case, if the early state of the universe were homogeneous, there would have to be a mechanism for separating matter and antimatter into galaxies.

In either case, one way of discovering that regions of antimatter exist elsewhere in the universe would be the unambiguous detection of antimatter nuclei in the primary cosmic rays. It is probable that a small flux of antiprotons (about 10^{-5} of the proton flux[6]) would be produced by cosmic-ray interactions with the interstellar gas, but these reactions should produce negligible fluxes of anti-helium and the more complex anti-elements. Therefore, observation of complex anti-nuclei such as helium, carbon or oxygen would almost certainly imply the existence of antimatter stars capable of nucleosynthesis.

Several teams have searched for antimatter[7-13]. The most stringent experimental upper limits for antimatter flux have been set by searches in large emulsion blocks that have been exposed at high altitudes to the incident cosmic rays. Examination of several thousand stopping cosmic-ray nuclei of charge greater than 2 showed no events with the characteristic annihilation products at the termination of the track. As Greenhill *et al.*[7] pointed out, however, these measurements apply only to low-energy antimatter flux, and it is reasonable to expect that antimatter might become more plentiful at higher energy. The observation of just one or two well verified, unambiguous anti-nuclei in the incident cosmic-ray flux would have profound astrophysical significance, because most present day cosmologies do not take account of significant amounts of antimatter.

In this article we report the first data from two flights of a new superconducting magnetic spectrometer, which was capable of a direct matter–antimatter separation in the cosmic rays. Antimatter events would appear in the spectrometer as trajectories which curve in the opposite direction to common matter, because of their negative charge. A detailed description of our equipment is given elsewhere[14]. Fig. 1 presents a schematic diagram and Table 1 summarizes some of the properties of the instrument.

Table 1 Specifications of the Magnetic Spectrometer

Flight dates	September 18, 1970, and May 8, 1971
Mean magnetic field integral	3.8 kG m (first flight) 4.3 kG m (second flight)
Spatial detectors	Four optically viewed spark chambers, four 1-cm gaps each
Optical reconstruction error	± 0.14 mm per spark chamber
Maximum detectable rigidity (set by optics and field integral)	75 GV/c (first flight) 90 GV/c (second flight)
Scattering material within the spectrometer	4.12 g/cm², 0.25 radiation lengths (first flight) 2.04 g/cm², 0.055 radiation lengths (second flight)
Geometry factor	660 cm² ster
Trigger counters	Pilot B and Y plastic scintillators
Time-of-flight measurement	With helium incident, timing error = ± 4 ns giving rejection of $>95\%$ against albedo events With $Z > 2$ incident, timing error = ± 2.5 ns giving rejection of $>99\%$ against albedo events
Charge resolution (helium through oxygen)	± 0.15 charge, by recording light levels in the three trigger scintillators
Trigger threshold	Selectable to gather data with charge Z greater than or equal to 1, 2, or 3
Residual atmosphere for flights	Roughly 5 g/cm² of air left above the instrument
Data-taking times	2.5 h with $Z \geq 2$ and 13 h with $Z \geq 3$

The spark locations and scintillator pulse height data were recorded on photographic film. All sparks were automatically measured on a cathode-ray tube encoding machine, and all pictures were visually scanned. Events showing occasional delta-ray activity in the spark chambers were redisplayed on a cathode-ray tube, and a scanner rejected the spurious tracks. Multiprong interactions in the apparatus and other back-

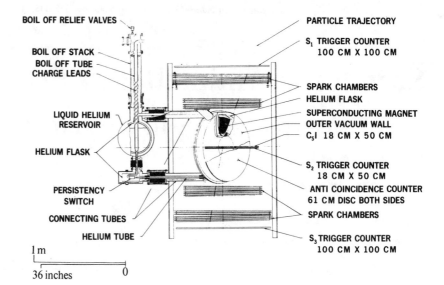

Fig. 1 Schematic SSCM detectors and cryogenics.

BOIL OFF RELIEF VALVES

BOIL OFF STACK
BOIL OFF TUBE
CHARGE LEADS

LIQUID HELIUM
RESERVOIR

HELIUM FLASK

PERSISTENCY
SWITCH

CONNECTING TUBES

HELIUM TUBE

PARTICLE TRAJECTORY

S₁ TRIGGER COUNTER
100 CM X 100 CM

SPARK CHAMBERS
HELIUM FLASK
SUPERCONDUCTING MAGNET
OUTER VACUUM WALL
C$_s$I 18 CM X 50 CM

S₂ TRIGGER COUNTER
18 CM X 50 CM

ANTI COINCIDENCE COUNTER
61 CM DISC BOTH SIDES

SPARK CHAMBERS

S₃ TRIGGER COUNTER
100 CM X 100 CM

1 m

36 inches   0

ground events were rejected after measurement. We fitted each good event to a curved trajectory through the magnetic field, and calculated its specific curvature K ($=1/$rigidity) and goodness of fit χ. Poorly fitting events were eliminated using a χ-cut, which should reject only 3% of the good data. Fig. 2 shows a specific curvature distribution for a sample of the fitted data. The maximum rigidity detectable with the automatic measuring machine was about 50 GV/c. A large body of common matter events starts at $K=0$ and extends beyond the end of the graph to geomagnetic cutoff at $K\approx0.25$ (c/GV). Any potential antimatter signal might be masked by some of the common matter events near $K=0$ spilling over into the region of negative K. Spillover could be due to measurement error or to occasional misinterpretation of an event by our automatic measuring and analysis sequence. Therefore, all events which had K less than zero were remeasured by hand on a machine which had three times the accuracy of the automatic machine. On remeasurement, most of these gave a positive specific curvature, but some remained with negative K and these we regarded as antimatter candidate events.

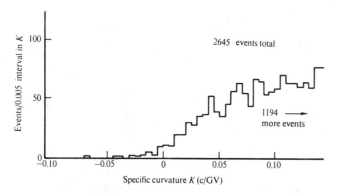

Fig. 2 A plot of specific curvature K distribution for a sample of data from second flight measured automatically. All events with $K<0$ to be remeasured on a more accurate machine.

Table 2 presents a list of possible explanations for these events. We are able to rule out source 2 as an explanation, because we observed no albedo time-of-flight (a difference in timing of about 10 ns) for any of the antimatter candidates. For sources 1 and 3, we made a Monte Carlo calculation which included the optical error, multiple coulomb scattering and nuclear scattering of our instrument. The optical error was obtained by a detailed study of several hundred

"straight-through" events which were taken for each flight with the superconducting magnet turned off. Fig. 3 shows the K-distribution for straight-through events in the second flight. The smooth curve is the Monte Carlo fit to these data. For the nuclear scattering, we used a diffraction shape for the angular distribution with a cutoff in total momentum transfer of $\Delta P_{max}=300$ MeV/c. Any greater momentum transfer than this in a collision should fragment the incident nucleus. Usually this would cause it to fail our spark chamber topology requirement and to give a much smaller pulse height reading in the bottom scintillator of the apparatus. Nuclear scattering should have contributed several antimatter candidates in the helium portion of our data, but the cutoff on momentum transfer becomes such a stringent limit for higher-Z elements (whose total momentum is Z times the rigidity) that in our $Z\geqslant3$ data there are no appreciable antimatter candidates to be expected from this source.

Table 2 Sources of Antimatter Candidates

(1) Optical reconstruction error moves an event over from $K>0$ region

(2) High-energy albedo nucleus simulates negative charge because it is incident from below

(3) Nuclear scattering in the material of the instrument caused an apparent negative curvature for a common matter event

(4) An actual antimatter signal

Table 3 presents the antimatter candidates and Monte Carlo predictions for all our data. We have presented helium (with about 9% admixture of higher-Z data) separately from our $Z\geqslant3$ data, because the nuclear scattering mechanism is unimportant for the latter. For the distribution of common matter events that could contribute in our Monte Carlo predictions we used also a power law distribution in rigidity, $dN/dR \propto R^{-j}$ where the value for $j=2.6$ was determined by a fit to all our data. A more detailed analysis of the rigidity spectra of our data is now being prepared. In terms of specific curvature, the power law becomes $dN/dK \propto K^{j-2}$. The Monte Carlo predictions of Table 3 are quite sensitive to the value chosen for j. Uncertainty in j causes the overall scale of the values to have an uncertainty of about $\pm20\%$. No significant enhancement is seen in any of the bins. In the high-Z data no events are seen for $R<33$ GV/c. This limit and the one set by statistics in the rigidity range between 33 and 100 GV/c are shown in Fig. 4. This figure also

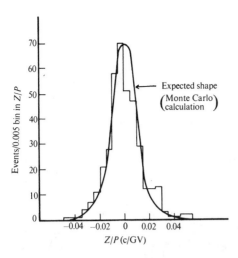

Fig. 3 377 events (second flight) straight-through (magnet off) data fit to curved trajectory.

presents results of previous experiments, all limits shown having 95% confidence level.

Because the antimatter candidates of Table 3 were potentially a greatly enriched sample, we inspected the spark chamber and pulse height records for each event. They all had a normal appearance, except for a single oxygen event which produced an interaction halfway through the last spark chamber and deposited 202 MeV (instead of the usual 124 MeV) in the bottom scintillator. The spark chambers showed that the interaction had at least three prongs at wide angles to the incident trajectory. Such a violent interaction must have fragmented the incident nucleus thoroughly, making it difficult for the event to produce a pulse of normal height in the bottom counter, let alone the abnormally large pulse we saw. The rigidity of this event was 40 GV/c, two standard deviations from a straight trajectory in our apparatus.

Table 3 Antimatter Search Statistics and Monte Carlo Predictions

Data sample	Rigidity range (GV/c)	Observed antimatter candidates	Monte Carlo predicted spillover
Flight 1	$R < 33$	5	2.6
Helium	$33 < R < 100$	7	6
(6,131 events)	$R > 100$	12	11
Flight 2	$R < 33$	0	1
Helium	$33 < R < 100$	0	4
(3,838 events)	$R > 100$	8	7
Flight 1	$R < 33$	0	0.4
Charge $Z \geq 3$	$33 < R < 100$	1	4
(3,152 events)	$R > 100$	5	7
Flight 2	$R < 33$	0	0.2
Charge $Z \geq 3$	$33 < R < 100$	4	9
(10,445 events)	$R > 100$	8	18
All data	$R < 33$	5	4
combined	$33 < R < 100$	12	23
(23,566 events)	$R > 100$	33	43

Common matter might occasionally make up such a large energy deposition in the lower scintillator, through the combined energy deposition of cascade protons (from the struck nucleus) and of inelastically produced mesons. The material of the spark chamber was sufficient to stop any evaporation products from the struck nucleus. Antimatter, of course, would have all of the energy deposition capability of common matter plus that of the extra mesons resulting from plentiful

annihilation processes. To evaluate the relative probabilities of these two processes, we made a Monte Carlo study that included cascade protons[15] from the struck nucleus in the spark chamber (taken to be aluminium), inelastic pion production[16], and the relativistic fragments of the incident nucleus. For the antimatter case, we added extra mesons to take account of annihilation processes. Averaging over all our data we found that an oxygen nucleus could interact in the fourth spark chamber; only 1% of the time could it make as large a pulse as the one we saw in the counter. On the other hand, we found that a 40 GV/c antimatter nucleus could produce as large a pulse 20–50% of the time. The difference between these two numbers (with even greater disparity with our original estimates) led us to make a detailed study of large pulse height probabilities for this event. We found that the chance of common matter producing the large pulse height is an increasing function of incident rigidity. Furthermore, only high-rigidity events could appear because of the optical distortions of our spectrometer as antimatter. Integrating over our known rigidity spectrum, we found that the chances were about 12% that this oxygen event was high-energy common matter that appeared as antimatter in the spectrometer because of optical distortion and made up the large pulse height by producing many inelastic mesons. Therefore, it finally appears only an insignificant three times easier to explain this event as true antimatter than it does as common matter spillover.

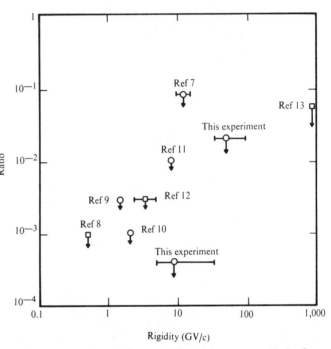

Fig. 4 Summary of present information on upper limits for antiparticles in the cosmic rays at 95% confidence limits. □, $\overline{P}/P$ ratio; ○, $\overline{A}/A$ ratio.

The limits set on antimatter flux by our experiment are considerably lower than for any previous experiment. We consider that we have done as well as possible in the search for antimatter events among helium with our present equipment, because of the contributions to antimatter candidates from nuclear scattering within the detector. A new apparatus, with much less material in the detector, could perhaps carry the helium search further. For a magnetic field integral of 3 kG m, however, the momentum transfer necessary to scatter a common matter event to negative curvature is greater than 90 Z MeV/c. As we noted before, total momentum transfer greater than 300 MeV/c is likely to fragment an incident nucleus. Therefore, on the higher charged elements, nuclear scattering becomes unimportant and our experiment is limited only by the statistics and by errors in optical reconstruction. We are planning to fly a second-generation magnetic spectrometer with improved optics, which will also have a thick Čerenkov counter following the momentum measurement. Antimatter candidates should interact in this counter about 50% of the time, and the Čerenkov threshold will diminish the response of the counter to the troublesome nonrelativistic cascade products which characterize interactions of common matter. With this innovation and multiple balloon flights, it should be possible to push the upper limit down by another decade or so, or hopefully to make the first positive unambiguous identification of complex anti-nuclei in the primary cosmic rays.

We thank the National Center for Atmospheric Research balloon launch crew at Palestine, Texas, for hospitality. We also thank Dr Charles Orth for help with the Monte Carlo program and Dr John Naugle, who called our attention to the possibility that our very small sample of very high-energy ordinary matter events had to be considered seriously as a cause for the one apparent anti-oxygen event. This work was supported by a NASA contract.

Received November 23, 1971.

[1] Alfvén, H., and Elvins, A., *Science*, **164**, 911 (1969).
[2] Hoyle, F., and Burbidge, G. R., *Nuovo Cimento*, **4**, 558 (1956).
[3] Morrison, P., *Nuovo Cimento*, **7**, 858 (1958).
[4] Kraushaar, W., Clark, G. W., Garmire, G., Helmken, H., Higbie, P., and Agogino, M., *Astrophys. J.*, **141**, 845 (1965).
[5] Alfvén, H., *Rev. Mod. Phys.*, **37**, 652 (1965).
[6] Rosen, S., *Phys. Rev.*, **158**, 1227 (1967).
[7] Greenhill, J. G., Clarke, A. R., and Elliot, H., *Nature*, **230**, 170 (1971).
[8] AppaRao, M. K. V., *Canad. J. Phys.*, **46**, S654 (1968).
[9] Aizu, H., Fujimoto, Y. H., Hasegawa, S., Koshiba, M., Mito, I., Nishimura, J., and Yokoi, K., *Phys. Rev.*, **121**, 1206 (1961).
[10] Evanson, P., and Meyer, P., *Twelfth Intern. Conf. on Cosmic Rays*, Hobart, Conference Papers, **1**, 138 (1971).
[11] Golden, R., *Twelfth Intern. Conf. on Cosmic Rays*, Hobart, **1**, 203 (1971).
[12] Bogomolov, E. A., Lubyanaya, N. D., and Romanov, V. A., *Twelfth Intern. Conf. on Cosmic Rays*, Hobart, Conference Papers, **1**, 144 (1971).
[13] Brooke, G., and Wolfendale, A. W., *Nature*, **202**, 480 (1964).
[14] Smith, L. H., Buffington, A., Wahlig, M. A., and Dauber, P., *Rev. Sci. Inst.* **43**, 1 (1972).
[15] Metropolis, N., Bivins, R., Storm, M., Turkevich, A., Miller, J. M., and Friedlander, G., *Phys. Rev.*, **110**, 185 (1958).
[16] Barashenkov, V. S., Maltsev, V. M., Patera, I., and Toneev, V. D., *Forschritte der Physik*, **14**, 372 (1966).

(*From left*), Glen Seaborg, Ed McMillan, Ernest Lawrence, Don Cooksey, Edward Teller, Herb York, and Alvarez (1951).

Alvarez after first flight at Mach 2, 1963.

Luis Alvarez receiving the National Medal of Science from
President Lyndon B. Johnson (1963).

Bubble chambers and their builders. (*From left*) Ron Rinta, Paul Hernandez, A. J. (Pete) Schwemin, Bob Watt, Alvarez, and Glen Eckman (1968).

Seventy-two-inch bubble chamber body and controls.

454A PST OCT 30 68 LA003

SPD260 SYA198 BC175   CDD257 VIA RCA ZCZC WUB0247 SWB5846 SOP9113

URNX CO SWSM 067

STOCKHOLM 67 30 1200 P1/50

1968 OCT 30 AM 6 28

PROFESSOR LUIS W ALVAREZ

4 NORTHAMPTON AVE BERKELEY CALIF

THE ROYAL ACADEMY OF SCIENCES TODAY

AWARDED YOU THE 1968 NOBEL PRIZE FOR PHYSICS FOR YOUR

DECISIVE CONTRIBUTIONS TO ELEMENTARY PARTICLE PHYSICS IN

PARTICULAR THE DISCOVERY OF A LARGE NUMBER OF RESONANCE

STATES MADE POSSIBLE THROUGH YOUR DEVELOPMENT OF THE TECHNIQUE

OF USING

COL 4 1968

SWB5846 PROFESSOR P2/17

HYDROGEN BUBBLE CHAMBER AND DATA ANALYSIS OUR VERY

WARM CONGRATULATIONS LETTER WILL FOLLOW ERIK RUDBERG PERMANENT SECRETARY

Telegram from the Nobel Prize Committee.

Alvarez being presented Nobel Prize for Physics by King Gustav VI,
with Princess Christina looking on (1968).

Nobel Banquet, December 10, 1968.

Alvarez being inducted into the National Inventors Hall of Fame (1978). (*From left*) wife Jan Alvarez, son Don, Juanita M. Kreps, Secretary of Commerce, daughter Helen, Alvarez, and John C. Dorfman, President of the Patent Lawyers Assocation.

Asteroid Impact research team (1979). (*From left*) Helen Michel, Frank Asaro, Walter Alvarez, and Luis Alvarez.

Walter Alvarez, Luis Alvarez,
and Richard Muller, at the
Bohemian Grove (1985).

Alvarez entertaining at Optical Society meeting. Photo by:
Richard Altman, © 1979 Applied Optics.

18

Examining the Kennedy Assassination Evidence

Richard L. Garwin

Even among outstanding physicists, Luis W. Alvarez combines a rare ingenuity and flair for innovation with a faculty for critical analysis. His broad contributions are marked by a drive to consider things that don't quite fit, to ask whether something new can be done, to draw conclusions from theories and hypotheses; and *then* to confront these misfits, needs, and predictions with analysis and facts.

I met Luie in the late 1940s and had some contact with him in the 1950s, mostly about physics applied to hydrogen bubble chambers. In 1957 we worked together intensively on a technical advisory committee dealing with national security matters, and after that I always valued Luie as a contributor to the panels on which I served for the President's Science Advisory Committee. These panels made technical contributions as often as program recommendations, and Luie contributed vastly and broadly to their effectiveness in the former.

John F. Kennedy, President of the United States, was killed by rifle shots in Dallas, Texas, on November 22, 1963. After extensive hearings into the facts of this crime, a commission chaired by Chief Justice Earl Warren concluded that the President was assassinated by a lone gunman, Lee Harvey Oswald, who was himself killed by Jack Ruby while in police custody.[1] The bizarre facts and the importance of the crime created intense and long-lasting interest in the assassination and raised continuing questions about conspiracy and the possibility of a second gunman.

At a colloquium I attended in Los Alamos, Luie presented his analysis of the Zapruder film of the shots actually striking the President. Later I served with Luie on a National Research Council panel which reviewed the acoustic evidence from the Dallas Police Department tapes relating to the Kennedy assassination.[2]

The best way to appreciate Luie's remarkable article reprinted here is to *read* it. It was intended as a

tutorial, as well as a presentation of new results in an important matter. The wealth of results derived from a minimum of data is typical of Luie's contributions. The Zapruder film, long available to official investigators, was first used by Luie only after frames appeared in *Life* magazine.[3]

From his analysis of the pictures, Luie deduced the timing of the shockwaves which disturbed the photographer's camera; he showed that when struck by a bullet ejecting a mass much greater than itself, it is possible and even normal for a body to recoil *toward* the direction of the gun; he showed that Zapruder's camera was running near 18 frames per second and specifically not at 48 and probably not at 24 frames per second; and he obtained internal evidence from a "featureless grass mound" to determine the precise speed of the President's limousine when the Warren Commission report stated that it ". . . is just a grassy plot. So there is no reference point on which we can reestablish the position of the car in the roadway."

Luie says, "I've long felt that the testimony of a physicist could have been of help to the Warren Commission." One can only agree that the testimony of a physicist could have been helpful, *if* that physicist were Luie and if even such a remarkable physicist had had the benefit of the informal (and at times destructive) arguments and hypotheses of skeptics of the Warren Commission Report. It is typical of Luie to have many hypotheses, most of which he rejects in the search for truth. In this case, he searched for refractive effects of a shock wave as the cause of the differing position of the flag seen in the *Life* pictures. He initially held that the President's head had recoiled because of unbalanced muscle forces. These preliminary views were abandoned when no calculations could make them plausible and when a more convincing explanation for the data was found.

Also typical of the best of physics (and physicists) is Luie's comment, "I solved the problem (to my own satisfaction, and in a one-dimensional fashion) on the back of an enevelope, as I sat in solitary splendor in the beautiful suite that the St. Louis hotel management supplied me in my capacity as president of the APS."

1. "Hearings Before the President's Commission on the Assassination of President Kennedy," (U.S. GPO, Washington DC, 1964).

2. N. F. Ramsey, L. W. Alvarez, H. Chernoff, R. H. Dicke, J. I. Elkind, J. C. Feggeler, R. L. Gawin, P. Horowitz, A. Johnson, R. A. Phinney, C. Rader, and F. W. Sarles, "Report of the Committee on Ballistic Acoustics" (Nat. Acad. Press, Washington DC, 1982), and summarized in Science **218**, 127 (1982).

3. Life, November 25, 1966.

It was not for Luie (or for Norman Ramsey, mentioned later in this article, or for many others who have served as president of the American Physical Society) to give up the practice of physics because of administrative burdens or high status.

The paper is rich in its informal advice for the physicist, or for anyone wishing to arrive at truth or to persuade others. Thus, "Mr. Dulles was an experienced intelligence agent, and his practiced eye caught an important clue, but he too quickly dismissed it as undecipherable, which of course, we now know it wasn't. The expert photoanalyst put the lid on the matter by his polite endorsement of Mr. Dulles's error." The physicist *always* asks, "Do I know this?" "How do I know this?" and "Is this still true?"

Luie makes simple tools do marvelous work, like a craftsman in wood or in optics for that matter. He is always willing to work and to explore: "As any physicist would do, I plotted the tabulated distance of the car . . . against frame number for these 14 selected frames." Unfortunately, *not* every physicist would do that. Luie's remarkable spirit is also revealed in his comment about the determination of film speed by the rate of hands clapping and hence the velocity of the President's limousine. "I did this work, and the analysis of the clapping, during the Christmas vacation following the publication of the November 26, 1966 issue of *Life*."

Finally, I note the importance of Luie's *breadth* of experience, ready and waiting to be applied to the problem at hand—from his memory of Fermi's measurement of the first atomic bomb yield to his experience in inventing image stabilizers for movie cameras, to the instructive measurement of neuromuscular response time using a gravity-driven timer (the dollar-bill parlor trick).

This paper provides a marvelous example of physics at its best applied to the analysis and clarification of important matters, where information of unrecognized value may be available.

Luie was a coauthor of the Report of the Committee on Ballistic Acoustics[4] published by the National Research Council—the mechanism used by the National Academy of Sciences, National Academy of Engineering, and the Institute of Medicine to do substantive studies for government sponsors. Professor Norman F. Ramsey chaired this committee, and the full list of authors is given in the reference.

The style of this report is doubly removed from that of Luie's paper reprinted here, in part because it is a product of a *group,* and the style of any one member would be inappropriate to the total group. Furthermore, it was addressed not to an audience of physicists, but to the House Committee on Assassinations and to the public as well. Also, it is a strict presentation of results, not a didactic, instructive presentation of thoughts, procedures, and wrong hunches followed. It also addresses only a very limited aspect of the assassination of President Kennedy.

The Ramsey Committee was directed to study only the "acoustic evidence" in the Kennedy assassination. The group had no responsibility to look at films or transcriptions of witnesses' testimony. The sum total of the acoustic evidence was the set of embossed plastic disk recordings of the few minutes before and after the gunshots which killed President John F. Kennedy supplied by the Dallas Police Department. The National Academy of Sciences was requested by the Department of Justice to do this study, in view of the previous analyses of magnetic recording copies of these original recordings, by workers at Bolt, Beranek and Newman (BBN) and at Queens College, New York.[5] Previous investigators had listened for many hours to these recordings, attempting to determine what happened in the Kennedy assassination and to find agreement or contradiction with hypotheses as to the murderer, or murderers, of the President. It had occurred to some that evidence might serendipitously be found on one of the two channels being recorded, because a microphone on one of the mobile radios was stuck in the "on" position for many minutes at about the time of the assassination. If the vehicle with the mobile radio had been in Dealey Plaza at the time of the assassination, it could well have recorded important evidence for constructing or refuting a hypothesis of the assassination.

Analogies are useful in understanding the nature of the evidence that might be available under those circumstances. Many of us recognize the aural "feel" of an accustomed room and would readily distinguish the sound of music, voices, or even the scraping of a chair from that of a similar sound in a great hall or outdoors. More technically, an impulse of sound at a particular point in space would produce a set of echoes which could be specific for the point of origin of the sound, the position of the sensing microphone, and the details of the space itself, even if the echoes are so close in time that they cannot be heard as separate sounds.

In the present instance, the question was whether the few seconds of recording identified by the previous investigators as containing imprints of the gunshots did indeed do so; whether the recording *could* have contained such imprints (because it was or was not made at the time of the assassination, or the microphone was or was not in Dealey Plaza); and whether the analyses of the recording supported (or negated) the hypothesis of a second gunman, especially one shooting from the

4. See n. 2.

5. Appendix to Hearings Before the Select Committee on Assassinations of the House of Representatives Ninety-Fifth Congress, Volume 8 (U.S. GPO, Washington DC, 1979).

"grassy knoll" in front of the Presidential limousine at the time of the assassination. Note that the question was not whether shots from a "second gunman" stuck or killed President Kennedy or wounded any of the other passengers; the *existence* of shots from the grassy knoll would vitiate the hypothesis that the President was killed by an individual gunman, acting alone, from the Texas Book Depository building.

The *House Assassination Committee* asserted that "scientific acoustical evidence establishes a high probability that two gunmen fired at President John F. Kennedy."

In contrast, the Ramsey Committee found that the recording could not possibly have contained imprints of shots because:

This identification of cross talk between Channels I and II shows conclusively that the previously analyzed sounds were recorded about one minute after the assassination and, therefore, too late to be attributed to assassination shots. A similar conclusion is reached independently by the analysis of the times of the acoustic impulses of intelligible cross talk between the two channels more than three minutes after the assassination. This analysis shows that the previously studied acoustic impulses were recorded after the motorcade was instructed to go to Parkland Hospital.

How was this identification of crosstalk established? Since I am writing about the contribution of Luie, and not presenting a history of the Committee on Ballistic Acoustics, I limit my comments to one aspect of the Report—testing the hypothesis that the few seconds of Channel I containing the "shots" was in fact recorded a good many seconds *after* the shots, at a time when the order was being broadcast to "Hold everything secure until the homicide and other investigators can get there."

Whether recognized formally or not, science and the establishment of truth (rather, the negation of falsehood) progress in two quite independent steps: the formulation of a hypothesis and the testing of that hypothesis. Progress is most rapid when formulation and testing are done by the same person or group, but it often happens that different talents or different personalities are better at one or the other aspect of scientific advance.

The Ramsey Committee had met with the previous investigators and had done enough analysis to question seriously the identification of signals on Channel I as the imprint of "shots" and to demonstrate that the stretch of noiselike recording identified by the BBN workers as containing the "shot from the grassy knoll" was displaced more than 200 milliseconds from that identified by the Queen's College workers. *Both* sets of workers could not be correct.

At this point, a communication was received from Steve Barber, a private citizen in Mansfield, Ohio, who had listened to copies of the tapes. Mr. Barber, noting his experience as a musician (percussionist), was persuaded that he could hear, adjacent to the recording of the "shots" on Channel I, an image of the very clear recording on Channel II containing the words "Hold everything. . . ." Because the section containing the shots was immediately adjacent to and *following* the "Hold everything . . .," a finding that the recording did indeed contain this broadcast (and that it was recorded by radio and not somehow later imprinted on the tape) would eliminate the possibility that the noise-like section contained any of the shots contributing to the assassination. Thus, in this case, the *hypothesis* is due to Steve Barber.

Most committee members, however, could not hear "Hold everything," but it was agreed that the possibility of identifiable crosstalk between Channels I and II should be studied further. If incontrovertible evidence could be attained that "Hold everything" was indeed recorded on the tape, it would be of vital importance in the working of the committee. Paul Horowitz, Charles Rader, and Norman Ramsey used the FBI facilities to make sound spectrograms ("voice prints") of the relevant portions of the two channels. Although visual inspection showed some similarities, such an inspection was not convincing one way or the other. Norman Ramsey recognized that not only the understandable words but *any* noise found on Channel II at this time and appearing on Channel I might be used to establish the coincidence of the few important seconds on Channel I with the "Hold everything . . ." on Channel II, which was obviously *after* the assassination. Accordingly, he looked at the sound spectrograms and was able to identify regions of one channel where the spectrogram had clearly describable changes in frequency with time. He observed that every identifiable feature common to both Channels I and II would have to occur at exactly the same (or proportional) spacing in time and at exactly the correct frequency. Cutting segments of the spectrogram containing 27 identifying features and sliding them along the spectrogram of the other channel, he located matches at the relative positions shown in Figure 1 and plotted in Figure 2. The straight line with a slope of 1.06 was consistent with a 6% difference in speed between the original embossing recorders (including any changes in recording speed involved in the transcription to the tape copies used by the committee). This interpretation was confirmed by the frequencies of the corresponding features on the two channels differing also by about 6%, as would be the case if the recording speeds differed. The match was indeed robust, as indicated by the fact that every member of the committee who tried the same experiment arrived at similar results.

Luie found this process too complex for his taste and sought a simple and convincing scheme to dem-

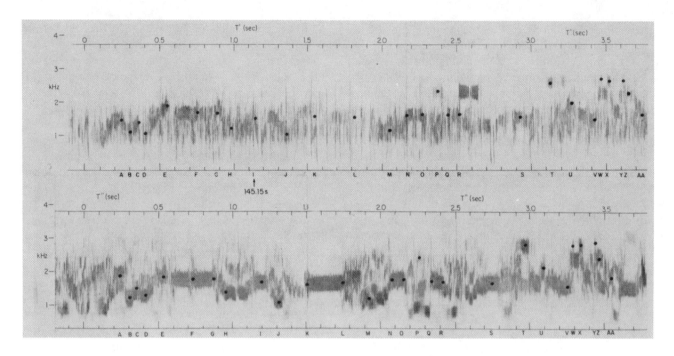

Fig. 1 Composite photograph of sound spectrograms on each of Channels I and II, with the apparent time on each being indicated by T' and T''. The audible ". . . hold everything . . ." phrase begins at approximately zero on both channels, but there is no special significance to the exact location of zero time on either channel. The impulses identified by the workers from Bolt, Beranek, and Newman, as arising from their conjectured grassy knoll shot occur above the arrow marked 145.15s; the proper location for this arrow was determined by comparing this sound spectrogram with that of Figure 5 in the Bolt, Beranek and Newman report. The letters designate the characteristic features listed in Table B-1 of Ramsey *et al.*, 1982, with the black dot identifying the characteristic point. The black dots were located by the procedure described in Appendix B of Ramsey *et al.*, 1982. (All figures are reproduced from Ramsey *et al.*, 1982).

onstrate the common origin of the two channels. He found five features on Channel I which were clearly identifiable by anyone who looked at the spectrograms—namely, sustained tones—and were also visible on Channel II at about the same time. It did not matter whether these tones originated in a human vocal tract enunciating "Hold everything. . . ." Just as in the Zapruder analysis, Luie did some simple measurements from copies of photographs, this time to determine the location of the center frequency of each tone on the individual graphs. He found the ratios of the frequencies thus obtained between the tones of Channel I and Channel II to be 1.054, 1.066, 1.065, 1.052, and 1.067. This easily checked measurement gave a frequency ratio of 1/1.062, compatible with a speed ratio of 1.06, since the number of cycles is conserved no matter what the recording speed. The Alvarez approach, involving only the five most easily made frequency measurements (four of which were among Ramsey's "features"), was easier to apply and easier to repeat than Ramsey's, which involved the measurement of the times and frequencies of some 28 points.

Other committee members played important roles in analyzing the likelihood of occurrence of these re-

sults if the two recordings had not been causally related, in the particularly difficult matter of timing events on the two channels, and in challenging these results. Occasional clearly audible "tie points" were available in the form of transmissions broadcast from Dallas Police Department Headquarters on both channels simultaneously. Because the recorders were, in principle, *sound* operated and because they were for the most part recording different transmissions from the field or from headquarters, the timing of any particular section of the recording is not easy to determine on an absolute basis or even relative to other portions of that recording or the other recording. It was further complicated by the fact that because Channel II was recorded by embossing grooves on a Gray Audiograph disk and the Gray Audiograph playback device incorporated a tracking stylus with rather high friction, the magnetic tape copy contained *repeats* of some of the recorded information.

The committee's work was further advanced by playing the original Gray Audiograph disk on a standard high-fidelity turntable. In this way, the repeats were eliminated and a more reliable estimate of relative timing on the two channels was possible. Contrary to Luie's expectation, informal communication

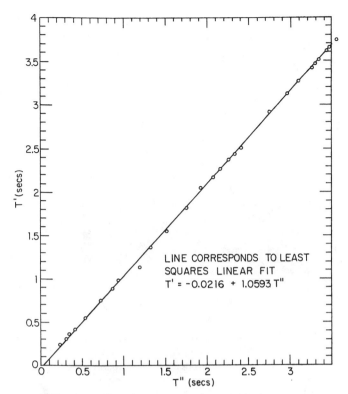

Fig. 2 Plot of T' versus T'' for corresponding characteristics with the values of Table B-1 of Ramsey *et al.*, 1982. The straight line in the figure is a plot of the given equation, which in turn is a robust linear regression fit to the plotted points. The analysis leading to this figure is given in Appendix B and Table B-1 of Ramsey *et al.*, 1982. The point farthest off the line is at $T'' = 1.195s$ and is for the incorrectly identified characteristic I, as discussed in Appendix B of that same publication.

of those results to the earlier investigators did not result in immediate agreement that the previously claimed ''shots'' were not accurate imprints of the assassination sounds.

A further measure of similarity and relative timing and speed of the two channels became available when my IBM colleagues, Ramesh Agarwal and Burn Lewis, and I made a direct comparison of the two-dimensional sound spectrograms of the two channels.[6] Looking at Figure 3, one is seized with the desire to cut out not a short feature, but the whole interesting segment of Channel I and to slide it over a long stretch of Channel II, taking the pattern cross-correlation of the two. The pattern cross-correlation is obtained from a sum over every frequency-time bin of the Channel I segment of the bin energy multiplied by the underlying frequency-time–bin energy of Channel II. Because of the possibility of dissimilar speeds on the two magnetic recordings, the search for a peak in this cross-correlation pattern has to be carried out not only in ''slide,'' but in stretch or ''warp.''

6. R. C. Agarwal, R. L. Garwin, and B. L. Lewis, IBM Research Report RC 9771 (1982).

This procedure was conducted blindly with the result shown in Figure 4. To test whether the procedure was reasonable, it was carried out in identical fashion in the neighborhood of a *known* and clearly identifiable (even to committee ears) voice recording on two channels, with the results shown in Figure 5.

In the case of the acoustic data, Luie did not think of the hypothesis (Steve Barber did), nor was he the first to verify the hypothesis (Norman Ramsey was). He simply presented what seemed to be the most accessible way to demonstrate the common origin (and hence synchrony) of the two channels. Other members of the committee not mentioned made major contributions to this aspect of the committee study and were the source of other findings presented in the Ramsey Report; but my topic is the work of Luie and not an assessment of the relative contributions of the members of the committee.

What can we learn from these accounts of Luie's contribution to the analyses of the Kennedy Assassination films and acoustic evidence? We add redundant evidence in support of the hypothesis that Luie has employed great faculties of innovation and criticism not only in his youth but in his eighth decade of life as well. More general, however, is the reminder that evidence and analysis cannot confirm a hypothesis, but they certainly can destroy one. In his Zapruder work, Luie confronted the assertion that one of the shots must have struck President Kennedy from the front, driving his head backward—evidence taken by many as proof of the existence of a second gunman. He showed, instead, that a backward impulse to the head is a common result of a bullet entering from the rear, and he has therefore negated the *negation* of the hypothesis of a single gunman in the Texas Book Depository.

Similarly, in the Ramsey Report, the assertion was confronted that the tape recording contained not only echoes of gunshots in Dealey Plaza, but a set of echoes which uniquely matched a bullet coming from the grassy knoll, which would thus vitiate the hypothesis of a single gunman. The results in the Ramsey Report demonstrate that the recording in question cannot contain imprints of the sounds of the assassination, and so it eliminates that particular criticism of the single-gunman hypothesis.

Another lesson to be drawn from both analyses is the value of publication. Contributions by different individuals in different places at different times must be brought together to provide the best technique and evidence. To paraphrase an oft-quoted statement, if Luie sees farther than others, it is because he is willing to stand on the shoulders of giants or for that matter of anyone who can provide a firm footing. And he willingly provides such footing for others.

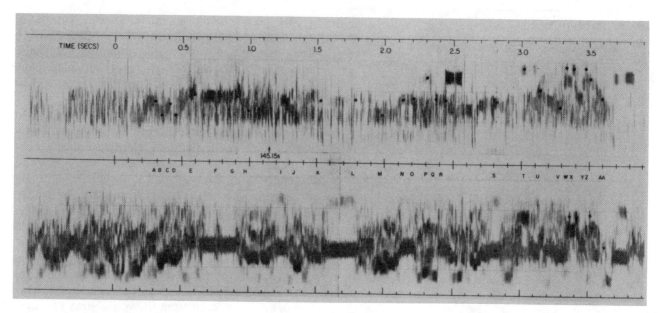

Fig. 3 Composite photograph of sound spectrograms on each of Channels I and II, with Channel I being the upper sound spectrogram and Channel II the lower. The audible ". . . hold everything . . ." phrase begins at approximately zero on both channels, but there is no special significance to the exact location of zero time on either channel. The impulses initially identified by the Bolt, Beranek and Newman group as arising from their conjectured grassy knoll shot occur above the arrow marked 145.15s, and those identified by the Queens College workers occur 0.2 second earlier; the proper location for this arrow was determined by comparing this sound spectrogram with that of Figure 5 in the Bolt, Beranek and Newman report. The letters and black dots designate corresponding characteristic features. One recording was speeded up by 6.7%. Since the interpretation of sound spectrograms depends on continuing gradations on darkness, copies in a printed report lose clarity. For this reason, photographs of the sound spectrograms have been retained in the National Research Council files.

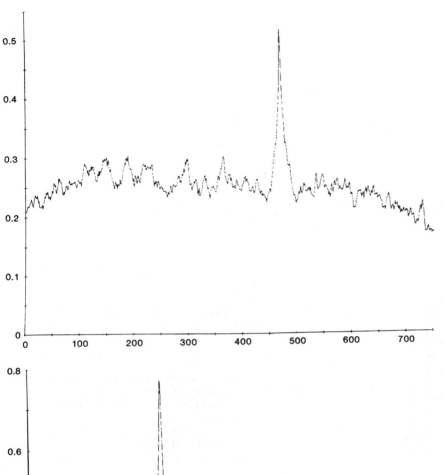

Fig. 4 Cross correlation between "Hold everything . . ." segments of Channels I and II sound spectrograms with time scale slightly compressed to produce the best correlation peak. The curve is produced by sliding 2.50 secs of Channel I along 10.00 secs of Channel II, 0.01 sec at a time, using frequencies in the band 600–3500 Hz. Much of the energy in Channel II spectrogram correlates precisely with that of Channel I at one and only one offset.

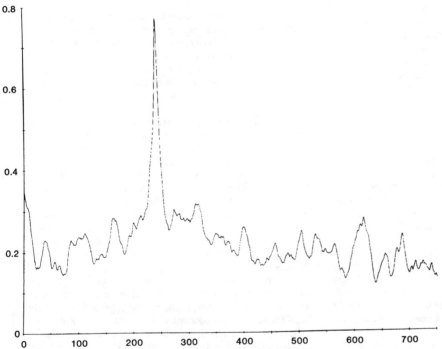

Fig. 5 Cross correlation between "You want . . . Stemmons" segments of Channels I and II sound spectrograms with time scale slightly compressed to produce the best correlation peak. The curve is produced by sliding 2.50 secs of Channel I along 10.00 secs Channel II, 0.01 sec at a time, using frequencies in the band 600–3500 Hz. The clear correlation peak between the two spectrograms demonstrates the causal origin of the peak of Fig. 4.

A physicist examines the Kennedy assassination film

Luis W. Alvarez

Lawrence Berkeley Laboratory, University of California, Berkeley, California 94720
(Received 26 January 1976)

The motion picture film of the Kennedy assassination taken by Abraham Zapruder was one of the most important exhibits examined by the Warren Commission. The author uses the tools of the physicist to draw some conclusions that escaped the notice of the Commission and its expert FBI photointerpreters. Among the subjects treated are (1) the timing of the gun shots, (2) a theoretical and experimental investigation of the "backward snap" of the President's head immediately after he was killed—yielding the surprising result that it was consistent with a shot fired from the rear, (3) the speed at which the camera was running, and (4) a previously undetected deceleration of the President's automobile just before the final shot. The emphasis throughout is not on the assassination but rather on the application of elementary physics principles to the solution of practical problems.

EDITOR'S NOTE

We publish this article by Luis Alvarez for its unique pedagogic usefulness. It brings to bear on a matter of public concern powerful and simple physical arguments that are within the reach of introductory physics students. It shows a physicist at work employing qualitative arguments, estimates, measurements, and calculations appropriate to the problem and to the accuracy of data available.

As always, we welcome readers' responses to this article and will select some for publication according to their appropriateness and the space available. We are interested in comments on procedures which Professor Alvarez uses to reach his conclusions and on the pedagogic uses to which the article can be put. We do not feel that this Journal is an appropriate forum for a discussion of alternative theories of the assassination.

I. INTRODUCTION

In the eleven years since the Warren Commission published its 26-volume report[1] on the assassination of President Kennedy, a controversy has continued over the validity of the Commission's findings. Dozens of books and countless articles have been written to show, for example, that Lee Harvey Oswald had nothing to do with the event, or that he was part of a conspiracy with the CIA or other parties in planning the assassination. Some of the books, such as Mark Lane's *Rush to Judgement,*[2] were best sellers. In December 1966 *Esquire* published an article[3] listing 35 different theories that had been advanced by as many authors, each suggesting a variation on the Warren Commission's official scenario of the assassination. And since then, many more theories have appeared.

In the light of such a long history of unsettled controversy, the reader might well wonder why yet another author would feel moved to write on the subject. The reasons are quite simple; in the first place, I continue to read, and to hear on radio and television that, "The laws of physics require that the President must have been shot from the front, whereas the Warren Commission places his assassin, Lee Harvey Oswald, behind him."

Such statements involve the backward snap of the President's head, immediately after the shot that killed him. I will show, both theoretically and experimentally, that such statements are simply incorrect; the laws of physics are more in accord with the conclusions of the Warren Commission than they are with the theories of the critics.

My second reason for writing this report is to show how an experienced physicist attacks a new problem. Textbooks tend to indicate that problem solving in physics is a straightforward matter; one proceeds step by step from the input data to the final answer. But in real life, as I will show, a physicist makes many mistakes, and backs up to correct them, one by one. (To those who feel the personalized style of this report is an uncorrected error, I apologize; the earliest version was intended only for a few friends, where the liberal use of personal pronouns wouldn't cause offense. When the report was finally finished, the task of squeezing all the first person singular pronouns out of the text seemed too formidable, so the author hopes the reader will accept his apology.)

After a decade of exposure to the various theories of the assassination, I have at least one advantage over the earlier writers. I've watched each new writer in turn criticize the earlier ones for speaking authoritatively in areas in which they weren't experts. I will, therefore, speak with authority only in areas in which a judge would most probably accept me as an "expert witness." For this reason, the reader will be spared any thoughts of mine on conspiracies, medical reports, the CIA, or ballistics. I haven't counted the number of times I have agreed with, or disagreed with the Commission's findings; I've done both in several different instances.

One of the aspects of physics that makes it appealing to those of us who practice it as a profession is that calculations and the results of experiments can be repeated at will. So all of the interesting observations I've made on the Zapruder assassination movie film can be repeated by anyone sufficiently interested in such matters. (And all of them have been duplicated at least once by others.) Most of the conclusions I reach will seem reasonable to physicists, but in one case I will simply give my "best guess," and not try to do any more persuading.

This report will cover my analysis of several events appearing in the assassination film, some theoretical calculations relating to the "head shot," and some firing range experiments that validated the theoretical conclusions based on the laws of physics as I have taught them for the past 40 years. My observations, analyses and conclusions also relate to the timing of the shots, the speed at which the camera was

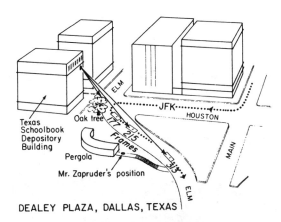

DEALEY PLAZA, DALLAS, TEXAS

Fig. 1. Dealey Plaza, Dallas, Texas, 22 November 1963. President Kennedy's route is shown down Main Street and Houston Street, turning onto Elm Street, in front of the Book Depository Building, where Lee Harvey Oswald was employed. Mr. Zapruder photographed the President's car throughout its passage along Elm Street, until it disappeared under an overpass. Physical evidence will be presented for three shots, at Zapruder frames 177, 215, and 313.

running—both matters of some dispute, and to a sharp deceleration of the President's car just before the President was killed. To the best of my knowledge, this strange behavior on the part of the President's driver has gone unnoticed by everyone else; I suggest a reason for it.

In pointing out some conclusions that seem persuasive to me as a physicist, I do not wish to give the impression that I think that a physicist's way of arriving at "the truth" is the best way or the only way. It works well in the world of physics and so long as I confine my attention to the physical evidence in the Kennedy assassination, I feel that my conclusions can be of help in elucidating what took place in Dealey Plaza, Dallas, on 22 November 1963 (see Fig. 1).

II. THE FILM, THE COMMISSION, AND THE CRITICS

A remarkable moving picture record of President John F. Kennedy's last living moments was taken by Abraham Zapruder in Dallas on 22 November 1963. The Zapruder film was viewed several times by the Warren Commission, and extensive testimony was presented to the Commission by FBI photoanalysts who had made detailed studies of the film, frame by frame. Nevertheless, a good many substantive observations were missed by the photoanalysts, and some of the information they gave to the Commission was incorrect.

With the publication of the 26-volume series containing the evidence presented to the Warren Commission,[1] together with a transcript of the hearings, a group of "Warren Commission Critics" came into being. These critics, or assassination buffs as they are sometimes called, have gone over the voluminous "exhibits" with fine-toothed combs, and have found many errors and contradictions. The assassination buffs attribute most of the errors to more than the sloppiness of a rapid publishing effort; they feel that the Warren Commission didn't do a thorough enough job in investigating many leads, and some of them take the position that the Commission actually ignored or suppressed evidence that Oswald was part of a conspiracy.

I was quite unaware of the strong criticism of the Warren Commission's actions when I first drew some conclusions from a study of the Zapruder film. A simplified and not too convincing report on my analysis of the timing of the shots was presented in a four-hour CBS documentary television program, "The Warren Report," 25–28 June 1967, the text of which is reproduced in Stephen White's book on that documentary.[4] It is difficult to explain a rather technical matter to a lay audience, and in a short space of time. I hope that the lifting of such limits in this report will permit me to explain the methods I used and the conclusions I drew.

III. HOW MANY SHOTS WERE FIRED, AND WHEN?

Publication of the Warren Commission Report and its supporting documentation initiated an intense controversy involving the timing of the shots. Witnesses testified that as few as two and as many as six shots were fired.

The Commission, noting among other bits of evidence, the presence of three spent cartridge cases on the sixth floor of the Book Depository Building near the abandoned Mannlicher-Carcano rifle, concluded that three shots had been fired by Oswald. They decided that one of the shots missed the car; this missing shot could have been either the first or second one fired, but the Commission favored the hypothesis that the second shot was the one that missed. The Commission decided that of these two early shots, the first one probably passed through the President's body before wounding Governor Connally of Texas, who was riding on a "jump seat" just ahead of the President, and the third one struck and killed the President in frame 313. Governor Connally stated quite positively (in the 25 November 1966 issue of *Life*) that he wasn't wounded by the first shot; his testimony was based on his recollection that he heard a shot, turned around, and was later wounded. His story agrees better with the shot timing to be developed in this section, which in turn is not in conflict with the Commission's "allowed but not favored" conclusions. My reasons for preferring physical evidence to the recollections of even the best witnesses are highlighted by noting that the Governor was not even aware that he had received bullet wounds in his wrist and in his thigh until after he had been admitted to the hospital and operated upon.

Several years after I wrote the previous sentence, I read a fascinating article in *Scientific American* by a man who qualified as an expert on the reliability of "eyewitness testimony." Robert Buckhout wrote[5]:

> "Eyewitness testimony is unreliable. Research and courtroom experience provide ample evidence that an eyewitness to a crime is being asked to be something and do something that a normal human being was not created to be or do. Human perception is sloppy and uneven, albeit remarkably effective in serving our need to create structure out of experience. In an investigation or in court, ... [the prosecution and the defense], and usually the witness, too, succumb to the fallacy that everything was recorded and can be played back later through questioning."

The above-mentioned issue of *Life* arrived on the day before Thanksgiving, and because of it I got very little sleep that long holiday weekend. It contained a set of reproductions in color of selected frames from the Zapruder film, illustrating the controversy between the Commission and the Governor. With my many years of experience in analyzing bubble chamber film, plus some moonlighting activities in photographic detective work as a background, I soon found myself completely engrossed in the Zapruder frames. My first observations and their subsequent "explanation" turned out, as I showed later, to be quite incorrect. But by the time I knew my first conclusions were wrong, I had devoted so many hours to a study of the pictures that I was subsequently able to see some things that I do believe have significance.

My attention was drawn to the way the flag, at the left front fender of the President's car, changed its shape from frame to frame in the *Life* photographs. I remembered that at Almagordo, Enrico Fermi had almost instantly measured the explosive yield of the first atomic bomb by observing how far small pieces of paper which he "dribbled" from his hand, were suddenly moved away from "ground zero" by the shock wave. (He had a precomputed table of numbers in his pocket, so he knew the explosive energy of the bomb long before any of the official measurements had been analyzed.) I thought I detected a deformation of the Presidential flag under the influence of the shock wave generated by a nearby bullet. From an elementary calculation involving the known properties of shock waves from bullets, and an assumption as to the surface density of the flag, it seemed to me reasonable to believe that the motions I detected were indeed due to the action of shockwaves. If such a conclusion could be confirmed, the vexing questions concerning the timing of the shots might be solved. (My knowledge of the strength of shock waves from bullets came from an experience I had in World War II, with W. K. H. Panofsky, who had built and was testing a "firing error indicator." This device was towed behind a plane, in a "sleeve," at which gunners fired for practice. It contained two microphones that recorded the shock waves from passing bullets.)

The frames reproduced in *Life* showed a total of only 1.3 sec of the critical moments in Dallas, so I had to wait until the following Monday to examine the sequence of 160 frames in the Law School Library's copy of the Warren Commission "exhibits."[6] When I saw the full set of frames, it was clear that the flag was simply flapping in the breeze. But the thought that effects of the individual bullets might show in the film was still very much in my mind. As I scanned the selected color photographs in *Life* and the full set of black and white copies in the exhibits, I noticed a striking phenomenon in frame 227 (Fig. 2). All of the innumerable pointlike highlights on the irregular shiny surface of the automobile were stretched out into parallel line segments, along the "8 o'clock–2 o'clock" direction. In the plane of the automobile, the parallel streaks appeared to be about 10 in. long.

To appreciate the significance of the streaks, one must remember that each frame of moving picture film is not an instantaneous snapshot, but a time exposure that lasts for about one-thirtieth of a second. For a point of light on the car to be spread out into a streak on the film, the optical axis of the camera must have an angular velocity relative to the line joining the camera and that point of light. If most of the frames had shown streaking, one would simply have concluded that Mr. Zapruder was a "sloppy tracker" who couldn't follow the motion of the President's car as it moved

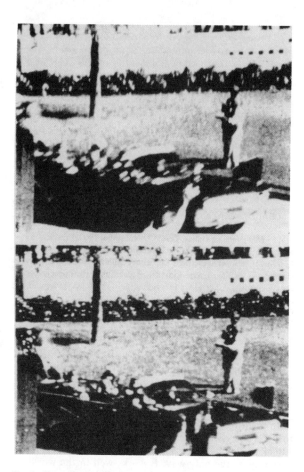

Fig. 2. Zapruder frames 227 (top) and 228 (bottom). Note that the highlights on the car which appear in frame 228 as points, are drawn out into streaks (along the 8 o'clock–2 o'clock direction) in frame 227.

past him, as he "panned" his camera to keep the President in his field of view. But the highlights showed as sharp points of light in most of the frames.

If we "transform" to a rotating coordinate system in which the car and the camera axis are at rest, we can better understand the significance of the streaks. In this system, a streak means that the camera axis has an angular velocity relative to the coordinate axis, and this means that a torque has been applied to the camera to produce the angular acceleration that gave rise to that angular velocity. Such a torque could be produced by a muscle spasm, or by a passing shock wave from a bullet. (I guessed that the frightening crack of a bullet in Dealey Plaza would set Zapruder's neuromuscular system into a temporary spasm. This phenomenon was demonstrated in the CBS documentary series, as we shall see.) For a long time, I thought that I had been the first person to attribute significance to the streaks I've just mentioned. But apparently Harold Weisberg did it first in his book *Whitewash*.[7]

My interest in moving picture camera jitter arose when I was photographing animals in Africa in the summer of 1962. I was bothered by my inability to suppress all visible jitter in a long focal length movie camera used without a tripod, and I started thinking of ways to build optical compensators so that hand-held movie shots would not exhibit the jitter that usually distinguishes amateur movies from those made on tripods by professionals. One night in Nairobi, I invented a solution to the problem. The Bell and Howell Company, which incidentally built Zapruder's camera, was supporting my development of working models of the movie camera stabilizer at the time the President was shot, and my U.S. camera stabilizer patents are owned by Bell and Howell. In the course of my work in movie camera stabilization, I learned that the jitter frequency of a hand-held optical device does not depend to first order upon the weight or the moment of inertia of the device, in spite of what a physicist's intuition would suggest, but instead depends mainly on the time constants of the neuromuscular-feedback system. Most people have a peak in their jitter power spectrum at about 3 cycles/sec. As we shall soon see, this frequency appeared in Zapruder's jitter spectrum when his neuromuscular system was set into oscillation—presumably by the sharp "crack" of the bullets.

Many people who have heard of my observation of "streaks" in the Zapruder film have concluded that the presence of such streaks is the important phenomenon, and that if someone tabulated the frames showing streaking, he would be repeating my observations. Even though CBS presented the data in this highly oversimplified manner, the presence of the streaks simply indicates that the angular velocity of the optical axis of Mr. Zapruder's camera (about a nearly vertical direction) did not match the angular velocity of the President's car, as it drove down Elm Street (Fig. 1). Such a mismatch in the two angular velocities would cause the image of the car on the 8-mm film to move relative to the edges of the "filmgate," during the roughly 30-msec exposure, and this motion would give rise to the streaking of the pointlike highlights. It is obvious that no information of any importance can be attached to such streaking, because no one can perform "hand tracking" accurately enough to avoid all streaking.

My observations involved the measurements of the streaking, but I didn't plot the meaningless streak length—proportional to the mismatch in angular velocity, $\Delta\omega$—but instead, the angular acceleration, α, averaged over two successive frames. Under normal conditions, when $\Delta\omega$ is large enough to give appreciable streaking, the angular acceleration—given by the *difference* in the lengths of the streaks in two successive pictures—is too small to be measured, since the streak lengths in successive frames are almost equal. The plot I made and showed to my friends at CBS is reproduced in Fig. 3. The frame number runs vertically, as on the film itself, and the angular acceleration of the camera axis is plotted horizontally. Since each measure of α involves the subtraction of streak lengths, $\Delta\omega_{n+1}$ and $\Delta\omega_n$ on two successive frames, the value of $\alpha_{n+1/2}$ is plotted at a "half integral frame number," midway between the two frames whose subtracted streak lengths are involved. In order to find α, one needs to know the "sign" of each of the two $\Delta\omega$'s to be subtracted. In other words, we must find out for each streaked frame whether the camera axis was moving toward the back or toward the front of the car. It turns out that the sign of $\Delta\omega_n$ can be found quite unambiguously, simply by observing where the camera was pointing on the $n-1$ and the $n+1$ frames. When I was assigning a plus or minus sign to each of the $\Delta\omega$'s by this technique, I found that the only place this technique didn't work was for frames 314 and 315. A closer examination showed that the numbering of these two frames had simply

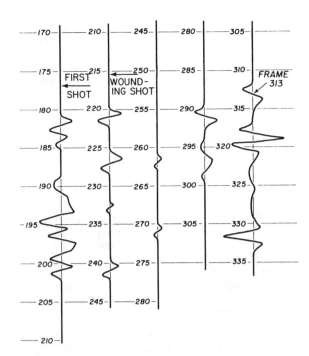

Fig. 3. Angular acceleration of Mr. Zapruder's camera, frame by frame. The frame numbers run vertically from 170 through 334. The angular acceleration for the $n + \frac{1}{2}$ frame is plotted as abscissa, in arbitrary units. Each such acceleration is determined by subtracting the length of the streak in the nth frame from that in the $n + 1$ frame, after assigning an algebraic sign to the streak length in each frame. (See text for details.) Accelerations plotted to the left are "clockwise looking down." Shots are associated (in the text) with pulse trains starting at about 182, 221, and 313.

been interchanged in the "exhibits," and when they were properly labeled, the signs of all $\Delta\omega$ could be determined without ambiguity. Although I later found that the interchange of these two frames was well known to the assassination buffs, the manner in which I detected it convinced me that my determination of the signs of the $\Delta\omega$'s, and therefore the signs and magnitudes of the α's were completely objective.

Figure 3 is a reproduction of my original graph of angular acceleration versus frame number. Angular accelerations plotted to the left correspond to motions of the camera axis that are "clockwise looking down." (The motion of the car and of bullets from the Book Depository are also clockwise looking down, as seen by Mr. Zapruder.) Thus the torque acting on the camera between frames 312 and 313 was "negative," meaning that it could have been caused by a direct interaction of the shock wave from the bullet that hit the President in frame 313, with the left hand side of Mr. Zapruder's camera. (This is important because the impact of the bullet can be seen in frame 313, and there isn't enough time available for the relatively sluggish neuromuscular system to have produced the observed torque on the camera axis.)

When I saw Fig. 3 for the first time, I felt confident that the trains of pulses of angular accelerations were largely the results of the excitation of Zapruder's neuromuscular system, by the sounds of bullets in Dealey Plaza. I had no experimental data to show that a camera would undergo such violent angular accelerations if held by a person who was startled by the sound of gunfire. But such a test was made for CBS by a firm well known to physicists—Edgerton, Germeshausen, and Greer—and films of the test were shown on the CBS program. While the audience watched, cameras held by two separate cameramen shook quite violently in response to gunfire, as Walter Cronkite was saying,[8]

> "Just as a rough check on [the Alvarez] theory, we decided to try it ourselves, using other cameramen holding similar cameras, standing on a rifle range, filming an automobile while a rifleman fired over their heads.

> "These two volunteers are aiming their cameras at a parked limousine. Their instructions: 'Hold the cameras as steady as possible, and keep filming no matter what happens.' The shots will come between them and the car. The cameramen are as far from the firing platform as Mr. Zapruder was from the sixth floor of the Book Depository. [Sound of gunfire in background.]

> "The reaction was obvious. The film taken by these cameramen showed the effect of the shots, despite instructions to hold steady. Even in steadier hands, motion was always noticeable. This frame shows highlight dots around the car's windshield. In reaction to a shot, the dots changed to crescents. And in the following frame they became streaks, comparable to streaks found in some frames from Mr. Zapruder's film."

In view of these tests, I feel that few persons would now dispute the cause and effect relationship between the shots in Dealey Plaza and at least some of the trains of streaks in Mr. Zapruder's otherwise well-tracked movies. If we accept this relationship, we can use the locations of the trains of streaks to shed useful light on the important question of the timing of the shots. No conclusions of the Warren Report have been so disputed as those concerning the timing of the shots, and the damage done by each bullet. Most observers remembered that three shots were fired, but the recollections embraced a range from two to six. Three spent cartridge cases lay on the floor by Oswald's Mannlicher-Carcano rifle abandoned near the sixth floor window of the Book Depository, overlooking Dealey Plaza. According to the Warren Commission Report, p. 110,

> ". . . the nearly whole bullet discovered at Parkland Hospital [to which the President was taken directly from Dealey Plaza] and the two larger fragments found in the Presidential automobile, which were identified as coming from the assassination rifle, came from at least two separate bullets and possibly from three."

One of the "boundary conditions" on the timing of the shots (assuming there were three—one from each ejected cartridge) was the FBI's finding that a skilled marksman could not space his shots more closely than 2.3 sec, or 42 frames of Mr. Zapruder's camera, with its measured frame rate of 18.3 per second. (I will discuss the frame rate later in this article.)

No problem was involved in deciding when the third and fatal bullet was fired; the gory photograph labeled frame 313 settled that question quite conclusively. The fates of the first and second bullets were debated at length by the Commission, and the following conclusion emerged: a bullet, fired in a one-second interval between frames 206 and 225, wounded the President by passing through his neck, and then wounded Governor Connally, who was seated just ahead of the President. This so-called "single bullet theory" as we have already learned, was later challenged by Governor and Mrs. Connally.

The Commission decided that the other bullet was never recovered, and after giving reasons to suggest that it could have been fired either before or after the shot that was identified as wounding the two men, the Commission favored the suggestion that the unrecovered bullet was fired after the one that wounded them.

If we now look at Fig. 3 in the light of this background material, we see that the obvious shot in frame 313 is accompanied immediately by an angular acceleration of the camera, in the proper sense of rotation to have been caused directly by shock-wave pressure on the camera body. The human nervous system cannot transmit signals fast enough for the angular acceleration between frames 312 and 313 to have been caused by Mr. Zapruder's muscles reacting to impulses from a brain that had been startled by the shot that killed the President. The expected neuromuscular reaction occurs about one-quarter to one-third of a second later, as shown by the large accelerations near 318. (I'll adopt five frames as Mr. Zapruder's experimentally determined reaction time, for reasons to be discussed later.) Another large acceleration peak occurs about two-thirds of a second after this group, so we observe three out of a possible four pulses spaced very nearly the canonical one-third of a second apart. For those readers who are surprised that the neuromuscular response time is so long, let me recall a common "parlor trick": A bets B that if A drops a vertically held dollar bill without any warning, B cannot stop its fall by pinching his fingers together, if his fingers are poised, ready to clamp together, at the bottom edge of the bill. The fact that the bill can almost never be stopped (unless A gives a precursor signal with his fingers) indicates that a nervous system "on hair trigger" takes more than one-sixth of a second (3.1 frames) to respond to an optical stimulus.

If we look between frames 206 and 225, the one-second interval in which the Commission suggested the "wounding shot" was fired, we see the start of a one-second-long train of pulses, spaced very nearly one-third of a second apart. We further note that the initial pulse of the series, at 221.5, is not in the proper direction to have been caused by a direct interaction of the shock wave with the camera; the camera turns toward, rather than away from the shock wave. The shock wave from a bullet fired from the Book Depository toward the car in its position at the time of frame 221 would have been considerably weaker at Mr. Zapruder's station than the shock wave in frame 313, so the lack of a direct physical interaction at the time of this earlier shot is not surprising. I therefore conclude that the accelerations at 220.5 and 221.5 were caused by Mr. Zapruder's neuromuscular response to an earlier stimulation. If we use Mr. Zapruder's thereby observed oscillation period of about five frames (which is close to the expected value), we place the "wounding shot" at about 215.5. I find it most interesting that although the determination of 215.5 as the frame number of this shot was derived directly from the appearance of the streaks, it is exactly halfway between two limits, only one second apart, set by the Warren Commission from very different data.

If we convert the Commission's language into the vernacular of the physicist, their conclusion could be stated: "The bullet that wounded the President and Governor Connally occurred at frame 215 ± 10." Although I would not have expected the conclusions of two such different studies to agree so closely, it is true that my estimated frame number for one of the two disputed shots agrees with the Commission's best estimate to within less than one-tenth of a second. The Commission based its findings largely on an examination of what the people in the car were doing; President Kennedy "seemed to be reacting (in frame 225) to his neck wound by raising his hands to his throat."[9]

I will ignore the two small accelerations between frames 245 and 280; each is caused by a single frame in which I judged that highlights might be smeared slightly more than the normal smearing caused by the imperfections of the half-tone process. I will return later to the short sequence of significant pulses starting at 290 since they require an explanation. They seemed to me to have less intensity, and to last a much shorter time than the three sets of pulses I identified as being triggered by bullets. I eventually found what I think is a reasonable explanation, not only for these angular accelerations, but also for a puzzling deceleration of the President's car at the same time—but that is getting a bit ahead of the story.

Because of the quietness of the acceleration graph between the pulse trains starting at 221 and 313 (except for the pulses which I feel have other explanations), and because of the obvious train of pulses starting at 182, I favor the view that the Commission's "missing shot" initiated this first train of pulses. My best estimate of the time of this shot is therefore 182 minus 5 (for Mr. Zapruder's calibrated time delay), or frame 177.

The Commission noted that about that time, the President's car was partially obscured from the sixth floor window, as it passed under a large tree. In a very thorough reenactment session in Dealey Plaza, photographs were taken by the FBI from the window near which the rifle and three spent cartridge cases were found. A limousine was moved along Elm Street, into positions corresponding to known frame numbers, and the Commission report reproduced sample groups of corresponding pictures: (1) from Mr. Zapruder's camera, (2) from the FBI camera in the sixth floor window showing the appearance of the limousine and a man sitting in the President's seat, and (3) from an FBI camera with a field of view equal to that of Mr. Zapruder's movie camera, located at the position from which he photographed the assassination. The FBI pictures corresponding to frames 166 and 186 are reproduced in the Commission's report, and both show that the President was clearly visible through the branches of the intervening tree in both views. It appears that the President had been unobscured before 186, during which time the gunman would have had a good opportunity to track him, and match the angular velocity and angular position of his gun with that of the President's body. The fact that the President's head might have been partially obscured by branches for one-half a second, at frame 177, would not, in my opinion, have had any appreciable effect on the gunman's tracking

ability, or feeling of confidence that his aim was good. Anyone who has ever driven a car in a heavy rainstorm, with a slow windshield wiper will realize that a partial loss of visual acuity for a half-second would not seriously affect a gunman's ability to perform good tracking, particularly when most of the car was still clearly visible through the holes in the trees. And if we remember that the decision to squeeze the trigger must have been made a few tenths of a second before the bullet was fired, the effect of the obscuring tree should have been negligible on the actions of the gunman, for a shot fired at frame 177.

I find it strange, on reading the testimony of experts on firearms (which I certainly am not), that they all looked at the photographs taken through the trees and testified whether or not a gunman could have fired at particular frame numbers. They treated the subject as though it was static—as though the gunman was presented with a stationary target behind a tree. They looked at the still photographs taken from the window in this static way, and decided that the gunman could have fired at certain frame numbers (when the President's body showed through a hole), but not at other times, when it was eclipsed. I can appreciate how they could have said such things under the stress of the investigation, when asked to comment on a set of still pictures, but I am surprised that no one mentioned what the real situation was like, with a large moving object containing a specific target fixed in its moving frame, that had a very nearly constant angular velocity with respect to the gunman. I don't believe a gunman would have been deterred from firing at frame 177, and I consider it most likely that the shot fired at that time was the one the Commission concluded missed the car and was unrecovered.

To return to the FBI's (assumed) minimum possible firing interval of 2.3 sec, we should compare this time with my best estimate of the time interval between what I identified as the first two shots. From frame 177 to frame 216 is 2.13 sec. To make this conform to the 2.3-sec limit, it is only necessary to change the timing of the two shots by one and a half frames each; if the first occurred at 175.5 and the second at 217.5, the time interval would be $42/18.3 = 2.3$ sec. Such a procedure of altering estimated numbers within their known errors is a standard technique in my own physics specialty of bubble chamber event analysis. We have complicated computer programs that alter measured angles and measured momenta of tracks (within the known errors) to match the constraints imposed by the laws of conservation of energy and momentum. Just as a bubble chamber physicist uses a "fitting routine" to make his events match a known constraint, I have shown that I can fit the 2.3-sec time interval constraint by two small adjustments in estimated frame number. Since the two changes of ± 1.5 frames are small compared to the extrapolation of five frames each, made to arrive at the two unfitted estimates, and since no one would really believe that such extrapolations were more accurate than 1.5 frames, I believe that the fitting procedure is justified. However, if the reader dislikes this fitting procedure, he can still accept my "unfitted estimates," by learning that the CBS tests turned up a "technician who had one hit and two misses" (at a moving car, in a three-dimensional mockup of the Dealey Plaza) "in 4.1 sec."[10] This is remarkably like the apparent performance of the marksman identified by the Commission as Lee Harvey Oswald and reduces the permissible time interval to 2.05 sec, which is within my unfitted estimate of 2.13 sec.

Let me now summarize the conclusions of this section. By an analysis of "streaks" in the Zapruder film, I identified the precise timing of two shots that had been pinpointed by other means by the Warren Commission. So far as I know, there is no real controversy concerning the timing of these two shots. I found evidence that convinced me that a third shot was fired at about frame 177. This firing time is allowed by the findings of the Warren Commission, even though they favored the idea that the "third shot" was fired between the two that they identified as surely hitting President Kennedy. And finally, this firing sequence is consistent with the memories of Governor and Mrs. Connally.

What limitations can be placed on these observations? If, as many people have suggested—and continue to suggest—two shots hit the President almost simultaneously from opposite directions, at frame 313 and very shortly thereafter, could I have detected this multiple firing? The answer to that question is "no." To be detected by the "streak method," two shots must be spaced by about 2 sec to be resolved as two separate shots, rather than a single shot followed by a slower than normal recovery time for Mr. Zapruder's neuromuscular system. But in the next section, I will be able to shed some light on the question of the "shot from the front."

I was bothered for some time by the weaker set of pulses lasting a shorter time, that show in Fig. 3, from frames 290 through 298. They don't look like the ones that seemed clearly associated with bullets. But obviously they required an explanation. I'll give my best explanation for them in the final section of this report, but I don't feel as certain about that explanation as I do about the other three cases.

IV. WHY DID THE PRESIDENT'S HEAD SNAP BACKWARD AFTER THE FATAL SHOT?

I must apologize for the tone of the following section, which may sound cold blooded and devoid of human feeling. My long delay in publishing this analysis derives largely from my feelings of inadequacy after many attempts to soften its impact. But I am finally convinced that the conclusions I reach in this section are important, and I have therefore done my best to make the text as free from emotional content as possible. John Kennedy was one of my personal heroes, and I had the pleasure of talking with him on two occasions. His death touched me deeply, and I hope the reader will bear that in mind as he studies this section.

Paul Hoch, who was then a graduate student at Berkeley, tried to interest me in one of the hottest and longest surviving controversies arising from a study of the Zapruder film. (It was the subject of several radio and television shows in April 1975, and testimony concerning it was taken during the Congressional Hearings on the CIA, in June 1975.) This controversy involves the unexpected behavior of the President's head immediately after it received the final and mortal shot. Everyone who studied the behavior of the people in the Zapruder film agreed that immediately after this shot, the President's head and body moved suddenly backward. The sixth floor window of the Texas Book Depository Building was behind the car, and the Warren Commission concluded that Lee Harvey Oswald shot the President from that window. Why then did the President's head recoil toward, rather than away from the gun as the

laws of physics would seem to demand? The assassination buffs argued at length about this action. I shall mention only three persons out of a great many who concluded in writing that the President was shot from the front. In his *Rush to Judgment*,[2] Mark Lane said, "So long as the Commission maintained the bullet came almost directly from the rear, it implied that the laws of physics vacated in this instance, for the President did not fall forward." Josiah Thompson, Professor of Philosophy at Haverford College, wrote a book that devoted a good deal of space to this problem.[11] He concluded that immediately after the President was wounded in the head from behind, another bullet fired from in front of the car hit his head and drove it back, by momentum conservation, toward the rear of the car. District Attorney James Garrison of New Orleans made similar claims in the highly publicized trial of Clay Shaw, in 1969. The thrust of all these arguments is that if the President was shot from two directions, almost simultaneously, there must have been a conspiracy, in contradiction to the Warren Commission's basic conclusion that Oswald acted as an independent agent.

Paul Hoch often pressed me for an explanation of the odd behavior of the President's head, and although I hadn't observed it myself, I usually suggested that the head had probably been held erect by muscles controlled by the brain, and that when the controls were suddenly damaged, the head fell back. I was finally convinced that this explanation was incorrect after Paul Hoch handed me a copy of Thompson's book as I was leaving Berkeley for the February 1969 meeting of the American Physical Society in St. Louis. On the plane I had time to study the book carefully. It is beautifully printed, with excellent photographs and carefully prepared graphs. When I studied the graph showing the changing position of the President's head relative to the moving car's coordinate system, I was finally convinced that the assassination buffs were right; there had to be a real explanation of the fact that the President's head did *not fall* back, but was *driven* back by some real force.

And the answer turned out to be simpler than I had expected. I solved the problem (to my own satisfaction, and in a one-dimensional fashion) on the back of an envelope, as I sat in solitary splendor in the beautiful suite that the St. Louis hotel management supplied me in my capacity as president of the APS.

I concluded that the retrograde motion of the President's head, in response to the rifle bullet shot, is consistent with the law of conservation of momentum, if one pays attention to the law of conservation of energy as well, and includes the momentum of *all* the material in the problem. The simplest way to see where I differ from most of the critics is to note that they treat the problem as though it involved only two interacting masses: the bullet and the head. My analysis involves three interacting masses, the bullet, the jet of brain matter observable in frame 313, and the remaining part of the head. It will turn out that the jet can carry forward more momentum than was brought in by the bullet, and the head recoils backward, as a rocket recoils when its jet fuel is ejected. (Col. William H. Hanson came to the same conclusion, independently.[12])

If a block of wood is suspended by strings from the ceiling, it is called a ballistic pendulum, and physicists or gunsmiths can calculate the velocity of a bullet shot into it to be

$$v_B = v_W M_W / M_B, \qquad (1)$$

where v_M is the velocity of the wooden block after it stops the bullet, and M_W and M_B are the masses of the wooden block and bullet. Equation (1) follows directly from the law of conservation of momentum:

$$v_B M_B = v_W M_W. \qquad (2)$$

In using a ballistic pendulum, we normally forget that the collision of bullet and wooden block is very inelastic. Of the incoming kinetic energy of the bullet, only a small fraction f appears as kinetic energy of the moving wooden block; the remaining fraction $(1 - f)$ goes into heating the wood. If $M_B \ll M_W$,

$$\text{KE}_W = f\,(\text{KE}_B),$$
$$M_W v_W^2 / 2 = f \times M_B v_B^2 / 2. \qquad (3)$$

From (3) and (2),

$$f = M_B / M_W. \qquad (4)$$

For the case of a 10-g bullet, and a block weighing 10 kg, it can be seen that 99.9% of the incoming kinetic energy goes into heating the block, and only 0.1% appears as mechanical energy. Ballistic pendulums are designed so that they contain the inelastically dissipated energy. Unfortunately, the human head is not able to contain the major fraction of the energy carried in by the bullet. This tragic aspect of the assassination is clearly visible in frame 313 of the Zapruder film, and is discussed in detail in the reports of the autopsy surgeons.

The mechanism of the retrograde recoil turns out to be rather simple, if one remembers that 99.9% of the incoming energy must be accounted for. The momentum associated with a given amount of kinetic energy varies as the square root of the mass of the object carrying that kinetic energy:

$$p = (2MK)^{1/2}, \qquad (5)$$

where p is the momentum, and K is the kinetic energy of the object with a mass M.

Figure 4 shows what happened when my friends and I fired bullets at melons that had been wrapped with Scotch glass filament tape, to mock up the tensile strength of the cranium. Under the influence of the bullet, some of the material making up the melon breaks through the reinforcement, and carries momentum in the forward direction. (Frame 313 of the Zapruder film shows this same phenomenon.) As we shall now see, the momentum carried forward in this way can be much larger than the momentum brought in by the bullet. For example, if the bullet weighed 0.1% of the melon weight, and if 10% of the incoming kinetic energy was used to propel 10% of the mass of the melon forward, then the momentum of the jet expelled forward would be $(10)^{1/2}$ times that of the incoming bullet. (I will use subscripts, b for bullet, j for forward moving jet, and m for melon.)

$$p_j = (2M_j K_j)^{1/2} = (2 \times 100 M_b \times 0.1 K_b)^{1/2}$$
$$= (10)^{1/2} (2M_b K_b)^{1/2} = (10)^{1/2} p_b, \quad (6)$$

since $M_j = 0.1 M_m = 100 M_b$, $K_j = 0.1 K_b$. The melon would then recoil backward with about twice the velocity it would have been expected to go forward, assuming it were made of wood. This is because the melon, acting at first as a ballistic pendulum, acquires a forward velocity equal to $v_m |_{\text{BP}}$

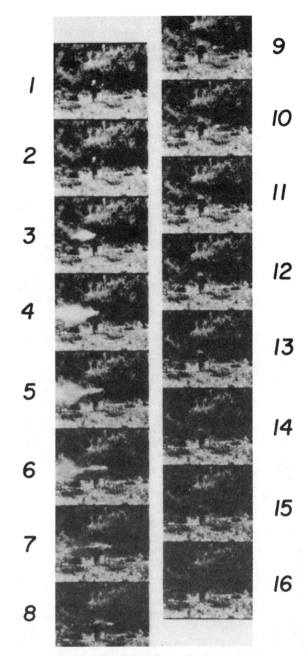

Fig. 4. Retrorecoil in a tape-reinforced melon hit by a high-velocity bullet. The bullet came from a rifle off the right-hand side of the frames. The forward jet (to the left) propelled the melon "backwards." (See text.)

$= p_b / M_m$. (The notation $v_m |_{\text{BP}}$ means the velocity one would expect the melon to have if it contained all the kinetic energy of the bullet, as a ballistic pendulum does.) But in the center of mass system of the melon, which is moving "forward" with the expected velocity, a jet moves forward with momentum equal to $(10)^{1/2} p_b$—as we have just seen. It gives the melon an equal and opposite momentum, in the moving (CM) system; in that system, $p_m = -(10)^{1/2} p_b$. If we neglect the 10% loss of mass by the melon to the jet, the recoil velocity of the melon (in the CM system) is

$-(10)^{1/2}$ times the "expected value." Since velocities add vectorally, the final velocity of the melon (in the laboratory system) is $[1 - (10)^{1/2}] v_m|_{BP}$. Since the square root of 10 is close to 3.16, the observed velocity of the melon is about $-2v_m|_{BP}$.

If one wants to know more about the details of the transfer mechanism of kinetic energy from the bullet to kinetic energy of the fragments thrown forward, he will have to ask someone more knowledgeable in the theory of fluid mechanics than I am. My intuitive feeling is that the conical shape of the interaction zone is the key to the nonnegligible efficiency of energy transfer. (It is clear that an appreciable mechanical energy transfer is only possible if the incoming energy can avoid "being thermalized.") The conical region is defined by the small entrance hole and the much larger exit hole in the melon. Transmission lines with tapered internal conductors are efficient transformers of electrical energy, and a tapered bullwhip can smoothly transform the energy given to a large mass, by the flick of the wrist, into roughly the same energy of a much smaller mass at the tip of the whip. The "crack" of the whip occurs when the tip of the whip goes supersonic. I believe that in a somewhat analogous manner, but of course in the opposite direction, the kinetic energy of the bullet is given in a "tapered region" to a progressively larger mass in the melon, to achieve the modestly efficient energy transfer that is demonstrated in our experiments.

Now that I've given the theory of the "jet recoil mechanism," I'll describe the experiments that gave rise to Fig. 4. When I showed my simple calculations to Paul Hoch, he said that no one would believe my conclusions (including himself) unless we could demonstrate the retrograde recoil on a rifle range, using a reasonable facsimile of a human head as a target. I discussed my theory with my longtime friend and associate at the Laboratory, Sharon "Buck" Buckingham. Buck is an enthusiastic deer hunter, and he offered his services if I would buy the melons into which he would fire the shots.

Buck did his first experiments in June 1969 at the San Leandro Municipal firing range. Before he started shooting, all the expert marksmen in attendance told him that he was wasting his time—one said, "I've been around guns all my life, and you must be out of your mind to believe something you hit with a bullet will come back toward you." Most of the targets were melons that Buck had reinforced by wrapping with 1-in. Scotch "filament tape," as mentioned earlier.

The results of the first test shootings were encouraging in that most of the reinforced melons were driven by their shots toward the gun as I expected, rather than away from the gun "as the laws of physics require."

Paul Hoch expressed an interest in the results of this test, but said that he wouldn't ask his fellow buffs to believe them unless he had photographic evidence to document the case. Paul enlisted the help of Don Olson, another physics graduate student and assassination buff, who had a remotely controlled Super 8 movie camera, and I was present as an observer. We were all impressed to find that Buck's early results could be duplicated before the camera. The performances were now more uniform, with six out of seven reinforced melons clearly recoiling in a retrograde manner toward the gun. (According to Paul Hoch, the other one "just rolled around a bit.")

Figure 4 is an enlargement of a section of the film showing shot number 4. The frame rate is 24 per second. The gun, a 30.06 rifle, is about 30 m out of sight on the right-hand side of the photographs. Its 150 grain hand-loaded soft-nosed[13] bullet hit the melon with a velocity of about 3000 ft/sec; the 6.5 Mannlicher–Carcano rifle found near the sixth floor window of the Book Depository building fired a 160-grain bullet at about 2165 ft/sec. (I am told that at a distance of 265 ft, the measured slant range from the Book Depository window to the President in frame 313, the bullet would have slowed down to about 1800 ft/sec.)

To relate these experiments to the melancholy affair in Dallas, we can use Thompson's[11] carefully measured velocity of the backward motion of the President's head. He finds that it was about 1.6 ft/sec, averaged over eight frames. In Fig. 4, the measured retrograde velocity of the melon is 4.5 ft/sec. It is obvious that if the melon had been hit by a slower bullet, and had been connected to a large mass, simulating a torso, rather than being free of restraint, it would also have moved back more slowly. But in spite of what appears to me to be a good semiquantitative match in velocities, we must remember that the important question at issue here is not the *magnitude* of the velocity, but its *direction!*

I believe that our experimental demonstration of retrograde recoil in head-like objects will convince most people that the laws of physics do not require a second assassin to have been firing at the President from the "grassy knoll," ahead of the car. It is important to stress the fact that a taped melon was our *a priori* best mockup of a head, and it showed retrograde recoil in the first test. If we had used the "Edison technique" and shot at a large collection of objects, and finally found one which gave retrograde recoil, then our firing experiments could reasonably be criticized. But as the tests were actually conducted, I believe they show it is most probable that the shot in frame 313 came from behind the car; after all, the jets visible in frame 313 were what suggested this mechanism to me.

Many of the assassination buffs wrote to Hoch to say that neither my "back of the envelope" numbers nor the experimental results agreed with Professor Thompson's measured head velocities. So, in case any readers of this article may be similarly bothered, I should point out that the three numbers I used in my analysis (two mass ratios and an efficiency) were each assumed to have the value of 10^i, where i is a positive or negative integer. In spite of this highly quantized nature of the input data, the calculated and observed velocities differ by only a factor of 3. The assassination buffs who argued with Paul Hoch in a quantitative way (neglecting the important sign of the velocity) usually suggested that I was assuming that the mass of the jet (10%) was too high. But they missed the fact that, if either this assumed mass ratio or the assumed efficiency of energy transfer were *reduced* by a combined factor of almost 10, the calculated and observed velocities would be equal. In addition, frame 313 shows that the event wasn't one dimensional, as the model was; the two jets visible in frame 313 have vertical components that would lower the longitudinal component of momentum, bringing the theory closer to the actual event. I don't want to be that quantitative; the theory wasn't designed to calculate the velocities to high accuracies—but to show qualitatively that the head could jerk backwards.

I will end this section by saying what I think can be concluded from our experiments. It is possible to disprove a theory, but never to prove one; no matter how often a theory has given correct predictions in the past, a single (repeatable) counterexample invalidates that particular theory. (Newton's theory of gravitation was disproved in this manner.) For these reasons, I believe that those arguments for a second assassin that derive from President Kennedy's head movements after frame 313 are now clearly invalid; a documented counterexample is now available to disprove the assertions of many writers concerning the consequences of Newton's laws of motion. I am convinced that everything that is known about the motion of the President's body in that short time interval is consistent with a shot from above and behind, where the sixth floor window of the Book Depository building was situated. But by the argument given earlier in this paragraph, I obviously can't prove that the bullet came from that window.

Dr. John K. Lattimer recently published an article[14] entitled "Observations Based on a Review of the Autopsy Photographs, X-rays and Related Materials of the Late President John F. Kennedy." Dr. Lattimer was apparently the first physician without governmental credentials to be given access to this material, which had been restricted for more than eight years, at the request of the President's family. Dr. Lattimer's article, published several years after the shooting experiments described above, says

> "These observations, made possible by actually seeing the autopsy photographs and the clothing, (and added to the previous laboratory and autopsy findings) have answered some of the questions that were in the mind of the author and have revealed no incompatibilities with the concept that two high-speed bullets hit the President, both fired downward and from the rear, as from the sixth floor of the Book Depository Building;—There were no signs of bullets or bullet wounds or bullet fragment tracks through the President's body running in any other location or direction, such as transversely, or from the front, to indicate bullet "hits" from any of these directions upon the President's head, body or limbs."

Several critics of the Warren report had predicted that when a "nonestablishment" expert on bullet wounds, such as Dr. Lattimer (with his "questions") was finally permitted to see the autopsy films, the "head shot from the front" would be confirmed. But Dr. Lattimer has ruled it out quite unequivocally.

Although Dr. Lattimer is now classified as a urologist, his biographical sketch[14] shows that he is an expert in the relevant fields:

> "In World War II, Dr. Lattimer was a military surgeon in the European Theater of Operations and had experience with military missile wounds of all types, almost always using X-rays for their localization. He served as a firearms range officer and also did experimental work on the wounding capabilities of various missiles on human tissues."

V. HOW FAST WAS THE CAMERA RUNNING?

Everyone who has watched football on TV knows that it is easy to distinguish a slow motion "instant replay" from the real thing, even when the play-back rate is not much slower than the normal rate. The clues come largely from our memorized knowledge of the oscillation frequency of the legs of runners moving at their fastest possible rates, and from our memory of the way objects fall in a "one g" gravitational environment.

But Mr. Zapruder's camera showed an automobile in which the occupants were for the most part sitting still, together with images of two motorcycle policemen who sat immobile on their seats all the while. The background comprised fixed structures, plus a few spectators who appeared to be standing still as the camera panned past them as it followed the President's car. So the clues we see in "instant replay football" on TV seem to be denied us in the Zapruder film.

If one accepted the FBI's subsequently measured frame rate of 18.3 per second for Mr. Zapruder's camera, the car was moving at a speed of approximately 12 mph. But an FBI report stated that, "The camera was set to take normal speed movie film or 24 frames/sec." Had the camera actually been operating at that rate, it would have been exceedingly difficult—if not impossible—to devise a sequence of Mannlicher-Carcano rifle shots that would have been within human capability, and therefore the multiple gunmen theories—so popular with many of the Warren Commission critics—could not have been ignored. (The higher the frame rate, the shorter is the time between any pair of numbered frames.) The Bell and Howell camera used by Mr. Zapruder had a "normal" button position, and a "slow motion" position, and I believe the intent of the FBI report was simply to answer the question, "Did Mr. Zapruder use normal or slow motion speed in taking his pictures?" Since the normal speed of 16- or 35-mm *sound* moving pictures is well known to be 24 frames/sec, I believe that the FBI was in turn saying, in effect, "He used normal speed." (I am now using my legally acceptable status as a "camera expert" to give an opinion outside the field of physics; I was for several years a salaried consultant to the Photoproducts Division of the Bell and Howell Company.) Actually the "slow motion frame rate" on the Zapruder camera was closer to 48 frames/sec.

I tried for some time to find a way to convince myself that the frame rate was 18.3 per second, and not the much higher "slow motion rate." But as I looked at the pictures again and again, I couldn't find a clue that could distinguish pictures of a car moving at 10 mph, together with some people who moved slowly, from pictures of a car moving at about 30 mph, with the same people still moving slowly, but not quite so slowly. I was about to give this problem up as hopeless when I noticed the action of a man standing beyond the car, as seen by the camera. He was clapping as the President drove by—a gesture that was common in the Kennedy era.

An elementary analysis of the muscle power involved in clapping shows that the power required, for a given maximum hand spacing, varies as the cube of the clapping frequency. The average velocity of the hands varies directly with the frequency, so the energy expended per cycle varies as the square of the frequency. Power is the time rate of expenditure of energy, so it involves an additional factor proportional to the frequency. It turns out that we can use the spectator's apparent clapping frequency, together with his observed and very natural maximum hand separation of about 1 ft, in the same way we use a running back's leg rate, to decide if we are watching live action, or slow motion "instant replay."

The spectator appears to move smoothly across the film from the right-hand edge, and about 1 (assumed) sec later (18 frames) disappears out of view beyond the left-hand edge. His apparent motion is of course due to Mr. Zapruder's panning action to follow the car. The clapping is shown in Frames 278 through 296 (Fig. 5), and even though the man's image is blurred because of the panning, it is evident that he has executed between 3½ and 4 full clapping cycles. I will assume that his apparent clapping frequency is 3.7 cycles/sec, and will ask how much greater this could be—due to a higher frame rate—and still be within reasonable human limits. The key to this particular analysis is the existence of the aforementioned cube law relating clapping frequency and muscle power. If a person doubles his clapping frequency, at constant amplitude, he must expend eight times as much power. The "steepness" of the cube law is what gives one the ability to distinguish film speeds by observations of clapping behavior, but only if normal clapping behavior is not too far from the "power barrier."

To answer this question, I clapped in synchronism with a metronome set at the assumed rate of 220 beats/min. I found I could clap quite comfortably at this rate of 3.7 per second, but I couldn't do so at twice the rate, with the same amplitude; to make 7.4 cycles/sec, which was an obviously unnaturally high rate, I had to reduce my amplitude considerably. I could just make it at 1.5 times 3.7 cycles/sec, but the effort felt quite unnatural. I am confident that anyone who repeats these experiments, as I have just done after a hiatus of several years, will be convinced that Mr. Zapruder's camera was running at very nearly 18 frames/sec. (It was certainly not running at 48 frames/sec, and I believe that 24 frames/sec can be ruled out, as well.) Although there is apparently no longer a serious controversy relative to frame rates, I wanted to share with my physicist readers the pleasure I had in discovering a "cube law clock" in the film.

VI. WHY DID THE PRESIDENT'S CAR SLOW DOWN ABRUPTLY JUST BEFORE THE FATAL SHOT?

The Commission was aided in its interpretation of the films by an FBI photoanalyst, Mr. Lyndal L. Shaneyfelt. My first disagreement with his testimony comes on p. 155 of Vol. V, where he was running the Zapruder film for Allen W. Dulles and John J. McCloy, members of the Commission. After the expert had made a comment relative to frame 222, the following conversation took place:

> *Mr. Dulles:* Jerky motion in Connally in the film.
>
> *Mr. Shaneyfelt:* There is—it may be merely where he stopped turning and started turning this way. It is hard to analyze.
>
> *Mr. Dulles:* What I wanted to get at—whether it was Connally who made the jerky motion or there was something in the film that was jerky. You can't tell.
>
> *Mr. Sheneyfelt:* You can't tell that.

Since Fig. 3 shows some "jerky motion" immediately after frame 222, it is a reasonable assumption that this is what had caught Mr. Dulles's attention. It was too bad that Mr. Dulles answered his own question concerning the possibility of distinguishing between the motion of a man in the

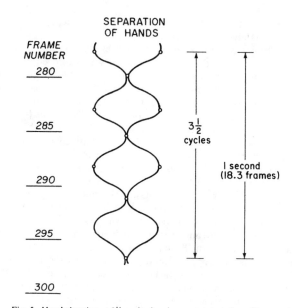

Fig. 5. Hand clapping at 3½ cycles/sec by a spectator allows film speed to be determined, within important limits. (See text.)

car, and a movement of the film (camera) as a whole. Mr. Dulles was an experienced intelligence agent, and his practiced eye caught an important clue, but he too quickly dismissed it as undecipherable, which, of course, we now know it wasn't. The expert photoanalyst put the lid on the matter by his polite endorsement of Mr. Dulles's error.

My second disagreement with this same FBI photoanalyst came when he testified concerning his inability to pinpoint the President's car, at frame 313, by examining the Zapruder film. He had this to say[15]:

> "Yes, I might state first that all of the other (reenactment) photographs were reestablished on the basis of the Zapruder film, using reference points in the background of the pictures.
>
> "As is apparent here from the photograph of the Zapruder frame 313, there are no reference points. There is just a grassy plot. So there is no reference point on which we can reestablish the position of the car in the roadway.
>
> "For this reason it was necessary to use the Nix film of the head shot and the Muchmore film of the head shot to establish this position in the road." [These films were shot from amateur movie cameras located on the opposite side of the street; one of them showed some identifiable background close to Mr. Zapruder's position, including Mr. Zapruder himself, instead of the plain grass that showed at that time in the Zapruder film.]

Mr. Shaneyfelt pinpointed the location of the car in 13 (or perhaps more) frames from 161 to 255, in which interval, there were architectural background features that were easily identifiable in the Zapruder frames. And as he said, the position of the car in frame 313 was determined from the two other films. These data were used in the FBI reenactment studies in Dealey Plaza. An open automobile, similar to the one in which the President rode, was moved in turn to the 14 (or more) positions as determined in the films. At each position, it was photographed (1) by a still camera with the same angular field as Mr. Zapruder's movie camera, from his original location, and (2) from the sixth floor window of the Book Depository building, through the rifle scope of the rifle found at that location immediately after the assassination. For each of these 14 selected frames, the Exhibits[16] show photographs (1) and (2), together with the original Zapruder frames; in the case of frame 313, the corresponding frames from the Nix and Muchmore films are shown, together with still shots of the stationary car from the Nix and Muchmore locations.

In addition to the several pictures corresponding to each of the 14 locations, the exhibits also tabulate various measurements made at the 14 locations. These include the distance of the car from a benchmark on Elm Street ("station C"), the distance between the rear seat of the car and the sixth floor window of the Book Depository building, and the angle of depression of the rifle sight in that window. The distances are given to the nearest tenth of a foot; they are probably accurate to somewhat better than 1 ft.

As any physicist would do, I plotted the tabulated distance of the car (from "station C") against frame number for these 14 selected frames. This graph is shown in Fig. 6, and all the points except that for frame 313 lie on a line with a slope equal to 11.8 mph. It is clear from the dispersion of the (Zapruder) points from a straight line that the final point (determined from the Nix and Muchmore films) does not lie on the extrapolated line. Two explanations are possible; the position of the car at frame 313 was incorrectly determined, or the car slowed down somewhere between frames 255 and 313. Neither of these possibilities seemed reasonable to me when I first saw Fig. 6, so I set myself the task of finding out which explanation was correct. (I did this work, and the analysis of the clapping, during the Christmas vacation following the publication of the November 26, 1966 issue of *Life*.)

The first relevant observation I made was that contrary to what Mr. Shaneyfelt said in his testimony, it was a trivial exercise to determine precisely where the car was at *each* of the 79 frames from where his "Zapruder data" stopped (at frame 255) to the final published frame, number 344. What he apparently failed to realize was that the approximately ten persons who were standing on the featureless background were "reference points" exactly as useful as if they were set in concrete. Their usefulness comes from two independent considerations. There is a linear relationship between any horizontal interval on the original film (or on the half-tone reproductions in the Exhibits) and the corresponding angular interval subtended at Mr. Zapruder's camera. In other words, every time the camera panned through an angle θ, a fixed object in the field of view moved to the left in the picture, a distance of $k\theta$. The value of the constant k (the focal length of the camera lens) could be determined with the aid of an accurate plan of Dealey Plaza, showing Mr. Zapruder's station. (The camera had a zoom lens of variable focal length, which I found had been used at very nearly its longest value.) From such a plan, one can measure the angles subtended by many architectural features, visible in the frames. Those angles, which can be measured with a high degree of precision, can be divided by the accurately measureable corresponding intervals on the film (or on the halftone reproduction) to give the corresponding value of k^{-1}. From then on, we can immediately tell through what angle the camera is being panned, frame

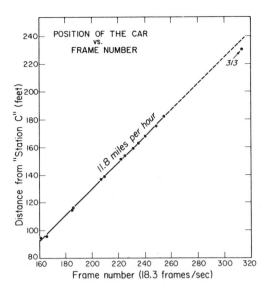

Fig. 6. Position of the President's car as determined by the FBI. Note that point 313 does not lie on extrapolated line.

by frame, by simply measuring the displacement of *any* stationary object in the field of view. That stationary object can be a concrete pole, or equally usefully, a person's foot that is temporarily bearing his weight, and is therefore fixed to the ground.

Since I didn't have an accurate enough plan of Dealey Plaza, I couldn't evaluate k with an absolute uncertainty as small as the relative uncertainty with which measurements could be made on the halftone reproductions. (The FBI could have done that with the theodolite they used in the reenactment session.) But that minor lack of absolute precision will have no effect on the very accurate measurements of the relative speed of the car before and after the strange and previously unseen deceleration I am about to describe. But before describing that event, I should mention that in one sequence, when no spectators are in the background, another interesting reference mark is available on the plain grass behind the car, in frames 313–334, the last ones reproduced in the exhibits. This mark is a white streak, whose position can be seen to move progressively across the film gate, in that sequence of 22 frames. It is clear that the white streak is really the image of a small shiny object that is reflecting sunlight into the camera lens. In this sense, it corresponds directly to one of the highlights on the car; it is "streaked" in every frame because the camera axis is moving relative to it in all frames.

Figure 7 shows the angular position of the car as a function of frame number, from frame 260 to the end of the sequence—a 4-sec interval of time in which the President was fatally wounded. This figure could have been drawn as an extension of the Commission-derived Fig. 6, which ends at frame 255, but I wanted the scale enlarged because the new individual points are now more precisely known. And all of this is in a region where the background

"... is just a grassy plot. So there is no reference point on which we can reestablish the position of the car in the roadway."[15][!]

The extreme smoothness of the curve comes from the fact

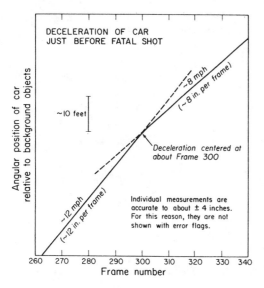

Fig. 7. Position of the President's car as determined in this paper. (See text.) Note the sudden deceleration of the car about one second before the President was fatally wounded (in frame 313). Error flags not shown; the 75 separate points have errors comparable to the width of the two straight lines.

that the smearing due to the camera accelerations (see in Fig. 3) cancels out; the measurements are made from a highlight on the car, to one of the reference points on the (featureless) "grassy plot" that I've just discussed. Any "jiggle" of the camera axis moves both of these reference points (on the car and on the ground) by the same distance on the film, leaving the distance between the two images on the film unchanged. These distances are plotted against frame number in Fig. 7, and I estimate that each point has a relative uncertainty of about 4 in. "in real space." The car had an average velocity of about 12 mph or about 12 in./ frame interval. I would normally show all the measured points on a curve such as this, but the scatter of the 75 points about the "best fit" two line segments is less than the width of the lines.

The car was moving almost exactly at 90° to the camera axis for these few seconds; one can easily check this by noting that the image of the horizontal strip separating the front and back compartments of the open car appears as a vertical stripe in one of these frames from Mr. Zapruder's downward-looking camera. For this reason we can translate relative positions of a car highlight and the background object on a frame-by-frame basis directly into the velocity of the car, simply by measuring the slope of the graph in Fig. 7.

The heavy car decelerated suddenly for about 0.5 sec (10 frames), centered at about frame 299, reducing its speed from about 12 mph to about 8 mph. Since the car was certainly being operated in some low gear ratio, the deceleration was no doubt caused by the driver reducing his foot pressure on the accelerator pedal. The question is then, "Why did the driver suddenly slow down at a time when a more natural reaction would be to speed up and weave to left and right, to avoid being hit again." I worried about this for some time, without finding any satisfactory answer. But then I found some testimony concerning a police siren that was remembered to have come just after the President was

killed (in frame 313). The many inconsistencies in the various witnesses' remembrances of exact times in this critical period made me feel that it was permissible to suggest that the siren, from an escorting police vehicle behind the President's car, had come a few seconds *before* the fatal shot. It would be most probable that an escorting officer, having heard one shot, and seeing the President wounded by a second shot, would hit the siren button when I'm suggesting he did. If the siren sound became apparent to Mr. Zapruder at frame 285, we would expect him to respond at frame 290, where we see the "unexplained and relatively weak angular accelerations" starting. We don't know the reaction time of the driver, but if it was 0.5 sec (9 frames), then he would lift his foot from the accelerator at frame 294, as Fig. 7 shows he did. Everyone will recognize that such a reaction on the part of the driver would be an unavoidable conditioned reflex; we all learn that when we hear a siren suddenly turned on, just behind our car, we lift our foot from the accelerator pedal. I haven't been able to think of any other reason why the driver of a car that has just stopped one or two high velocity rifle bullets would suddenly *slow down* his rate of travel.

The driver of the car, Agent William R. Greer, recalls that he speeded up the car in this period[17]:

> *Mr. Arlen Specter:* Do you recollect whether you accelerated before or at the same time or after the third shot?
>
> *Mr. Greer:* I couldn't really say. Just as soon as I turned my head back from the second shot, right away, I accelerated right then. It was a matter of my reflexes to the accelerator.
>
> *Mr. Specter:* Was it at about that time that you heard the third shot?
>
> *Mr. Greer:* Yes, sir; just as soon as I turned my head.
>
> *Mr. Specter:* What is your best estimate of the speed of the car at the time of the first, second, or third shots?
>
> *Mr. Greer:* I would estimate my speed was between 12 and 15 mph.
>
> *Mr. Specter:* At the time all of the shots occurred?
>
> *Mr. Greer:* At the time the shots occurred.

But since Fig. 7 shows that the car was still moving at the slower rate through the last of the published Zapruder frame—number 334—it is apparent that Mr. Greer's memory doesn't jibe with the recorded facts. This is what Professor Buckhout pointed out in his article on the reliability of eyewitness testimony[5]; all past events aren't recorded in a person's memory as on a magnetic tape, to be recalled later. That is why I find the photographic record so interesting; it doesn't have the normal human failings.

Certainly, the car eventually speeded up, and this is doubtless what Agent Greer recalled. In view of the disparity of several seconds between what the agent remembered of this terrible event and what actually happened, the reader may come to accept my conclusion that memories of the siren were similarly off by a few seconds. That's all it takes to turn the otherwise fantastically absurd deceleration of the car into a reasonable conditioned reflex on the part of the driver to the sound of a siren going off in his ear, and to shake up Mr. Zapruder at the same time. But as I said in the introduction, I can't prove that this is the way it happened.

As stated earlier, the streaks in the "grassy plot" were doubtless made by a small object reflecting light from the sun into the lens of Mr. Zapruder's camera. Figure 8 shows how this streak moved across the film gate in the camera (frames 313–334). This particular interval of just more than 1 sec coincides exactly with the climax of the events in Dealey Plaza. The President has just been fatally shot as the streak appears in the background, labeled 313. In the following second, Mr. Zapruder experiences great difficulty in continuing his earlier smooth tracking. He sees clearly in his view finder what has happened to his President, and it is a traumatic experience for him:

> Mr. Zapruder: . . . I heard a second shot and then I saw his head opened up and the blood and everything came out and I started—I can hardly talk about it. [The witness crying.][18]

But to return to the streaks in Fig. 8, let us first realize what that figure would have looked like if the shots had not been fired. Mr. Zapruder's tracking ability has been checked during the quiet periods of Fig. 3; a given highlight on the car, in those periods, stays pointlike, and at a fixed location in the film gate. Under such circumstances, a point of light in the background, such as that shown in Fig. 8, would move across the film gate on a straight line, at constant velocity. But because the camera shutter closes between exposures, while the film is being "pulled down," the straight line just mentioned would appear as a "dashed line" drawn by a draftsman using a straightedge.

Contrast the evenly spaced dashes on a straight line that Zapruder was capable of "drawing," with the dashes of Fig. 8 which appear to have been drawn by a spastic; that might even be the correct word to describe Mr. Zapruder's condition in that ghastly second after frame 313. (Until I realized that the labels on frames 314 and 315 had been interchanged in the exhibits, I thought Mr. Zapruder had lost even more control of his muscles than he actually had.)

Starting at frame 331, we see the streaks move up to the right and then back quite rapidly to the left. This phenomenon might be related to the "crescent"-like streaks seen in the CBS tests.[8] In Fig. 3, I couldn't plot this two-dimensional excursion of the camera axis, but one can see from that figure, at frame 332, that something pretty violent is happening. If I'd had access to the enlarged color prints that Governor Connally is shown viewing in *Life,* it would have been worthwhile plotting tracking curves like Fig. 8, for the whole sequence of frames. My reason for saying this is that such a curve complements an acceleration graph, such as Fig. 3. Ideally, the two should yield the same information, but in practice, the tracking curve shows more. This can be seen by comparing Fig. 8 with Fig. 3, in the vicinity of frame 325. From Mr. Zapruder's measured oscillation time of five frames, I expected to see an acceleration peak in Fig. 3, near this frame. But I've already mentioned the fact that of all the expected ones, a third of a second apart, only this peak was missing. However, a glance at Fig. 8 shows that there was quite a space in Mr. Zapruder's relatively smooth tracking curves at this point. This

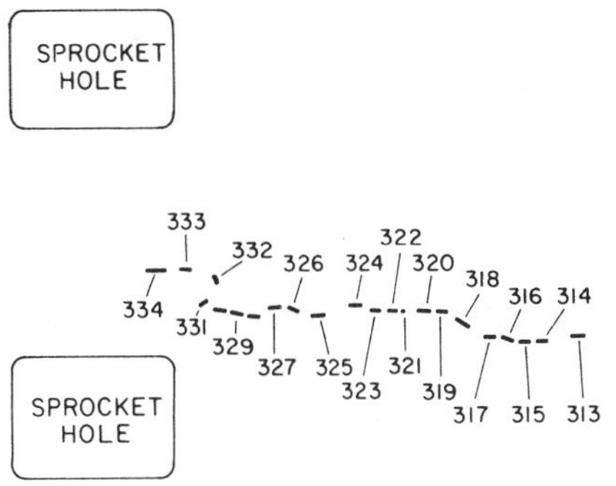

Fig. 8. "Streak" in the grass. The motion of Mr. Zapruder's camera axis is shown in two dimensions during the 1.2-sec period immediately following the fatal shot. (See text.)

example illustrates the fact that tracking curves are more sensitive than the angular acceleration graphs that derive from subtracted streak lengths.

I'll close this section by recalling that the wealth of data shown in Fig. 8, encompassing the climactic second in Dealey Plaza, involves a time period when an FBI photointerpreter told the members of the Warren Commission that from those pictures alone, there was no way to tell where the car was. I hope that this section will demonstrate what I've long felt—that the testimony of a physicist could have been of help to the Warren Commission, as it searched for the truth in early 1964.

ACKNOWLEDGMENTS

As I've indicated in the body of this paper, I've had help from several friends in the shooting experiments, particularly "Buck" Buckingham and Don Olson. Paul Hoch was for a long time my most knowledgeable and severest critic. In the absence of his always friendly but persistent criticism, this report of my study of the Zapruder film would have been much less convincing that I now hope it is. His vast store of knowledge concerning all aspects of the assassination was of great help to me, in my position of having read almost nothing of the literature critical of the Warren Commission's Report. His help in such matters is clearly evident in the text, where I acknowledge the work of others who anticipated conclusions that I reached later, but independently. Paul made many suggestions for improvements in a 1970 draft of this report, almost every one of which I incorporated. I don't believe we've discussed the assassination more than once or twice since then, and I haven't talked to Paul since I started the final rewriting a few months ago. He is now writing a book on the assassination, and we are agreed that although we each learn from the other, when final versions are being written, only one person can be responsible. And finally, I would like to acknowledge a great deal of constructive editorial criticism and help from Richard A. Muller.

This work was done with support from the U.S. Energy Research and Development Administration. Any conclusions or opinions expressed in this report represent solely those of the author and not necessarily those of the Lawrence Berkeley Laboratory nor of the U.S. Energy Research and Development Administration.

[1]*Hearings Before the President's Commission on the Assassination of President Kennedy* (U.S. GPO, Washington, DC, 1964).

[2]M. Lane, *Rush to Judgement* (Holt, Rinehart and Winston, New York, 1966), p. 55.

[3]Esquire, 205 (December 1966).

[4]S. White, *Shall We Now Believe the Warren Report?* (Macmillan, New York, 1968).

[5]R. Buckhout, Sci. Am. **231**, 23 (December 1974).

[6]Reference 1, Vol. XVIII, pp. 1–80.

[7]H. Weisberg, *Whitewash* (Weisberg, Hyattsville, MD, 1966; reissued by Dell, New York, 1966).

[8]Reference 4, p. 228.

[9]Reference 1, Report, p. 98.

[10]Reference 4, p. 82.

[11]J. Thompson, *Six Seconds in Dallas* (Geis, 1967).

[12]W. H. Hanson, *The Shooting of John F. Kennedy: One Assassin; Three Hits, No Misses* (Naylor, San Antonio, 1969).

[13]The fact that the bullets were soft nosed, rather than fully jacketed (as the Mannlicher–Carcano bullets were), was apparently important in intensifying the explosive jet effect. Dr. John K. Lattimer tells me that when he repeated our melon experiments with a duplicate of Oswald's weapon and jacketed cartridges, the effect was not as vigorous nor as dependable, even though it did occur about half the time. When he repeated it on skull models, however, the effect was still *more* violent than ours, even with the jacketed bullets.

[14]Dr. J. K. Lattimer, Resident and Staff Physician, 34 (May 1972).

[15]Reference 1, Vol. V, p. 159.

[16]Reference 1, Vol. XVIII, pp. 86–313.

[17]Reference 1, Vol. II, p. 119.

[18]Reference 1, Vol. VII, p. 571.

19

Submarines, Quarks, and Radioisotope Dating

Richard A. Muller

What do submarines, quarks, and radioisotope dating have in common? Among other things, they are subjects that I worked on largely because of the influence of Luis W. Alvarez. As different as they appear, these three subjects were directly linked to each other in my research, along with other diverse regions of science and technology including laser fluorescence, cosmic radiation, the Big Bang, superheavy elements, and thermonuclear fusion. While working with Luie, I often felt like Odysseus, tossed between distant shores by capricious gods.

The saga began in 1974, shortly after Luie nominated me to join a group of academic scientists called "Jason," whose members spent two months every summer working on United States national security problems. (The group had been created in 1960, when names from mythology were popular for government projects.) Luie had been a Jason for many years. The members applied their abilities and knowledge in science to such questions as strategic arms limitations verification, vulnerability of United States missiles to surprise attack, and the security of our fleet of nuclear submarines.

While working in Jason, I became particularly interested in the security of the United States nuclear fleet, because our confidence in the ability of our submarines to elude detection allowed the United States to adopt a relatively sane and stable defense policy. As long as we believed that our submarines would survive a surprise attack, it was unnecessary to "launch on warning"; instead we could absorb such a strike and respond in a measured way. The ability of our submarines to hide had been investigated many times before, but with the advance of technology it was necessary to reevaluate this security frequently. With my background in elementary particle physics, it seemed appropriate for me to reexamine the question of whether a nuclear submarine could be detected from the minute radioactivity left in its wake. Also working on the problem was Will Happer, a professor of physics at Columbia. After a few weeks of study, we tentatively came to the surprising conclusion that there was a new technology that could threaten the security of our submarines.

The new technology was "laser resonance fluorescence." Happer was an expert in the use of lasers, and he knew that intense lasers could be tuned to excite particular atoms and even particular isotopes. The excited levels decay with a characteristic radiation that allows the presence of the atom to be detected. Happer told me how laser resonance fluorescence had been used to detect single atoms of cesium vapor in the presence of large backgrounds of other atoms. Unlike radioactive decay, which allows detection of only those atoms which disintegrate during a finite counting period, the laser method could, in principle, detect each atom in the sample. Happer suggested that a variant of this technique could be used to detect the small number of radioactive atoms in the wake of a nuclear submarine.

After a great deal of calculation and many attempts at invention, we finally concluded that the number of radioactive atoms was too small and the ocean too big to allow for a practical implementation of the method. The radioisotopes induced in seawater by a passing nuclear submarine are too short-lived to give the laser method a substantial advantage over detection of the radioactive decays. For most physics projects failure of a new idea would be a disappointment, but on this project failure meant that our submarines were still secure, and Happer and I were delighted. Nonetheless, I was sufficiently intrigued with the idea of detecting radioactive atoms with a laser that when I returned to Berkeley I described it to Luie, who had been my thesis advisor and mentor.

In one of the great leaps for which he is famous, Luie immediately suggested an application in a totally different field: radiocarbon dating. He told me about a proposal by Michael Anbar, then at SRI International, in Stanford, California, to attempt radiocarbon dating with a mass spectrometer. I had difficulty understanding it because I knew almost nothing about the subject, but in the next few days I read extensively about Willard Libby's invention of radiocarbon dating in the late 1950s and began to appreciate the potential of Anbar's scheme.

Libby had recognized that the atmosphere contains a constant level of the radioactive isotope of carbon, carbon-14. Cosmic rays constantly bombard the upper atmosphere, creating free neutrons. These neutrons are absorbed by nitrogen nuclei which, after proton emission, become carbon-14. This new carbon-14 replenishes that lost through radioactive decay with a half-life of 5700 years; at equilibrium there is one atom of carbon-14 for every 10^{12} atoms of stable carbon. All plants and animals have this level until they die and the carbon-14 in their cells is no longer being replenished. One gram of carbon from a living organism has 14 decays per minute of carbon-14; from the smaller fraction of carbon-14 in a dead sample one can deduce the "age" of a sample, the length of time since it went out of equilibrium with the atmosphere. Although the decay rate is low, the absolute number of carbon-14 atoms in a gram of carbon is huge: 5×10^{10}. Anbar planned to detect and count these atoms with a mass spectrometer. If he succeeded at counting the large number of atoms rather than the infrequent decays, he could greatly extend the sensitivity of radiocarbon dating. Much smaller samples could then be used, and older ages could be measured.

Anbar's method failed, largely due to the inability of the mass spectrometer technique to suppress all sources of background that could simulate a carbon-14 atom. Luie suggested that Happer's scheme of laser resonance fluorescence might be used for single-atom detection of carbon-14. I spent the next several months talking to experts, reading, and learning everything I could about laser fluorescence. I discovered that the problem was much more difficult than I had anticipated. This method also had problems with background, from pressure-broadened carbon-13 lines and from continuum emission of trace contaminants. I realized that if one is searching for a signal at the 10^{-12} level, one must consider every conceivable stable atom or isotope to be a potential background. (Our blood contains arsenic at 3000 times this level and gold at 10 times this level.) I finally concluded that the laser method would not work, at least not until it became much more highly developed. I temporarily forgot about carbon-14 and went back to my main basic research project at the time, an experiment to study the 3° Kelvin cosmic microwave background radiation.

Luie soon interrupted my peace with a new and brilliant idea. He had been thinking about quarks, the particles hypothesized by Gell-Mann and Zweig to make up the proton and neutron but which had never been seen as separate entities. The standard explanation for their absence was that they were "confined" by forces which increased indefinitely as one tried to pull a quark from a nucleus. Luie considered this explanation unphysical and suggested instead that quarks hadn't been found because experimenters had looked for the wrong signature. The charge of the quark was predicted to be a fraction ($\frac{1}{3}$ or $\frac{2}{3}$) of the proton charge, and most experiments depended on this unique characteristic. If instead the quarks had integral charge, as first suggested by theorists Han and Nambu, they might have been seen but mistaken for other singly charged particles. The recent introduction of the quantum number called "color" had made integral quarks an attractive alternative to fractional quarks, and Luie had devised a method to search for integral quarks with the incredible sensitivity of one part in 10^{18}.

We would use the Lawrence Berkeley Laboratory 88 inch cyclotron as a large mass spectrometer. Luie had used a cyclotron for this purpose once before, in 1939, in his discovery of the natural existence of helium-3 and the radioactivity of tritium. He used a cyclotron then because he had one (and didn't have a mass spectrometer), but he learned of its remarkable resolution even under conditions of high-beam current. This was just the property needed for an integral quark search, so for the first time in 35 years the cyclotron would be again used in this mode. We would tune the cyclotron through various mass regions and look for a singly charged object with a mass different from that of known particles. If they are stable, quarks would have accumulated in air from cosmic-ray production and could be found in atmospheric hydrogen gas (which has the same chemistry as singly charged quarks). We would be able to identify individual quarks even in a background of 10^{18} hydrogen nuclei (the number we could accelerate in a reasonable counting period), for when the cyclotron is detuned from the hydrogen resonance frequency, no hydrogen is accelerated. Luie thought that we could complete the experiment in a few months.

By "we," I knew Luie meant me. He had recognized the importance of the measurement and figured out how to do it, and it would be my job to do the detailed experiment design and to make the measurements. I didn't resent this division of labor; I was grateful that Luie thought highly enough of me that I was the one he chose for this collaboration. The conception of such a project is the difficult part; carrying it through is relatively straightforward, although time-consuming. Yet I would be a coauthor of the discovery papers, if any. It seemed like a very good deal. It would also be an opportunity to learn how to use the cyclotron. But I was worried that I would have to neglect my principal research project, the cosmic microwave background measurements. I guessed as best I could the probability of a major discovery in the quark search and somehow came up with the figure of 10%. I weighed this probability against the importance of such a discovery and decided to take the risk. I turned more and more of the effort of the cosmic microwave experiment over to my colleague George Smoot and graduate student Marc Gorenstein.

Soon after we began the quark search, we read a paper by Zel'dovich and Okun that showed that stable heavy particles in the atmosphere should exist at levels much higher than the 10^{-18} level we had initially calculated. The primary source would not be cosmic rays but the cataclysmic explosion at the creation of the universe, the "Big Bang." Quarks, if stable, would be found in ordinary matter at the 10^{-10} level as remnants of nuclear reactions during the explosion, just as the cosmic microwave background is a remnant of the tremendous heat released at that time. Measurement of the density of quarks would tell us about the temperature and density of the universe during the first second of its existence. Our project suddenly transformed from a study of elementary particles to one which also could be the most fundamental measurement in cosmology since the discovery of the background radiation a decade earlier.

As if research in fundamental particles and cosmology weren't enough, we soon realized that discovery of negatively charged quarks might provide a new source of energy. Such quarks could be absorbed on the nuclei of hydrogen atoms to form neutral particles which could fuse with protons without requiring the high temperature normally required to overcome the Coulomb repulsion. If the quark were ejected during the fusion, it could catalyze another fusion. Similar catalysis of fusion had been observed with negative μ mesons, but the μ meson was too readily captured by the fused nucleus to assure continued catalysis. With the heavier quarks, it might prove practical. I envisioned quark separation plants to distill the quarks from seawater and power plants which would mix the quarks with deuterium.

The search took two years. Soon after we began, we were joined by Edward Stephenson, who taught us about low-energy particle identification, and by William Holley, who taught us about the complexities and subtleties of modern cyclotron operations. We tuned the cyclotron continuously over the mass range $1/3$–12 amu (the mass of a carbon-12 atom). As the accompanying paper shows, we found no integrally charged quarks. We were able to demonstrate that if such particles do exist in nature, their abundance is less than one part in 10^{18}. (Recall that the predicted level from the Big Bang was one part in 10^{10}.) Despite the null result, I didn't regret my decision to become involved in the search. I had taken a calculated risk and had lost. One more paper would join the enormous literature of null results.

While we were making the search, Ed Stephenson told me about an interesting new idea he had been discussing with Arvand Jain, a visiting physicist at the 88 inch cyclotron. We had both read in the newspapers about the reported discovery of "superheavy elements" in rock samples. (Superheavy elements are elements higher in mass than those in the known periodic table which are predicted to be stable or semistable by nuclear shell structure theory.) The new idea was to verify the existence of these particles using the cyclotron as a mass spectrometer, just as we were doing in the quark search.

It was a excellent idea, and I was disturbed that I hadn't thought of it. I had been working with the cyclotron for over a year and had heard about the superheavy report, but hadn't taken the trouble to put the two together. I realized that I had been lazy and had become too narrowly focused on the details of one experiment. I had been using the cyclotron in a virtually unexploited mode and achieved a sensitivity far greater than most people knew was possible, but I hadn't even bothered to think about other potential uses. The idea of looking for superheavy elements led nowhere, for the original report was mistaken and was eventually retracted, but the moment I heard from Stephenson about the possibility of such a search I promised myself that I would take time to look for other possible applications. Almost immediately I thought of radiocarbon dating.

I became excited as I realized that the cyclotron mass-spectrometer technique might be just what was needed for direct detection of carbon-14 atoms. It took a day of calculation and of checking cyclotron parameters (such as the current of carbon ions that the cyclotron could accelerate) before I was certain that the method would work. The key was the high energy of the emerging beam, which allowed particle-identification techniques to eliminate the backgrounds that had plagued Anbar and his mass spectrometer.

Luie walked into my office just as I finished the calculations, and I announced to him that I had solved the carbon-14 problem (which we hadn't discussed for over a year) and that I now knew how to count carbon-14 atoms directly, without having to wait for their decay. As was customary between Luie and myself, I didn't tell him my solution immediately; I gave him a moment to try to invent the method himself. Luie paused and I worried that he would reinvent my idea on the spot. Fortunately, after a few seconds he looked at me skeptically and asked for my solution. "Use the cyclotron as a mass spectrometer, just as we did in the quark search," I said. He paused again and I guessed that he was doing in his mind the same calculations that I had been doing for the last day. Then he smiled, put out his hand and said simply, "Congratulations!"

I wrote an internal report on the method in July 1976; an expanded paper with the first experimental results (the age of a deuterium sample was obtained by measuring the radioisotope tritium) was widely circulated in preprint form a few months later and was published.[1]

The highest compliment that a scientist can receive is praise from his scientific heroes. One of my proudest

1. L. W. Alvarez, Phys. Today **35**, January, 25 (1982); T. S. Mast and R. A. Muller, Nuclear Science Applications **1**, 7 (1981); R. A. Muller, Phys. Today **32**, February, 23 (1979); and R. A. Muller, Science **196**, 489 (1977).

moments came when Luie showed me part of a letter that he had written about my work:

> Dr. Muller solved a problem which is two decades old (increased sensitivity for radioisotope dating) with a technique which is three decades old. I am particularly appreciative of Dr. Muller's work because I was aware of all the necessary ingredients, including the importance of the problem, yet I failed to bring them together.

The method was immediately adopted by several groups around the world, in particular by Grant Raisbeck in France, who used the Orsay cyclotron, and by D. Erle Nelson and collaborators in Canada, who adapted the method to the tandem electrostatic accelerator in place of a cyclotron. In one of those accidents that seem to occur frequently in science, about nine months after our first successful demonstration of the method, the idea of using an accelerator for radioisotope dating was invented independently by a collaboration including Ken Purser (General Ionex Corporation), Harry Gove (University of Rochester), and Ted Litherland (University of Toronto). This group discussed the subject among themselves for the first time at the very American Physical Society meeting at which we were presenting a description of our measurements with tritium and our attempts with carbon-14. They not only missed our talk, but they also failed to see my paper in *Science* until they had already made plans for their first experiment. Purser and his colleagues realized, as had Nelson, that the tandem accelerator offered a particular advantage over the cyclotron for carbon-14 detection, because the main background expected, nitrogen-14, does not form negative ions and hence is not accelerated in a tandem machine. The high sensitivity achieved in the first experiments of this General Ionex/Rochester/Toronto collaboration was particularly potent in convincing scientists around the world that the method had promise.

The use of an accelerator permits the dating of samples thousands of times smaller than had been useful previously and allows the detection of other radioisotopes (such as beryllium-10 and chlorine-36) whose decay rate is too slow to permit easy detection. The new method is now in active use by nearly two dozen research groups around the world, and accelerators totally dedicated to radioisotope dating are planned or in use at laboratories in the United States, Canada, England, France, Switzerland, and Japan. Important new discoveries have been made in many fields, including archeology, atmospheric and earth sciences, astrophysics, and oceanography. There have been a host of review articles on the new method, several sessions of conferences, and three topical conferences totally dedicated to the method. Our research at Berkeley has concentrated on improvement of the technique, and we are now developing a small tabletop cyclotron ("cyclotrino") which we hope can make the accelerator technique more widely available to scientists.

I have learned several lessons from this adventure. The clearest is the interrelation of apparently different fields of science. I have had a great deal of fun talking to scientists in unfamiliar fields about the most important problems they were pursuing. I continue to hear from scientists around the world who are wondering if the accelerator method could help them, and I continue to learn about new areas of science in this way.

Many apparently wild jumps in research are really the revival of old ideas (integral quarks, fusion catalysis, direct detection of carbon-14 atoms) not forgotten and given new life by developments in theory or technology. Yet it is easy to miss the interconnections, simply by not bothering to look for them. I worked with the cyclotron for over a year before I made any effort to think about other applications, and I did force myself to think only when shamed into doing so by a colleague who had an idea that I thought I should have had. The surprising lesson is how easy it is to be lazy.

I feel certain that there is more to be learned from this story, but I am not really sure what it is. What is the optimal strategy for productive research? How much time should be spent thinking, exploring new ideas, and how much should be spent concentrating on one project? How can one estimate the risk of a jump to a new area of research? In retrospect I believe that I came close to having nothing of importance to show for several years of effort. I worried at the time that I had become involved in too many projects, a worry encouraged by some of my colleagues who tried to be helpful by advising me not to spread my efforts too thin. During the period of this story I was also the principal investigator on a project in adaptive optics called the "rubber mirror," a participant in the American Physical Society study on nuclear reactor safety, and lecturer for an upper division course in the Physics Department at Berkeley. My spare time was spent trying to get financial support for my research.[2] Was I a dilettante or an interdisciplinary scientist? The difference is substantial but subtle. Was I chasing chimeras? An element of self-doubt and uncertainty may have helped keep me on the track. Most important of all was the model and encouragement of Luie, who never seemed to think for a moment that I was wasting my time, or doing anything wrong. Perhaps this is why Luie has caught more chimeras than anyone else I know.

2. R. A. Muller, Science **209**, 880 (1980).

Quarks with Unit Charge: A Search for Anomalous Hydrogen

Richard A. Muller, Luis W. Alvarez, William R. Holley
and Edward J. Stephenson

Abstract. *Quarks of charge +1 and other anomalous hydrogen have been sought by using the 88-inch cyclotron at Berkeley as a high-energy mass spectrometer, with natural hydrogen and deuterium as the sources of ions. No quarks were observed, and limits were placed on their ratio to protons on the earth that vary from $< 2 \times 10^{-19}$ for high masses (3 to 8.2 atomic mass units) to 10^{-13} for the lowest masses ($< 1/3$ atomic mass unit).*

Quarks, the components of the nucleon proposed by Gell-Mann (*1*) and Zweig (*2*), have been the subject of extensive searches by both elementary-particle and cosmic-ray physicists (*3*). In most of these experiments the quarks' fractional charge (1/3 or 2/3 that of the proton) was the distinctive signature that would have indicated their presence. But fractional charge is not required by all quark models; in particular, theories proposed by Lee (*4*) and by Han and Nambu (*5*) have quarks of integral charge ($Z = 1$). These theories are more complex, however, in that they require more than the three quarks of Gell-Mann and Zweig; the Han-Nambu theory, for example, has nine. When the classical quark theory was improved by the addition of the new quantum number "color," the number of fractionally charged quarks had to be increased from three to nine (*6*). The integrally charged quarks of Han and Nambu, on the other hand, accepted color in a very natural way and did not require an increase in number (*7*). It seems reasonable to speculate that the quark searches failed because they assumed the wrong signature; perhaps quarks have integral charge.

Experimental limits on the existence of quarks with integral charge are not nearly as good as those for quarks with fractional charge. Experiments at accelerators would have observed $Z = 1$ quarks only if they were produced with a cross section at least 10^{-3} of that for proton-antiproton pair production and if their mass were no more than 4 atomic mass units (amu) (*8, 9*). If quarks are stable against decay, as they are in several versions of the theories, then a search in terrestrial material is potentially more sensitive than a search at accelerators since the natural production of quarks had literally aeons of time to take place. Quarks with $Z = +1$ would be equivalent chemically to a new isotope of hydrogen, and would be found on the earth mixed with that element. With this in mind, we decided to search in terrestrial hydrogen for $Z = +1$ quarks, using the same technique that was used in the discovery of stable (and natural) ^{3}He by Alvarez and Cornog (*10*); that is, we used a cyclotron as a high-energy mass spectrometer. After we began this work, we found that Zeldovich and co-workers (*8, 11*) had shown that if such particles existed they would be found in natural hydrogen not only from cosmic-ray production, but also as remnants of the big bang, in much the same way that protons and the cosmic microwave radiation are thought to be remnants. The ratio q/p of quarks to protons predicted by Zeldovich, Okun, and Pikelner depends somewhat on the assumed values for the quark mass and the cross sections, but is of the order of $q/p = 10^{-10}$ to 10^{-12}. They also calculated that q/p from cosmic-ray production is $\simeq 10^{-17}$.

The most sensitive search before ours was by Alvager and Naumann (*12*), who introduced deuterium into a mass separator and scanned in the charge-to-mass ratio (Z/m) to look for new isotopes of hydrogen. Ions with $Z > 1$ were eliminated by sending the accelerated beam through a nickel foil, which stopped molecules and particles of high Z. Unfortunately, this technique works only for high values of mass, where background from scattered beams is low; for quark masses $m_q > 6$ amu, they were able to set the limit $q/p < 3 \times 10^{-18}$. For the low-mass region, the best limit was obtained by Kukavadze et al. (*13*), who used a mass spectrometer to find $q/p < 10^{-8}$ for $m_q > 2$ amu. Thus, previous experiments do not rule out $Z = +1$ quarks with $m_q < 6$ amu at the levels predicted by Zeldovich and Okun.

We used the 88-inch sector-focused cyclotron at the Lawrence Berkeley Laboratory for our search. Both hydrogen and deuterium gases were introduced as sample materials into the filament ion source. Only ions with the appropriate value of Z/m, given by the cyclotron resonance equation, were accelerated. In their original discovery of stable ^{3}He, Alvarez and Cornog scanned in Z/m by varying the cyclotron's magnetic field; in our search we kept the magnetic field constant and scanned in frequency.

For the 88-inch cyclotron to be "in tune"—that is, for it to accelerate particles into our detectors—several criteria must be met. The magnetic field profile (the field as a function of radius) must be appropriate to compensate the relativistic increase in mass as the particle is accelerated, and the voltages applied to the electrostatic deflector and dee must be at the correct values to assure efficient beam extraction. To eliminate the need to change the magnetic field profile during a scan, we operated the cyclotron in the third harmonic mode, where the particle revo-

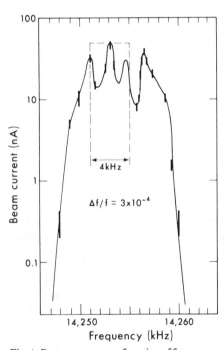

Fig. 1. Beam current as a function of frequency for D$^+$ ions, as measured by a Faraday cup. The smooth curve is a guide to the eye. The effective width of a "flattop" curve with the same area is 4 khz, as shown, giving a fractional width $\Delta f/f$ of 3×10^{-4}. For these measurements, we reduced the beam current to prevent heat damage to the cyclotron deflector electrodes. The fine structure at the top of the resonance curve reflects changes in the number of turns made by the beam in the cyclotron before extraction.

In figure: 100 · 10 · 1 · 0.1 (Beam current (nA)); 4 kHz; $\Delta f/f = 3 \times 10^{-4}$; 14,250 · 14,260 (Frequency (kHz))

lution frequency is one-third of the oscillator frequency and the maximum particle velocity is only $0.1c$, where c is the speed of light. During scans we changed the deflector and dee voltages in proportion to the change in frequency. The ultimate proof that we were in tune in our scans came from the observation of expected beams; for example, when we set the cyclotron to accelerate $Z/m = 1/2$ (D^+) and scanned to $Z/m = 1/3$, we found that DH^+ molecular ions were being efficiently accelerated and extracted.

The emerging beam from the cyclotron was focused onto two silicon detectors: a thin one (20 to 60 μm) to measure the ionization rate dE/dx, followed by a thick one (1 mm) to measure the total energy. By requiring a coincident signal between the two detectors, and that the signals in them correspond to a single $Z = 1$ particle with the correct energy, we were able to discriminate against heavier particles, molecules, and background radiation from the cyclotron. Except when the cyclotron was tuned near the resonant frequency for known species (integral values of Z and m) the number of background counts was essentially zero; during one run we operated for over an hour with no background counts. The rapid falloff of the tails of the cyclotron reso-nance allowed us to search quite close to values of Z/m for which known particles exist. Figure 1 shows the beam current plotted against frequency for a D^+ beam. By the time we had moved off a main peak by about 1 percent, the rate of particles emerging from the cyclotron was reduced to less than one per minute.

The advantage of the cyclotron over an ordinary mass spectrometer does not come from its high resolution (see Fig. 1) or from its beam current (10 to 50 μa), but from the high energy of the emerging beam: several million electron volts per nucleon. This high energy allowed us to send the beam into the particle identification detectors, and to get useful information on a particle-by-particle basis. The cyclotron has, of course, one additional advantage. Had we discovered quarks in natural material, we would have had a high-energy beam of them, and study of their properties would have been straightforward.

For quark masses above 2 amu, we used deuterium gas in the cyclotron ion source. The deuterium had been separated from water at the Savannah River Heavy Water Plant, using the GS process followed by vacuum distillation and electrolysis (14). The efficiency for each of these concentration processes increases as the mass of the isotope increases; thus for masses just above 2 amu quarks would have been concentrated by at least a factor of 6600, the ratio of ^1H to ^2H in river water. For masses >> 2 amu we know that the concentration of quarks could not have been increased by more than a factor of $5 \times 6600 = 33 \times 10^3$, since 20 percent of the original deuterium in the water was recovered, and obviously no more than 100 percent of the original quarks could have been recovered. Kaufman and Libby (15) have shown that for electrolytic separation, this transition to maximum concentrations takes place below a mass of 3 amu. Thus, the sensitivity of a search in deuterium is 6600 times greater than that of a search in hydrogen for quark masses near 2 amu, and 33,000 times greater for quark masses above 3 amu. In a typical run a 13-μa deuterium beam was observed as we tuned through the $^2H^+$ frequency, corresponding to 8×10^{13} ions per second. The deuterium beam was measured with a Faraday cup, preceded by collimators which simulated the acceptance geometry of the silicon detectors. Our typical dwell time for a particular value of Z/m as we manually scanned in frequency was 2 seconds, during which we would have recorded as significant the observation of even a single particle with the proper signature in the dE/dx and total E detectors; no quarks were observed. Thus the ratio of quarks to deuterons was less than 6×10^{-15}. The ratio of quarks to ^1H is then less than 10^{-18} near mass 2 amu, and less than 2×10^{-19} for masses ≥ 3 amu.

For quark masses in the range 1 to 2 amu we used ordinary hydrogen in the source rather than deuterium; our sensitivity was corresponding 6600 times poorer. We repeated the scan in the mass range 1.5 to 2 amu with deuterium, thinking that it would give us some additional sensitivity in the range just below 2 amu. We saw no quark events in this scan. It is difficult to estimate the sensitivity of this scan without studying the detailed physical chemistry of the isotope separation process; thus we claim no limit obtained from this scan. For the mass range 0 to 1, we looked for quark-hydrogn molecules Hq, assuming that their chemistry would be similar to that of H_2. The mass range was covered in two runs; for quark masses of 1/3 to 1 amu the H_2^+ current we obtained was 1.7 μa, and for the mass range 0 to 1/3 amu the H_2^+ current was 0.16 μa. The corresponding sensitivities are in the range $q/H = 10^{-14}$ to 10^{-13}.

We have observed no quarks of $Z = +1$, nor other anomalous isotopes of hydrogen. The limits from our search and

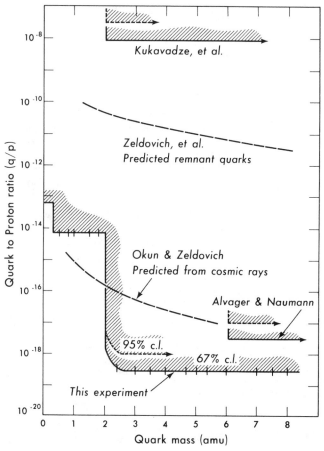

Fig. 2. Quark-to-proton ratio as a function of quark mass. Shown are the experimental limits, as well as predictions of the levels expected from cosmic-ray and big bang production if $Z = +1$ quarks exist. The predictions were scaled with mass to take into account changing cross sections and cosmic-ray fluxes. The short vertical bars indicate the narrow regions (generally less than 1 percent wide) where we could not search because of the existence of beams of common ions with integral values of Z and m. The experimental limits plotted as solid lines are for the 67 percent confidence level, corresponding to approximately 1 standard deviation. Each limit must be raised by a factor of 3 (as indicated with dashed lines) for a 95 percent confidence level.

previous experimental results are plotted in Fig. 2. These limits correspond to 1 standard deviation, that is, a confidence level of 67 percent. For 95 percent confidence levels, each limit must be adjusted upward by a factor of 3, as indicated on the plot. In addition, we have taken the calculations of Zeldovich *et al.* (*8*) and, after scaling them appropriately with energy, plotted the predicted q/p ratios. For several regions corresponding to integral values of Z and m, the intensity from known ions (such as $^2H_2^+$ and $^{14}N^{2+}$) was sufficiently high to require the removal of the silicon detectors from the beam. As a result there are regions, generally less than 1 percent wide in mass, in which we were unable to search for quarks. These regions are indicated in Fig. 2 by short vertical bars. The existence of these beams served the useful purpose of proving that the cyclotron was still in tune and that if there were quarks at concentrations greater than the limits we have placed, they would have been observed.

RICHARD A. MULLER
LUIS W. ALVAREZ
WILLIAM R. HOLLEY
EDWARD J. STEPHENSON

*Lawrence Berkeley Laboratory,
Berkeley, California 94720*

11 January 1977

References and Notes

1. M. Gell-Mann, *Phys. Lett.* **8**, 214 (1964).
2. G. Zweig, preprint CERN 8182 TH/401 (1964).
3. L. W. Jones, "A review of quark search experiments," preprint UM-HE76-42, Department of Physics, University of Michigan, Ann Arbor, 1976; *Rev. Mod. Phys.*, in press.
4. T. D. Lee, *Nuovo Cimento* **35**, 933 (1965).
5. M. Han and Y. Nambu, *Phys. Rev. B* **139**, 1006 (1965).
6. The idea that is now called color was introduced by O. W. Greenberg, *Phys. Rev. Lett.* **13**, 598 (1964). For a more recent review, see S. L. Glashow, *Sci. Am.* **233**, 38 (October 1975).
7. Y. Nambu and M. Han, *Phys. Rev. D* **10**, 674 (1974).
8. Ya. Zeldovich, L. Okun, S. Pikelner, *Usp. Fiz. Nauk* **87**, 113 (1965); *Sov. Phys. Usp.* **8**, 702 (1966).
9. P. Franzini, B. Leontić, D. Rahm, N. Samios, M. Schwartz, *Phys. Rev. Lett.* **14**, 196 (1965); V. Cocconi, T. Fazzini, G. Fidecargo, M. Legros, N. H. Lipman, A. W. Merrison, *ibid.* **5**, 19 (1960); W. Baker *et al.*, *ibid.* **7**, 101 (1961); A. Zaitsev and L. Landsberg, *Sov. J. Nucl. Phys.* **15**, 656 (1972).
10. L. Alvarez and R. Cornog, *Phys. Rev.* **56**, 379 (1939); *ibid.*, p. 613.
11. L. Okun and Ya. Zeldovich, *Comments Nucl. Part. Phys.* **6**, 69 (1976).
12. T. Alvager and R. Naumann, *Phys. Lett. B* **24**, 647 (1967).
13. G. Kukavadze, L. Memelova, L. Suvorov, *Zh. Eksp. Teor. Fiz.* **49**, 389 (1965); *Sov. Phys. JETP* **22**, 272 (1966).
14. See, for example, A. Standen, Ed., *Kirk-Othmer Encyclopedia of Chemical Technology* (Wiley, New York, ed. 2, 1968), vol. 6, pp. 815–918.
15. S. Kaufman and W. F. Libby, *Phys. Rev.* **93**, 1337 (1954).
16. We are grateful to D. L. Hendrie, H. E. Conzett, R. A. Gough, and J. T. Walton for their cooperation and suggestions and to G. F. Smoot for help during the experimental runs. This work was supported by the Energy Research and Development Administration and the National Aeronautics and Space Administration.

20

Ivory Towers and Smokestacks

William Humphrey

Ivory towers are white and smokestacks are black. Somehow this seems appropriate, because in many ways industry is as different from academia as black is from white. An industrial company expects to generate income at the rate of $100,000 per employee per year. A university research group consumes money at a comparable rate. In industry, trade secrets and know-how are regarded as confidential property, as edges over the competition. In the university setting, "publish or perish" is the rule. I could expand these dissimilarities to include tenure, exchange programs, and so forth, only to establish the obvious point that academia and industry abide by different rules. Few can successfully function in both environments. Luis W. Alvarez is one of those rare people who can.

My perspective of Luie's participation in the business world is but a snapshot of his involvement in the evolution of one venture. Luie has had a long history of business experiences, including his long-held seat on the Hewlett Packard board and consulting with IBM. He also shared the sad experience of his mentor, Ernest Lawrence, who attempted to commercialize an improved color television tube. When I came into the picture his extensive contacts built up over the years with business and financial leaders established the foundations for the business experience I now describe.

Luie and Pete Schwemin worked together beautifully for years exploiting Luie's ideas in the university, so it was inevitable that they form the technical nucleus of a start-up venture in the early 1960s. At this time I received my doctorate in Luie's bubble chamber group. I was also interested in vision, optics, and instrumentation, so I fit in well with one of their first areas of interest, a thin lens altered in power by sliding one optical element across another. I became a consultant with the embryonic firm when it was a small storefront operation in downtown Oakland with only one full-time machinist, Pete's wife as a part-time secretary-accountant-receptionist, and lots of attention from Pete and Luie in the evenings. It wasn't long before full attention to the technical aspects of the business became critical. Then Luie presented me with the possibility of a career change from basic research to applied physics in his start-up company. It was a surprise to everyone, myself included, when I accepted his offer. Soon I was president of a company with three other

regular employees where, as president, I helped carry the garbage cans out for collection every Wednesday.

When I joined Luie's start-up company its name was ORDCO, Optical Research Development Corporation. This is the first of six names that were associated with it, since over time its name was changed either to label technical assets that were split off into separate business entities, or to create a more descriptive name for our changing business goals.

After some tentative attempts at manufacturing an innovative, inexpensive optical range finder which Luie had conceived and Pete had executed, ORDCO began to evolved into a licensing company. Our technical concepts were patent-protected, modeled to show feasibility, and licensed to major manufacturers. Royalty income would form the base for further growth. Our first major technical asset, the variable-focus lens, was the subject of the accompanying article. We thought that it had such applications as the focusing mechanism for cameras and cosmetically acceptable spectacle lenses adjustable in power from infinity to reading distance.

When Luie and his wife Jan traveled to Africa, he found that his home movies, often taken with a telephoto lens, bounced around, detracting from their quality. So he invented a stabilizer that could be incorporated into a movie camera to produce a solid image, free of the annoying vibrations and lurches seen in movies taken with a hand-held camera or from a moving vehicle. This led to a series of inventions and patents by Luie and myself. Several schemes to produce aircraft collision avoidance systems followed, along with other innovative developments which added to the technical resources of the company.

Our technical creativity was no problem, but successful licensing was. Access to big names in the business world was easy with Luie's extensive network of contacts. We negotiated with American Optical and Polaroid in our attempt to exploit the variable-focus lens and with Bausch and Lomb, Bell & Howell, and Mark Systems to exploit stabilized optics. But these licensing experiences were frustrating. When a licensed company cut back or management changed, the outside licensed technology suffered most. Marketing and production decisions were beyond our control and counter to our best interests. However, these arrangements kept a trickle of funds flowing in which was sufficient to keep the doors open, thanks to Hugh Davy, Luie's long-time friend. As ORDCO's financial guide, Hugh kept the banks from locking our doors during our long period of dismal financial performance.

In 1972 the company split with ORDCO, retaining the stabilized optics patent portfolio then under license, while the remaining technology was assigned to a new company, Humphrey Research Associates. Humphrey was to search out an attractive business area and create a full-blown business. This marked the end of our licensing company experiment and the beginning of an exciting period of growth.

In one way, the licensing experiment was a success. It kept the company alive, barely, while we developed an extensive library of both licensed and unlicensed technology. With the unlicensed technology available to us, we tried to find the potential markets in which we could create a successful business. We settled on the ophthalmic instrument market, since our professional customers were easy to identify, and we had strength in optics. Our first product was an ambitious project using Luie's variable-power lenses and an offshoot which I developed for correcting astigmatism. The instrument could also "refract" patients (produce a complete spectacle correction) without any mechanisms in front of the patient's face. It was magic— something you had to experience to believe. However, the instrument and development costs escalated, and we needed more money. In 1974, the country was in a recession and venture capitalists didn't venture much. Raising money took innovation, plus lots of personal contacts from Luie and his friends. We eventually gave a group of local businessmen a rare opportunity to invest in a high-tech start-up at an early stage, and they gave us the funds we needed so badly for our growth. By now Jack Lloyd, who we knew from previous Rad Lab projects, had just sold his small start-up business. Jack came to be president of Humphrey

Instruments, and from then on things moved quickly. We moved to larger quarters in Berkeley, added engineering, purchasing, and marketing people, and brought out our first product, the Vision Analyzer. At its initial showing we had a spectacular draw and got our unknown company the attention it so badly needed to gain a toehold in the field. We sold instruments and somehow managed to circumvent a series of financial problems without going bankrupt. We moved again to new and much larger facilities in San Leandro and introduced a successful automatic lens-measuring machine. We were making money!

Throughout the start-up phase of the company the relationship among the founders was a close one, with continuous contact not only between the full-time people, who now included Pete Schwemin, but also with Luie, whose primary association was still with the university. Success, however, diluted these relationships as more people of diverse backgrounds became involved in the company. Then too, the daily threat of impending bankruptcy, which had been such an effective glue in unifying our actions, was now a thing of the past. By the time we sold the company in 1979, much of the original interpersonal magic had vanished, and within a few years all the founders were off doing their own things. Jack Lloyd had started a noncompeting medical instrument company, I stayed with Humphrey Instruments, and Luie and Pete had rejuvenated the ORDCO technology into a company called Schwem Technology.

Luie planted the seed and nurtured the growth of our commercial endeavor, which now provides a livelihood for hundreds of employees. Humphrey Instruments manufacturers and markets top-quality, innovative instruments, and in the process, has provided a healthy profit for every investor who ever participated in the company. This must be close to the definition of success in business.

Development of variable-focus lenses and a new refractor

LUIS W. ALVAREZ, Ph.D., Sc.D.

ABSTRACT—This paper traces the events that led to the development of the variable-focus lens, the variable-power astigmatic lens, and the principle of remote refraction and "phantom lenses." These developments formed the basis for invention of a new subjective refractor–the Vision Analyzer–whose novel design and rapid operation are described.

KEY WORDS—ophthalmic optics, variable focus, lenses, refraction, refractors, refractive error, cylindrical lenses.

Introduction

You might wonder how I came to be talking here today, since I am a professional physicist who normally deals with the smallest and shortest-lived particles of matter—a field with absolutely no known or even imagined practical applications. People in the ophthalmic professions frequently ask me how I became interested in ophthalmic optics—a very practical science. They ask specifically what was behind the invention of variable-focus lenses, the principle of "remote refraction" and "phantom lenses," and the development of the new subjective

"I looked at bifocals and decided I didn't like what I saw."

refractor (the Vision Analyzer). I will try to answer these questions.

About 20 years ago my interest in such matters was kindled when I started to develop presbyopia; that will immediately tell most of you how old I am. I didn't have to wear bifocals for a while, but I did obtain some 1 D clipons that I used for reading at home in the evenings. I looked at bifocals and decided I didn't like what I saw. It seemed to me that someone should have invented something better in the 200 years since Benjamin Franklin made his first pair of bifocal spectacles.

As many people had done, I thought first about distorting a flexible envelope of plastic material by internal liquid pressure to duplicate, more or less, the action of the crystalline lens of the human eye. I didn't really like that idea so I next turned my attention to lenses that could slide relative to each other at right angles to the optic axis. My friend and associate, Mr. A. J. Schwemin, built such a "lens" that used 4 sliding elements. But it was too complicated, even though it did act as a variable-focus thin lens.

It didn't occur to me then that it was possible to make a continuously variable power thin lens from only two elements that could slide relative to each other at right angles to the optic axis. I later found that many people had patented such objects, but none of them worked because they didn't incorporate the proper curve. It turns out that there is only one correct curve, and when I discov-

ered it I was able to patent it because it was different from the earlier inventions and, most importantly, it worked.

The morning after I found the equation for the variable-focus lens, I saw William Humphrey —who was then a graduate student in my research group. I told him what I had done and then asked, "Would you like to see the equation of the surface?" Instead of saying "Yes," as I had expected, he said, "No, I'd rather derive it myself." He came back in less than an hour with exactly the same equation it had taken me several years to discover. He helped me recover from my shock by saying, "After all, I knew there *was* a solution, and you didn't."

The variable-focus lens

Many people are intrigued by the variable-focus lens (VFL) and are interested in how it works. Fig. 1 shows a schematic drawing of the VFL with its two elements in two different positions along the horizontal (x) direction. One can tell from the general appearance of the elements that they have a variation in spherical power along

the horizontal direction. At the top of the figure we see that the upper left-hand element has its spherical power increasing toward the left; the average spherical power in each square is labeled by the first of the two numbers in that square. Of course, the power changes smoothly, and it is only shown here in steps for illustration. The figure shows that in the right-hand vertical column all the spherical powers are −2 D. In neighboring columns, moving to the left, the spherical power changes to −1 D, to zero D, and finally to +2D at the left edge. It will be clear from looking at the numbers in the upper figure that when the two elements are placed close to each other with all four edges lined up, the sum of the spherical powers is zero at all points—so the VFL in its "neutral position" acts like a plano piece of glass.

The right-hand number in each square tells how much diagonal cylindrical power there is (along the 45° and 135° meridians) in addition to the spherical power. The astigmatism for the pair of lens elements cancels out so long as the top and bottom edges of the elements are aligned.

If we slide one element horizontally relative to the other, as shown at the bottom of Fig. 1, we find that the sum of the spherical power is directly proportional to the amount of motion along the X axis and that the *power is the same over the whole area of overlap*. In fact, the center of the lens stays fixed at the center of the region of overlap, and in terms of performance the lens is almost indistinguishable from a lens in a trial set. The lower part of Fig. 1 shows a pencil of light going through a square in the left-hand element with the numbers 0, 2 and through the other element in a square labeled 1, −2. The sum of 0 and 1

Fig. 2. The optical effects of two variable-power astigmatic lenses (VAL) are illustrated. Each "lens" consists of a pair of optical elements which when completely aligned produces no optical power (center illustrations). Sliding the elements horizontally in opposite directions produces smoothly increasing cross-cylinder power at fixed axes—90° and 180° for the left VAL and 45° and 135° for the right VAL.

indicates a sphere of +1 D; the sum of 2 and −2 indicates a cylinder of 0 D power. By adding the powers in the two elements above, below, to the left, and to the right of the cells just examined, one finds the sums are always +1 D of spherical power and 0 D of cylindrical power. One can also see that if the elements are displaced by varying distances along the X axis, the new spherical power is proportional to the displacement and constant over the whole area of overlap.

My colleague, Dr. William Humphrey, pointed out something I hadn't realized: if the two elements were moved relative to each other at right angles to the direction that gave them varying spherical power, they generated varying amounts of cylindrical power in addition to the spherical power that had been produced by the first displacement. So a pair of such elements could generate any combination of sphere and cylinder power by motions along two axes at right angles to each other.

It was clear to both Dr. Humphrey and me that if the pair of lens elements could also be rotated together about the optic axis, any possible spectacle lens prescription could thus be produced. We thought for a few days about the possibility of building a replacement for the standard refractor, but we decided against it; the new technology was very different, but we couldn't convince ourselves that it had any great advantage over the technology embodied in the existing standard refractors.

The variable-power astigmatic lens (rotation free)

Several years later, Dr. Humphrey invented the variable-power astigmatic lens (VAL). He was apparently the first person to make use of the fact that any possible cylindrical lens power can be generated as the optical sum of a spherical lens plus two variable-power crossed cylinder lenses— one with axes at 0 and 90 degrees, the other with axes at 45 and 135 degrees. With the aid of our friend, Mr. Schwemin, we soon had vari-

able astigmatism lenses to play with, and we found that Dr. Humphrey's idea was correct: we could generate a cylindrical lens having any power and axis without rotating anything. The VAL involves only sliding motions along the horizontal direction (Fig. 2). Soon afterwards we put together the first "one-eyed" refractor that could correct both spherical and astigmatic errors without requiring any rotatable optical elements. We thought this instrument had some promise as a replacement for the standard refractor. However, the astigmatic readout wasn't convenient; it involved a pointer that moved over a piece of polar coordinate graph paper. Microcomputers, that eventually solved the readout problem, hadn't yet appeared on the market. So we went back to our other activities, knowing that we were getting very close to inventing a new clinical refractor.

Principles of remote refraction

In a search for ways to eliminate instrument myopia and remove the obtrusiveness of an instrument between patient and doctor, William Humphrey invented the concept of "remote refraction." Imagine a 10-D myope wearing his glasses and looking at a distant *concave mirror* on whose surface he sees a sharply focused projected letter (Fig. 3). If the myope takes off his glasses, the images will be hopelessly blurred. However, if he places his glasses between the mirror and the projector at a place near the center of curvature of the mirror, he can see the target letters just as well as he could when he wore the glasses, provided only that he also is near the mirror's center of curvature.

There are two ways of understanding how this happens. The first is to say that the mirror forms a real image of the spectacle lens at the standard vertex distance in front of the myope's eye. One could say that the concave mirror produces "phantom lenses" in front of the patient's eyes, and these phantom lenses have the same optical properties as the physical lenses. A second explanation is particularly easy to demonstrate. If the myope is completely presbyopic, there is only one position in front of his eye where he can see an object clearly without his glasses—and that is at 10 cm. Since he *can* see the distant test letter clearly when his glasses are removed and placed in front of and close to the projection lens, the light from the projector must pass through the spectacle lens and be reflected

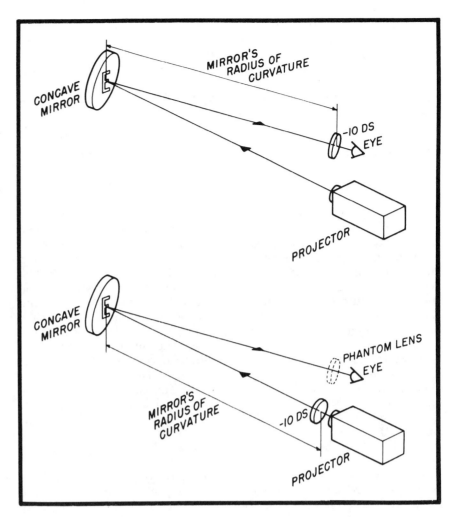

Fig. 3. The concept of "remote refraction" and "phantom lenses" is illustrated for a 10-D myope who, with glasses on, clearly sees in-focus letters projected onto a concave mirror (top). With the −10 D lenses removed and placed in the projecting beam near the center of curvature of the mirror, the myope can again see the letters clearly, provided that the eyes are also near the mirror's center of curvature (bottom). This latter remote refraction can be explained by saying the concave mirror produces a phantom lens in the spectacle plane.

from the concave mirror in such a way as to produce a sharp aerial image of the test letter that is 10 cm in front of the myope's eye. That this is really happening can be demonstrated by placing a piece of paper in the light path between the mirror and the myope's eye; a sharp image of the letter will appear on the paper when it is placed 10 cm from the eye, and the image becomes blurred as the paper is moved closer or farther away. This principle was found to work for any prescription, either plus or minus, with or without any amount of astigmatism at any axis.

A new subjective refractor; the vision analyzer

Once the principle of remote refraction was developed, we had the major elements for creating a novel, rapid, and accurate instrument for performing *clinical subjective refraction*. Designing the actual instrument involved many considerations such as human factors pertaining to doctors and patients and knowing what they both expect for a vision examina-

tion. The instrument we developed is called the Vision Analyzer. Functionally, it replaces the standard refractor. Operationally, it is very different and piques one's curiosity to say the least.

Physically, the Vision Analyzer looks more like a modern table with some controls than like the optical instrument it really is (Fig. 4). Contained within the table are projectors which create separate test targets for each eye and variable-focus spherical and cylindrical lenses (Fig. 5). A computer keeps track of the patient's responses and prints out the measurement of the refractive error and other information.

Fig. 4. A view of the Vision Analyzer from the doctor's side. Optical controls are near the doctor's right hand, and measured refraction parameters are displayed on the console in the center. At the doctor's option and under his control, the patient may make adjustments of lens and prism powers using the knob located near his right hand. A permanent record is made by the printer at the doctor's far right.

The Vision Analyzer works without using lenses before the eyes. The patient is seated at the instrument with forehead cushioned against a rest. Through electrical control of the seat and headrest, the eyes are quickly positioned with respect to light beams that measure the interpupillary distance and set the vertex distance at any value, includ-

ing zero (to measure the refractive error at the cornea, if desired). The patient can be wearing glasses, no glasses, or contact lenses. The "phantom lenses" of the instrument can be superimposed in space with a patient's glasses to provide an exact overrefraction without resorting to calculations of lens effectivity.

To measure visual acuity with or without corrective lenses, the examiner sets the variable focus lenses to zero and projects either monocularly or binocularly whatever acuity chart he chooses from the array displayed on the console in front of him. Separate controls permit the examiner to set room and target luminance independently over a wide range of values. Thus, acuity can be measured under "standard" conditions and also under conditions simulating those for which the patient may be having difficulty (for example, night driving).

The refractive error is measured subjectively by changing the power of the variable focus lenses in the beam of the projectors so that the targets seen separately by each eye in the concave mirror are brought into sharpest possible focus. With the variable focus

lenses, the power is changed smoothly and continuously (much like prism power is changed with a Risley prism) without interruption by dark intervals or movement of the visual field. Thus, determination of the optimum spherical power can be made quickly by whatever method the practitioner prefers. At the examiner's option, the *patient* can turn a single knob to make, for example, the letters on the red and green backgrounds equally black.

Measurement of astigmatism with the Vision Analyzer is extremely rapid and precise, and is one of its most remarkable features. The objective was to avoid hunting for cylinder power and axis separately, having to go back and change one after changing the other, and having to change the spherical component 0.25 D for each 0.50 D change in cylinder power. The basic idea is akin to the way phorias are measured with prisms. For a patient whose left eye deviates, when fusion is disrupted, by $7^\triangle$ up and out along a 45 degree meridian, one doesn't look for the power and axis of the neutralizing prism (that is, $7^\triangle$ base down and in at axis 135 degrees). Instead, examiners use a test

(such as the Maddox rod test, von Graefe prism-dissociation test, or cover test) that measures the components of the patient's phoria in the horizontal and vertical meridians. In the present example the phoria components would be $5^\triangle$ exo (base in) and $5^\triangle$ left hyper (base down left eye). (Most examiners are so used to thinking only of the two independent components of a phoria—$5^\triangle$ in each direction, in this case—that it requires some effort to realize that the components can be combined into the true total phoria of $7^\triangle$ at $45°$.)

The Vision Analyzer measures the *components* of ocular astigmatism. (As far as I have been able to learn, no other instrument has ever done that.) The patient simply looks at an obliquely oriented line and adjusts or reports when it appears sharpest; a similar setting is made for a vertical line. A pair of slightly tilted lines, one to either side of the "test line," improves the sensitivity of the measurement (Fig. 6). A small computer in the instrument quickly converts the patient's astigmatism in component form to the conventional polar coordinate form. (This is mathematically identical to the addition of the two $5^\triangle$ components of the phoria to give a $7^\triangle$ at 45 degrees). A unique and valuable feature of measuring the components of astigmatism with variable-focus crossed cylinders is the *independence* of measurements of the spherical and astigmatic portions of the refractive error. There is no need to change the sphere after changing the cylinder. This feature combined with continuously variable lens powers makes refracting with the Vision Analyzer extremely fast.

Refraction at different light and adaptation levels is easily performed with the Vision Analyzer. Hence, it is a simple matter to measure a patient's night myopia, for example. The light from the projector is all reflected back from the concave mirror in a small bundle about the size of the eye, rather than scattered back from the usual screen to cover most of the wall

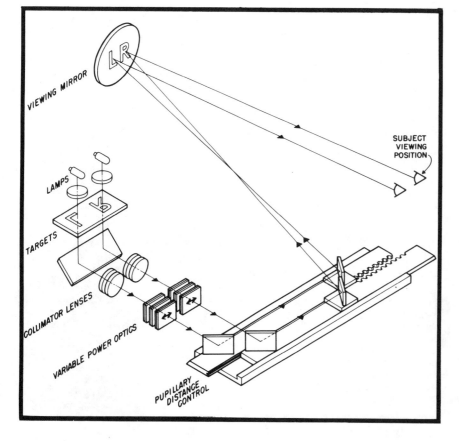

VIEWING MIRROR

SUBJECT VIEWING POSITION

LAMPS

TARGETS

COLLIMATOR LENSES

VARIABLE POWER OPTICS

PUPILLARY DISTANCE CONTROL

Fig. 5. Schematic of the optics in the Vision Analyzer.

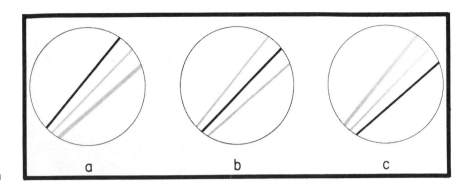

Fig. 6. An obliquely oriented line target is used with the Vision Analyzer to measure one component of the astigmatism. The surrounding pair of lines increases the sensitivity of the test. The correct setting is indicated when the patient reports neither *a* nor *c,* but *b.* The other component is measured with a vertical line target.

behind the patient. For this reason, the target brightness can be made enormously greater than that normally available in clinical refraction. This feature of the Vision Analyzer makes refraction possible even with fairly dense ocular opacities.

The Vision Analyzer has the capability of measuring other important vision functions. The projectors' narrow beams reflected by the mirror into each pupil permit binocular testing without need for polarizing filters or a septum. Fixation disparity can be quickly neutralized both horizontally and vertically to indicate the prism necessary to eliminate binocular stress created by a phoria. Stereopsis can be assessed qualitatively or measured using contour targets or random-dot stereograms. Fusional vergences are measured conventionally with variable-power prisms. And, near-point testing is performed with a swing-down unit to simulate a wide range of working distances.

I'll conclude by stating what I think is the most significant of the five innovations introduced in the Vision Analyzer. These innovations are 1) continuously variable spherical lenses, 2) continuously variable crossed-cylinder lenses, 3) measurement of two components of astigmatism, rather than of astigmatic power and axis, 4) remote refraction, with nothing in front of the eyes, and 5) separate binocular channels, without the need for Polaroids or septums.

I believe that within a few years, it will be generally agreed that the determination of astigmatic error is most quickly and accurately made through the measurement of two independent components of astigmatism, rather than by the Jackson flip-crossed-cylinder test. To appreciate what the "2-component method" does for astigmatism, it is useful to imagine what a standard refractor would look like if we didn't have Risley prisms to measure the components of a patient's phoria. Suppose we had to measure "phoria power and axis" as examiners now measure "cylinder power and axis." That would require another ring in the refractor, with a whole set of prisms of different powers mounted so that when the orientation of one prism changed, all the others would follow it. We would need a "Jackson flip prism," and the examiner would ask again and again, "Is A better (closer) than B," as he rotated the small prism to a new angle, and flipped it.

This illustration shows what a two-component system does to simplify and to make more accurate the measurement of phorias.

That same two-component system has exactly the same effect on measuring astigmatism, it allows the designer to eliminate the ring of "geared-together cylinders," and it lets the examiner or the patient "dial in" his precise correcting cylinder with as little effort as is customarily expended in measuring the "two phorias," using a Maddox rod and Risley prisms. It is for this reason that I believe the cylinder measuring technique of the Vision Analyzer will be listed in future textbooks as the most significant of the several new features I've described today.   **AOA**

Submitted for publication in the JAOA in April, 1977.
*University of California, Berkeley
Department of Physics
Berkeley, CA 94720*

ABOUT OUR AUTHOR—Luis W. Alvarez received his B.Sc. from the University of Chicago in 1932, and M.Sc. in 1934, and his Ph.D. in 1936. Dr. Alvarez joined the Radiation Laboratory of the University of California, where he is now a professor, as a research fellow in 1936.

Early in his scientific career, Dr. Alvarez worked concurrently in the fields of optics and cosmic rays. He is co-discoverer of the "East-West effect" in cosmic rays. For several years he concentrated his work in the field of nuclear physics.

During the war he was responsible for three important radar systems; the best known of these is a blind landing system of civilian as well as military value (GCA, or Ground-Controlled Approach). He also worked with Fermi at the first nuclear reactor, and later at Los Alamos.

In 1960 he was named "California Scientist of the Year" for his research work on high-energy physics. In 1961

he was awarded the Einstein Medal for his contribution to the physical sciences.

In 1964 he was awarded the National Medal of Science for contributions to high-energy physics; in 1965 he received the Michelson Award, and in 1968 the Nobel Prize in Physics.

21

The Cretaceous-Tertiary Iridium Anomaly and the Asteroid Impact Theory

Frank Asaro

In early October 1977, Luis W. Alvarez and his son Walter dropped into my office at the Lawrence Berkeley Laboratory (LBL). Luie introduced Walt who had recently come to Berkeley from Columbia's Lamont Doherty Geological Observatory. The Alvarezes had developed a project which melded their disciplines, physics and geology, and they now needed some chemistry help.

Walt, while studying the geology of the Apennines near Gubbio, Italy, had found an interesting 1 cm thick clay layer that marked the 65-million-year-old boundary between the Cretaceous and Tertiary geological periods. This thin clay which was sandwiched between two enormous limestone formations, had Cretaceous fossils just below it and Tertiary just above. This change was part of a world-wide extinction pattern, of species and other taxa, including dinosaurs, that had taken place abruptly at the end of the Cretaceous. However, what is geologically abrupt can take a long time by human standards, thousands of years or more. Luie and Walt thought that if they could determine how long it took to deposit this clay and thus how long it took for the Cretaceous-Tertiary (K-T) extinctions, it would shed light on the extinction mechanisms. Walt, with William Lowrie from the Institute for Geophysics in Zurich, Switzerland, had correlated the Upper Cretaceous geomagnetic zones at Gubbio with those on the Atlantic, Pacific, and Indian Ocean bottoms. They associated the K-T boundary with a particular paleomagnetic zone (Anomaly 29-reversed) and from that estimated the average sedimentation rate at Gubbio.

Luie thought that the time to deposit the 1 cm of clay from Gubbio could be estimated from the extraterrestrial material it contained, as meteoritic debris is continuously deposited on the Earth: about ten million kilograms a year. Luie had calculated that the element iridium would be the best monitor for this extraterrestrial material. The Earth's crust is depleted in iridium, which was carried to its core by molten iron when the Earth was young. Although in the last 4 billion years the Earth has accumulated considerable extraterrestrial material containing iridium, its crust still only has one ten-thousandth the iridium abundance in meteorites.

When Luie and Walt approached me to make the neutron activation measurements for iridium, I was not enthusiastic. My group had measured about 15,000 ancient pottery sherds by neutron activation to determine their place of origin and had seldom detected iridium. Also, we had unsuccessfully tried to measure iridium in samples from Dr. Andrei Sarna-Wojcicki of the U.S.G.S., and I was concerned about a conflict of interest. Discussions with Andrei established there was no conflict, but I still had reservations.

When I explained my doubts, Luie and Walt countered that the limestone would dissolve in acid, and enormous enhancements of iridium would be obtained. Although I still didn't think the experiment would work, there were three reasons to try it. First, our group had recently obtained a much larger germanium gamma-ray detector, so our measurement sensitivity would increase significantly. Second, and most important, the abundance of the 25 other elements we would measure in Walt's samples would be useful for him even if we detected no iridium. Finally, I thought it would be interesting to work with Luie, as I had been intrigued by his cosmic-ray search for undiscovered chambers in the Egyptian pyramids.

The two stable iridium isotopes, iridium-191 and iridium-193, upon neutron irradiation are partially transformed into the radioactive isotopes 74 day iridium-192 and 19 hour iridium-194, respectively. The characteristic gamma radiation emitted by these radioisotopes is used to measure the iridium abundance in the samples. Irradiated rock samples, however, produce enormous radiations from the more abundant elements which mask those of iridium. By waiting several weeks for many of the radiations to die away, particularly those of sodium and rare earths, iridium at the level of a fraction of a part per billion (ppb) can be detected. Without chemically separating the iridium, only iridium-192 is useful; its abundant 468 keV gamma ray is relatively free from interferences.

Walt brought us twelve rock samples from the Gubbio region which he had collected that summer. There was no funding for these measurements, so we had to squeeze them in with our other work. This situation was aggravated by the failure of one of our gamma-ray detectors.

Because of this failure an irradiation backlog of 300 samples had built up, and these had to be measured before the new work could proceed. It was not until

April that the Gubbio samples received their 8 hour irradiation in the Berkeley reactor. We detected no 468 keV iridium-192 gamma rays with our regular detector and decided to try the new one which was 18 times larger. At 1:45 PM on June 21, 1978, we put sample 1010-U, the acid-insoluble residue from the red clay in the upper half of the K-T boundary from the Contessa section, on the new detector for 224 minutes. We saw a 468 keV iridium-192 line that measured 3 ppb of iridium. We found later that iridium was lost in the acid dissolution, and the true value was 9 ppb.

We were happy to have found iridium but disturbed to have found so much. From a sedimentation rate of 10 millimeters per 1000 years, we expected 0.004 ppb, which in the acid-insoluble fraction would be 0.08 ppb since we found the limestones were 5% clay. Something was wrong with our ideas of how the iridium got into the clay.

Because iridium was in the K-T boundary layer and this layer was distributed world-wide, we concluded that iridium should be distributed world-wide. Since we knew of no terrestrial source which would produce and deliver that much iridium, the source had to be extraterrestrial.

In July Luie was checking out explanations for the iridium anomaly, and I was checking the validity of our experimental procedures. I now had more work than I could handle, so we asked Helen Michel, my nuclear chemist associate for 25 years, to join our team. Helen, who was in charge of the neutron activation facility at LBL, had measured most of the Gubbio samples. She also had a hobby in plant biology, developing new orchid hybrids.

Walt was planning his 1978 summer in Italy and Denmark to collect more samples to confirm our discovery. We had found the iridium distribution so sharp that a 10 centimeter uncertainty in the sample position above and below the K-T boundary obscured the pattern. We needed higher precision in sampling than had been previously necessary in geological collections. Walt then collected samples across the Gubbio K-T boundary and determined their stratigraphic positions to better than 1 cm. He also collected from five other Apennine locations and from the two center sections of the K-T boundary at Stevns Klint in Denmark. We submitted an abstract to the Spring American Geophysical Union meeting claiming the discovery of extraterrestrial iridium at Gubbio.

Luie considered whether Earth's passage through giant nebular clouds would provide the K-T iridium anomaly. These clouds would have also provided hydrogen, which would have reacted with the oxygen in Earth's atmosphere to suffocate the dinosaurs and other species. The process, however, was so slow that the plant life on Earth should have been able to maintain the oxygen supply; Luie discarded the idea.

Another source of the iridium anomaly could be a supernova, which others had considered as the cause of the K-T extinctions. Luie calculated the probability that another star would be close enough to our Sun to provide the iridium anomaly on the Earth—one in a billion in a hundred million years. Though unlikely, it was an easily testable hypothesis. A supernova which produced the iridium would also have produced plutonium-244, whose 80 million year half-life would assure that most of it would still be around. Further, the relative abundance of the two iridium isotopes in the K-T boundary produced by a special supernova would probably be different than found for the Solar System. We used neutron activation analyses to check both predications.

Ten-hour plutonium-245 is produced by neutron irradiation of plutonium-244, and its gamma rays and those of its daughter, americium-245, are detectable. To purify the irradiated plutonium fraction before measurement required us to move quickly before the plutonium-245 decayed to an undetectable level. For 36 hours Helen and I continuously ran chemical separations, while Luie and Walt fed us pizza, snacks, and chili. We didn't find any plutonium-244 in our Italian samples (except for one that contained laboratory plutonium-244), and in March 1979 we concluded that not more than 10% of the iridium could have been produced by a supernova.

We deduced the iridium-191/iridium-193 isotopic ratio by measuring the gamma-ray ratio of iridium-192/iridium-194 after neutron irradiation of the K-T iridium. As the iridium-194 half-life was short and complete chemical purifications were required, more pizza and chili sessions were necessary. Walt's wife Milly even dropped by with strawberry ice cream and chocolate chip cookies during one long night. But we ran into all sorts of problems. The relative neutron capture cross sections of the iridium isotopes in two research reactors were so position-sensitive that we had to place standards symmetrically about the unknown.

We also discovered that the measured gamma-ray ratios depended on the sample's distance from the detector. Following a suggestion by Luie, we suspended an optically flat glass plate on three sapphire ball bearings and placed our sample at its center. In June 1979 we knew there was less than a 1.5% difference between the isotopic ratios of the K-T iridium and ordinary terrestrial iridium. We then sent in an abstract to the Geological Society of America describing our experiment and claiming that a supernova was not responsible for the K-T iridium.

We had had enough of checking low-probability scenarios, no matter how testable. Also, there had been a half-dozen major extinctions in the last half-billion years, so we were looking for a repetitive process. Therefore, our mechanism should provide a world-wide iridium anomaly, it should be able to selectively cause enormous extinctions of life, and it should repeat every hundred million years or so.

Early on, Luie thought that the extinctions might be caused by tidal waves which selectively drowned animals. A friend, Chris McKee, had suggested that the iridium could have come from the impact of a 10 km asteroidal body in the ocean. But drowning just couldn't cause the extinction of 75% of the species and 50% of the genera which existed 65 million years ago, so we discarded that idea.

Luie also considered other sources of hydrogen which would use up part of Earth's atmospheric oxygen and cause suffocation. The Sun was a promising prospect, but nothing coming from it would bear closer scrutiny as an extinction cause. Neither novas, super solar flares, nor any known solar activity repeated five times in 500 million years. Then Luie tried to get hydrogen (and iridium) from Jupiter, dislodging material with impacting comets and asteroids, but that didn't work out either. On July 14, 1979, Luie wrote a detailed letter to Helen, Walt, and me describing how a comet or Apollo asteroid grazing Earth would fragment in the atmosphere, its pieces burning with a dense smoke as they fell to Earth. The smoke would obscure the Sun, which would turn off photosynthesis, and the animals would die of starvation. We abandoned this idea because grazing collisions happen rarely, the large Apollo objects do not fragment easily, and the smoke particles would settle quickly. But Luie was getting closer. When he had the object hit the Earth, he struck paydirt, and the asteroid impact theory was born. We were all extremely critical, but the theory survived and by August 4 we had a rough draft of our Science paper circulating.

There was, however, a fly in the ointment. The iridium from an asteroid impact should be distributed worldwide, and we had found it only in Gubbio. So when Walt arrived with his fine K-T samples from Italy and Denmark, we immediately ground up a kilogram of Danish boundary in preparation for neutron activation. Because we had an irradiation suitable for Ir analysis already scheduled in a few days hence, we quickly prepared one sample for that irradiation and a second for our normal neutron activation procedures.

On November 4, 1978, we began measuring the first sample and were discouraged when we detected no iridium. But Luie, a constant source of inspiration, said we just had to take a ten thousand times larger sample and do chemistry before and after the irradiation. For-

tunately other experiments delayed us. Meanwhile, the second sample, which was being processed automatically, was completely ignored because we had not found iridium in the first. By June 1979, the activation profile of the second sample was complete except for the iridium measurements on the big detector. A cursory check of the iron abundances in the two samples, however, did not agree, nor did those of many other elements. Our neutron activation analysis precision for finely ground homogenized samples is 1% or 2%. So something was wrong, and top priority was *always* given to solving data discrepancies. In a short while we were able to prove that the person preparing sample 1 broke it because of the hurry to get it ready. Working next to that individual was another sample preparer who was processing oil shales from the western United States. When the first sample 1 broke, a new sample 1 was inadvertently made from the oil shale rather than the Danish K-T boundary. As the agreement between the chemical abundances of sample 1 and the particular oil shale (NAA sample 1047—N) was the same as the average of the counting errors, there was no doubt about the substitution.

We were sitting in Helen's office when it occurred to someone that it might be useful to see if there was iridium in the second Danish K-T boundary sample. When we looked at its gamma-ray spectrum, which had already been taken with our small germanium detector, the two iridium gamma-ray peaks looked like mountains! Of all the time that we spent on this project, that was my happiest day. I was sure then that we were right.

In the second draft of the paper, on August 25, 1979, the Danish iridium data were included. Our final manuscript was sent to Science the day before Thanksgiving, when all measurements had been confirmed.

Some reviewers commented that since Italy and Denmark were geographically close, we had not proven that the iridium deposition was worldwide. By the time the paper appeared in June 1980, we had found the iridium anomaly in New Zealand samples generously provided by Dale Russell from the National Museums of Canada. Since you cannot get further from Europe than New Zealand and still be on Earth, we felt we had proven that the iridium deposition was worldwide.

We still do not know the answer to our first question, "How long did it take to deposit the K-T boundary clay in Italy?"

6 June 1980, Volume 208, Number 4448

SCIENCE

Extraterrestrial Cause for the Cretaceous-Tertiary Extinction

Experimental results and theoretical interpretation

Luis W. Alvarez, Walter Alvarez, Frank Asaro, Helen V. Michel

In the 570-million-year period for which abundant fossil remains are available, there have been five great biological crises, during which many groups of organisms died out. The most recent of the great extinctions is used to define the boundary between the Cretaceous and Tertiary periods, about 65 million years

microscopic floating animals and plants; both the calcareous planktonic foraminifera and the calcareous nannoplankton were nearly exterminated, with only a few species surviving the crisis. On the other hand, some groups were little affected, including the land plants, crocodiles, snakes, mammals, and many kinds

Summary. Platinum metals are depleted in the earth's crust relative to their cosmic abundance; concentrations of these elements in deep-sea sediments may thus indicate influxes of extraterrestrial material. Deep-sea limestones exposed in Italy, Denmark, and New Zealand show iridium increases of about 30, 160, and 20 times, respectively, above the background level at precisely the time of the Cretaceous-Tertiary extinctions, 65 million years ago. Reasons are given to indicate that this iridium is of extraterrestrial origin, but did not come from a nearby supernova. A hypothesis is suggested which accounts for the extinctions and the iridium observations. Impact of a large earth-crossing asteroid would inject about 60 times the object's mass into the atmosphere as pulverized rock; a fraction of this dust would stay in the stratosphere for several years and be distributed worldwide. The resulting darkness would suppress photosynthesis, and the expected biological consequences match quite closely the extinctions observed in the paleontological record. One prediction of this hypothesis has been verified: the chemical composition of the boundary clay, which is thought to come from the stratospheric dust, is markedly different from that of clay mixed with the Cretaceous and Tertiary limestones, which are chemically similar to each other. Four different independent estimates of the diameter of the asteroid give values that lie in the range 10 ± 4 kilometers.

ago. At this time, the marine reptiles, the flying reptiles, and both orders of dinosaurs died out (1), and extinctions occurred at various taxonomic levels among the marine invertebrates. Dramatic extinctions occurred among the

of invertebrates. Russell (2) concludes that about half of the genera living at that time perished during the extinction event.

Many hypotheses have been proposed to explain the Cretaceous-Tertiary (C-T)

extinctions (3, 4), and two recent meetings on the topic (5, 6) produced no sign of a consensus. Suggested causes include gradual or rapid changes in oceanographic, atmospheric, or climatic conditions (7) due to a random (8) or a cyclical (9) coincidence of causative factors; a magnetic reversal (10); a nearby supernova (11); and the flooding of the ocean surface by fresh water from a postulated arctic lake (12).

A major obstacle to determining the cause of the extinction is that virtually all the available information on events at the time of the crisis deals with biological changes seen in the paleontological record and is therefore inherently indirect. Little physical evidence is available, and it also is indirect. This includes variations in stable oxygen and carbon isotopic ratios across the boundary in pelagic sediments, which may reflect changes in temperature, salinity, oxygenation, and organic productivity of the ocean water, and which are not easy to interpret (13, 14). These isotopic changes are not particularly striking and, taken by themselves, would not suggest a dramatic crisis. Small changes in minor and trace element levels at the C-T boundary have been noted from limestone sections in Denmark and Italy (15), but these data also present interpretational difficulties. It is noteworthy that in pelagic marine sequences, where nearly continuous deposition is to be expected, the C-T boundary is commonly marked by a hiatus (3, 16).

In this article we present direct physical evidence for an unusual event at exactly the time of the extinctions in the planktonic realm. None of the current hypotheses adequately accounts for this evidence, but we have developed a hypothesis that appears to offer a satisfactory explanation for nearly all the available paleontological and physical evidence.

Luis Alvarez is professor emeritus of physics at Lawrence Berkeley Laboratory, University of California, Berkeley 94720. Walter Alvarez is an associate professor in the Department of Geology and Geophysics, University of California, Berkeley. Frank Asaro is a senior scientist and Helen Michel is a staff scientist in the Energy and Environment Division of Lawrence Berkeley Laboratory.

Identification of Extraterrestrial Platinum Metals in Deep-Sea Sediments

This study began with the realization that the platinum group elements (platinum, iridium, osmium, and rhodium) are much less abundant in the earth's crust and upper mantle than they are in chondritic meteorites and average solar system material. Depletion of the platinum group elements in the earth's crust and upper mantle is probably the result of concentration of these elements in the earth's core.

Pettersson and Rotschi (17) and Goldschmidt (18) suggested that the low concentrations of platinum group elements in sedimentary rocks might come largely from meteoritic dust formed by ablation when meteorites passed through the atmosphere. Barker and Anders (19) showed that there was a correlation between sedimentation rate and iridium concentration, confirming the earlier suggestions. Subsequently, the method was used by Ganapathy, Brownlee, and Hodge (20) to demonstrate an extraterrestrial origin for silicate spherules in deep-sea sediments. Sarna-Wojcicki *et al.* (21) suggested that meteoritic dust accumulation in soil layers might enhance the abundance of iridium sufficiently to permit its use as a dating tool. Recently, Crocket and Kuo (22) reported iridium abundances in deep-sea sediments

and summarized other previous work.

Considerations of this type (23) prompted us to measure the iridium concentration in the 1-centimeter-thick clay layer that marks the C-T boundary in some sections in the Umbrian Apennines, in the hope of determining the length of time represented by that layer. Iridium can easily be determined at low levels by neutron activation analysis (NAA) (24) because of its large capture cross section for slow neutrons, and because some of the gamma rays given off during de-excitation of the decay product are not masked by other gamma rays. The other platinum group elements are more difficult to determine by NAA.

Italian Stratigraphic Sections

Many aspects of earth history are best recorded in pelagic sedimentary rocks, which gradually accumulate in the relatively quiet waters of the deep sea as individual grains settle to the bottom. In the Umbrian Apennines of northern peninsular Italy there are exposures of pelagic sedimentary rocks representing the time from Early Jurassic to Oligocene, around 185 to 30 million years ago (25). The C-T boundary occurs within a portion of the sequence formed by pink limestone containing a variable amount of clay. This limestone, the *Scaglia ros-*

sa, has a matrix of coccoliths and coccolith fragments (calcite platelets, on the order of 1 micrometer in size, secreted by algae living in the surface waters) and a rich assemblage of foraminiferal tests (calcite shells, generally in the size range 0.1 to 2.0 millimeters, produced by single-celled animals that float in the surface waters).

In some Umbrian sections there is a hiatus in the sedimentary record across the C-T boundary, sometimes with signs of soft-sediment slumping. Where the sequence is apparently complete, foraminifera typical of the Upper Cretaceous (notably the genus *Globotruncana*) disappear abruptly and are replaced by the basal Tertiary foraminifer *Globigerina eugubina* (16, 26). This change is easy to recognize because *G. eugubina*, unlike the globotruncanids, is too small to see with the naked eye or the hand lens (Fig. 1). The coccoliths also show an abrupt change, with disappearance of Cretaceous forms, at exactly the same level as the foraminiferal change, although this was not recognized until more recently (27).

In well-exposed, complete sections there is a bed of clay about 1 cm thick between the highest Cretaceous and the lowest Tertiary limestone beds (28). This bed is free of primary $CaCO_3$, so there is no record of the biological changes during the time interval represented by the clay. The boundary is further marked by a zone in the uppermost Cretaceous in which the normally pink limestone is white in color. This zone is 0.3 to 1.0 meter thick, varying from section to section. Its lower boundary is a gradational color change; its upper boundary is abrupt and coincides with the faunal and floral extinctions. In one section (Contessa) we can see that the lower 5 mm of the boundary clay is gray and the upper 5 mm is red, thus placing the upper boundary of the zone in the middle of the clay layer.

The best known of the Umbrian sections is in the Bottaccione Gorge near Gubbio. Here some of the first work on the identification of foraminifera in thin section was carried out (29); the oldest known Tertiary foraminifer, *G. eugubina*, was recognized, named, and used to define the basal Tertiary biozone (16, 26); the geomagnetic reversal stratigraphy of the Upper Cretaceous and Paleocene was established, correlated to the marine magnetic anomaly sequence, and dated with foraminifera (30); and the extinction of most of the nannoplankton was shown to be synchronous with the disappearance of the genus *Globotruncana* (27).

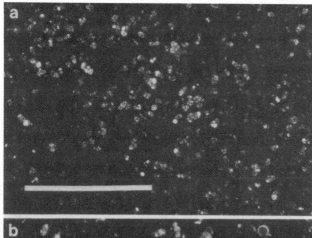
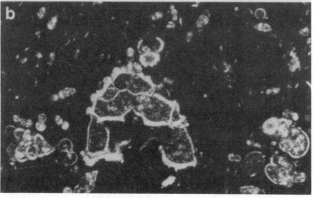

Fig. 1. Photomicrographs of (a) the basal bed of the Tertiary, showing *Globigerina eugubina*, and (b) the top bed of the Cretaceous, in which the largest foraminifer is *Globotruncana contusa*. Both sections are from the Bottaccione section at Gubbio; they are shown at the same scale and the bar in (a) is 1 mm long.

Results from the Italian Sections

Our first experiments involved NAA of nine samples from the Bottaccione section (two limestone samples from immediately above and below the boundary plus seven limestone samples spaced over 325 m of the Cretaceous). This was supplemented by three samples from the nearby Contessa section (two from the boundary clay and one from the basal Tertiary bed). Stratigraphic positions of these samples are shown in Fig. 2.

Twenty-eight elements were selected for study because of their favorable nuclear properties, especially neutron capture cross sections, half-lives, and gamma-ray energies. The results of these analyses are presented in Fig. 3 on a logarithmic plot to facilitate comparison of the relative changes in elements over a wide range of concentrations. The only preparation given to these samples was removal of the $CaCO_3$ fraction by dissolution in dilute nitric acid. Figure 3 shows elemental abundances as gram of element per gram of insoluble clay residue. The limestones generally contain about 5 percent clay. The boundary clay layer contains about 50 percent $CaCO_3$, but this is coarse-grained calcite that probably crystallized during deformation long after deposition. Chemical yields of iridium in the acid-insoluble fraction averaged 44 percent for the red and gray Contessa boundary clays, and this value was assumed for all the other samples.

Twenty-seven of the 28 elements show very similar patterns of abundance variation, but iridium shows a grossly different behavior; it increases by a factor of about 30 in coincidence with the C-T boundary, whereas none of the other elements as much as doubles with respect to an "average behavior" shown in the lower right panel of Fig. 3. Figure 4 shows a typical gamma-ray spectrum used to measure the Ir abundance, 5.5 parts per billion (ppb).

In follow-up experiments we analyzed five more samples from the Bottaccione section, eight from Gorgo a Cerbara (28 kilometers north of Gubbio), and four large samples of the boundary clay from the two sections near Gubbio and two sections about 30 km to the north (31). The chemical yield of iridium in the acid-insoluble fraction was 95 ± 5 percent for the Contessa boundary clay, and a 100 percent yield was assumed for all the other samples.

Figure 5 shows the results of 29 Ir analyses completed on Italian samples. Note that the section is enlarged and that the scale is linear in the vicinity of the C-T boundary, where details are important,

but changes to logarithmic to show results from 350 m below to 50 m above the boundary. It is also important to note that analyses from five stratigraphic sections are plotted on the same diagram on the basis of their stratigraphic position above or below the boundary. Because slight differences in sedimentation rate probably exist from one section to the next, the chronologic sequence of samples from different sections may not be exactly correct. Nevertheless, Fig. 5 gives a clear picture of the general trend of iridium concentrations as a function of stratigraphic level.

The pattern, based especially on the samples from the Bottaccione Gorge and Gorgo a Cerbara, shows a steady background level of ~ 0.3 ppb throughout the Upper Cretaceous, continuing into the uppermost bed of the Cretaceous. The background level in the acid-insoluble residues is roughly comparable to the iridium abundance measured by other

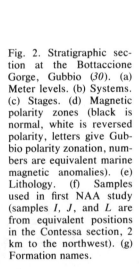

Fig. 2. Stratigraphic section at the Bottaccione Gorge, Gubbio (30). (a) Meter levels. (b) Systems. (c) Stages. (d) Magnetic polarity zones (black is normal, white is reversed polarity, letters give Gubbio polarity zonation, numbers are equivalent marine magnetic anomalies). (e) Lithology. (f) Samples used in first NAA study (samples I, J, and L are from equivalent positions in the Contessa section, 2 km to the northwest). (g) Formation names.

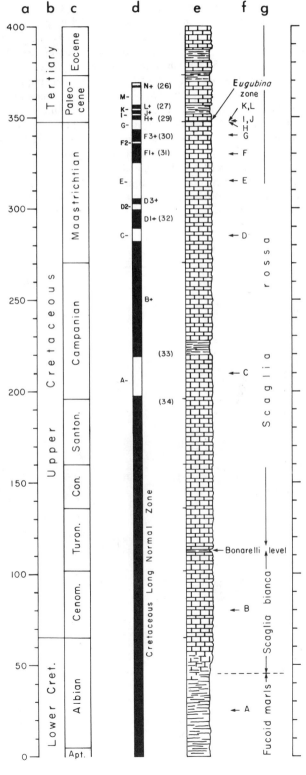

workers (*19, 22, 32*) in deep-sea clay sediments. This level increases abruptly, by a factor of more than 30, to 9.1 ppb, the Ir abundance in the red clay from the Contessa section. Iridium levels are high in clay residues from the first few beds of Tertiary limestone, but fall off to background levels by 1 m above the boundary. For comparison, the upper dashed line in Fig. 5 shows an exponential decay from the boundary clay Ir level with a half-height of 4.6 cm.

To test the possibility that iridium might somehow be concentrated in clay layers, we subsequently analyzed two red clay samples from a short distance below the C-T boundary in the Bottaccione section. One is from a distinctive clay layer 5 to 6 mm thick, 1.73 m below the boundary; the other is from a 1- to 2-mm bedding-plane clay seam 0.85 m below the boundary. The whole-rock analyses of these clays showed no detectable Ir with limits of 0.5 and 0.24 ppb, respectively. Thus neither clay layers from below the C-T boundary nor clay components in the limestone show evidence of Ir above the background level.

The Danish Section

To test whether the iridium anomaly is a local Italian feature, it was desirable to analyze sediments of similar age from another region. The sea cliff of Stevns Klint, about 50 km south of Copenhagen, is a classical area for the C-T boundary and for the Danian or basal stage of the Tertiary. A collection of up-to-date papers on this and nearby areas has recently been published, which includes a full bibliography of earlier works (*6*, vol. 1).

Our samples were taken at Højerup Church (*33*). At this locality the Maastrichtian, or uppermost Cretaceous, is represented by white chalk containing black chert nodules in undulating layers with amplitudes of a few meters and wavelengths of 10 to 50 m (*14*). These undulations are considered to represent bryozoan banks (*34*). The C-T boundary is marked by the *Fiskeler*, or fish clay, which is up to 35 cm thick in the deepest parts of the basins between bryozoan banks (*14*) but commonly only a few centimeters thick, thinning or disappearing, over the tops of the banks. The fish clay

at Højerup Church was studied in detail by Christensen *et al.* (*14*), who subdivided it into four thin layers; we analyzed a sample mixing the two internal layers (units III and IV of Christensen *et al.*). These layers are black or dark gray, and the lower one contains pyrite concretions; the layers below and above (II and V) are light gray in color. Undisturbed lamination in bed IV suggests that no bottom fauna was present during its deposition (*14*). Above the fish clay, the Cerithium limestone is present to a thickness of about 50 cm in the small basins, disappearing over the banks. It is hard, yellowish in color, and cut by abundant burrows. Above this is a thick bryozoan limestone.

The presence of a thin clay layer at the C-T boundary in both the Italian and Danish sections is quite striking. However, there are notable differences as well. The Danish sequence was clearly deposited in shallower water (*35*), and the Danish limestones preserve an extensive bottom-dwelling fauna of bivalves (*36*), echinoderms (*37*), bryozoans (*38*), and corals (*39*).

Foraminiferal (*40*) and coccolith (*41*)

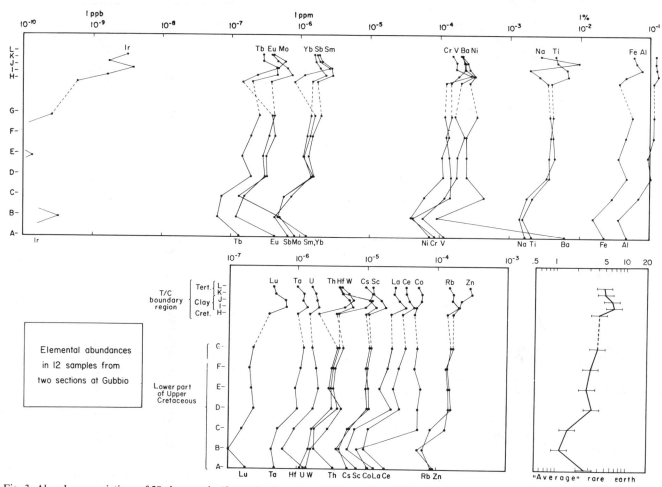

Fig. 3. Abundance variations of 28 elements in 12 samples from two Gubbio sections. Flags on "average rare earth" diagram are ± 30 percent and include all rare earth data.

zonation indicates that the C-T boundaries at Gubbio and Stevns Klint are at least approximately contemporaneous, and they may well be exactly synchronous. However, no paleomagnetic results are available from Stevns Klint, so synchroneity cannot be tested by reversal stratigraphy.

Results from the Danish Section

Seven samples were taken from near the C-T boundary (Fig. 6). Fractions of each sample were treated with dilute nitric acid, and the residues were filtered, washed, and heated to 800°C. The yield of acid-insoluble residue was 44.5 percent for the boundary fish clay and varied from 0.62 to 3.3 percent for the pelagic limestones (Table 1).

Neutron activation analysis (24) and x-ray fluorescence (XRF) (42, 43) measurements were made on all seven samples both before and after the acid treatment. This measurement regime was more sophisticated than that used for the Italian sections studied earlier, and 48 elements were determined.

The Cretaceous and Tertiary acid-insoluble residues were each rather homogeneous in all of the measured elements, and the two groups were only slightly different from each other. The residue from the clay boundary layer was much different in composition (Figs. 7 and 8 and Table 2), and this suggested a different source for the boundary clay.

As shown in Table 1, the Ir in the boundary layer residue rises by about a factor of 160 over the background level (~ 0.26 ppb). A 1-cm thickness of this layer would have about 72×10^{-9} gram of Ir per square centimeter. To test whether there is enough Ir in the seawater to contribute to this value, we made a measurement of the Ir in the ocean off the central California coast. In water passed through a 0.45-μm filter Ir was undetected, giving an upper limit of 4×10^{-13} g of Ir per gram of seawater. If the depth of the shallow ancient Danish sea is assumed to be less than 100 m and our limit for Ir in seawater is applicable, then the maximum Ir in the 100-m column of water should be 4×10^{-9} g/cm^2, almost a factor of 20 lower than the observed value. So there was probably not enough Ir stored in the seawater to explain the amount observed in the Danish boundary. Iridium has apparently not been detected in seawater. One tabulated result (44) contains a typographical error that places the value for indium in

the atomic number position of iridium. Iridium has been detected (45) in a warm spring on Mount Hood in northern California at a level of 7×10^{-12} g per gram of water, and in two cold-water sources at levels of 3×10^{-13} to 4×10^{-13} g per gram of water. Many other cold-water sources in this area had Ir levels less than 1×10^{-13} g/g.

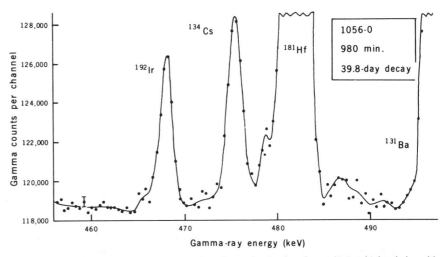

Fig. 4. Typical gamma-ray spectrum used to determine Ir abundance (5.5 ppb) in nitric acid-insoluble residues without further chemistry. Note that the entire spectrum rests on a background of 118,000 counts. Detector volume was 128 cm^3; length of count was 980 minutes. Count began 39.8 days after the end of the irradiation. Residue is from a Tertiary limestone sample taken 2.5 cm above the boundary at Gorgo a Cerbara (see Fig. 5).

The Boundary Layers

The whole-rock composition of the Contessa boundary layer (a mixture of red and gray clay) is shown in Table 3. There are two recognizable sublayers, each about 0.5 cm thick, the upper being red in color and the lower gray. The elemental iron content, which may explain

Fig. 5. Iridium abundances per unit weight of $2N$ HNO$_3$ acid-insoluble residues from Italian limestones near the Tertiary - Cretaceous boundary. Error bars on abundances are the standard deviations in counting radioactivity. Error bars on stratigraphic position indicate the stratigraphic thickness of the sample. The dashed line above the boundary is an "eyeball fit" exponential with a half-height of 4.6 cm. The dashed line below the boundary is a best fit exponential (two points) with a half-height of 0.43 cm. The filled circle and error bar are the mean and standard deviation of Ir abundances in four large samples of boundary clay from different locations.

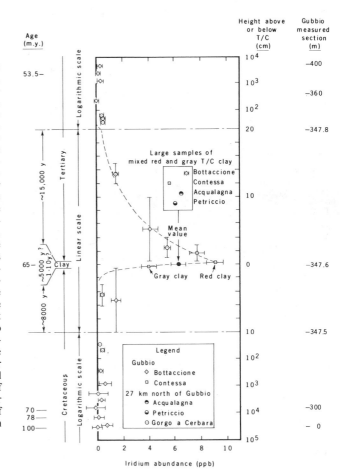

the color, is significantly higher in the residue of the red layer (7.7 versus 4.5 percent) than in the gray, and so is the Ir (9.1 ± 0.6 versus 4.0 ± 0.6 ppb). Boundary samples were analyzed from the Bottaccione Gorge nearby and two other areas about 30 km to the north.

In samples taken near the Italian boundary layer, the chemical compositions of all clay fractions were roughly the same except for the element Ir. However, there are discernible differences, as shown in Fig. 9, which suggest that at least part of the boundary layer clay had a different origin than the Cretaceous and Tertiary clays.

The Danish C-T boundary clay is somewhat thicker than 1 cm and is divided into four layers, as mentioned earlier. Only a single mixed sample from the two middle layers was measured, so no information is available on the chemical variations within the boundary. The average Ir abundance is 29 ppb in the whole rock or 65 ppb based on the weight of acid-insoluble residue.

The whole-rock abundances and mineral composition of the Danish boundary clay are shown in Table 3, and the abundances of pertinent trace elements are shown in Table 2. The major silicate minerals that must be present were not detected, so the other mineral abundances were normalized to give the amount of

Table 1. Abundance of iridium in acid-insoluble residues in the Danish section.

Sample*	Abundance of iridium (ppb)	Abundance† of acid-insoluble residues (%)
SK, +2.7 m	< 0.3	3.27
SK, +1.2 m	< 0.3	1.08
SK, +0.7 m	0.36 ± 0.06	0.836
Boundary	41.6 ± 1.8‡	44.5
SK, −0.5 m	0.73 ± 0.08	0.654
SK, −2.2 m	0.25 ± 0.08	0.621
SK, −5.4 m	0.30 ± 0.16	0.774

*Numerical values are the distances above (+) or below (−) the boundary layer; SK, Stevns Klint. †The boundary layer has a much higher proportion of clay than the pelagic limestones above and below. ‡Some iridium dissolved in the nitric acid. The whole-rock abundance was 28.6 ± 1.3 ppb.

calcite expected from the calcium measurement. The boundary clay fraction is far different chemically from the limestone clay fractions above and below, which are similar to each other. Pyrite is present in the boundary clay, and elements that form water-insoluble sulfides are greatly enhanced in this layer. The trace elements that are depleted are those that often appear as clay components. The element magnesium is an

exception. Its enhancement may be due to replacement of iron in the clay lattices in the sulfide environment or to a different, more mafic source for the boundary layer clay than for the Tertiary and Cretaceous clays.

Recent unpublished work by D. A. Russell of the National Museums of Canada and by the present authors has shown that the boundary layer whole-rock concentration of Ir in a section near Woodside Creek, New Zealand, is approximately 20 times the average concentration in the adjacent Cretaceous and Tertiary limestones.

A Sudden Influx of Extraterrestrial Material

To test whether the anomalous iridium at the C-T boundary in the Gubbio sections is of extraterrestrial origin, we considered the increases in 27 of the 28 elements measured by NAA that would be expected if the iridium in excess of the background level came from a source with the average composition of the earth's crust. The crustal Ir abundance, less than 0.1 ppb (19, 22), is too small to be a worldwide source for material with an Ir abundance of 6.3 ppb, as found near Gubbio. Extraterrestrial sources with Ir levels of hundreds of parts per

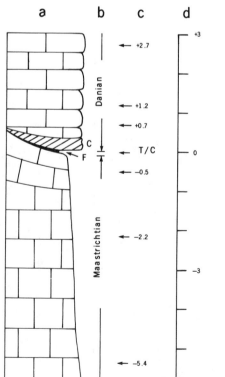

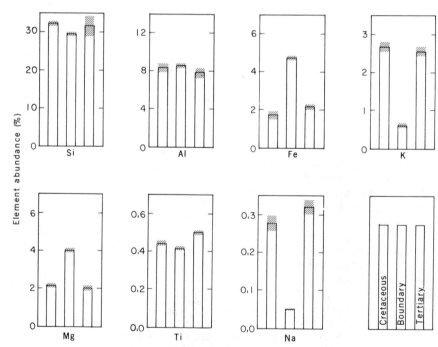

Fig. 6 (left). Stratigraphic section at Højerup Church, Stevns Klint, Denmark. (a) Lithology (C, Cerithium limestone; F, fish clay). (b) Stages. (c) Samples analyzed in this study. (d) Meter levels. Analytical results are given in Tables 1 to 3.   Fig. 7 (right). Major element abundances in acid-insoluble fractions from Danish rocks near the Tertiary-Cretaceous boundary. The crosshatched areas for the Cretaceous and Tertiary values each represent root-mean-square deviations for three samples. (Only two measurements of magnesium and silicon were included in the Cretaceous values.) For the boundary sample the crosshatched areas are the standard deviations associated with counting errors. Measurements of silicon and magnesium were done by XRF (42), all others were by NAA.

billion or higher are more likely to have produced the Ir anomaly. Figure 10 shows that if the source had an average earth's crust composition (46), increases significantly above those observed would be expected in all 27 elements. However, for a source with average carbonaceous chondrite composition (46), only nickel should show an elemental increase greater than that observed. As shown in Fig. 11, such an increase in nickel was not observed, but the predicted effect is small and, given appropriate conditions, nickel oxide would dissolve in seawater (47). We conclude that the pattern of elemental abundances in the Gubbio sections is compatible with an extraterrestrial source for the anomalous iridium and incompatible with a crustal source.

The Danish boundary layer, which has much more Ir than the Italian C-T clay, is even less likely to have had a crustal origin. Rocks from the upper mantle (which has more Ir than the crust) have less than 20 ppb (48) and are therefore an unlikely worldwide source. There are, however, localized terrestrial sources with much higher Ir abundances; for example, nickel sulfide and chromite ores (48) have Ir levels of hundreds and thousands of parts per billion, respectively. The Danish boundary layer, however, does not have enough nickel [506 parts per million (ppm)] or chromium (165

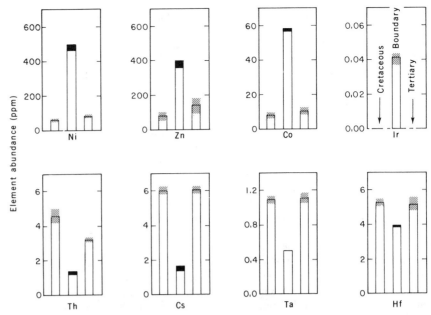

Fig. 8. Selected trace element abundances of Danish acid-insoluble residues. First bar is the mean value [root-mean-square deviation (RMSD) is shown by the crosshatched areas] for the given element in the three Cretaceous residues. Second bar is the abundance (counting error is shown by solid areas) for the given element in the boundary layer residue. Third bar is the mean value and RMSD for the given element in the three Tertiary residues. Measurements were by NAA, except for zinc, which was measured by XRF (43). Significant amounts of nickel, zinc, cobalt, iridium, and thorium in all samples dissolved in the $2N$ HNO_3. Very little cesium, tantalum, or hafnium in any of the samples dissolved in the acid.

ppm) to explain its Ir in this fashion, unless the marine chemistry concentrated Ir preferentially (even in the sulfide environment) and disposed of the other elements elsewhere. The probability of these effects occurring worldwide seems less likely than an extraterrestrial origin for the Ir.

We next consider whether the Ir anomaly is due to an abnormal influx of

Table 2. Abundance of trace elements in the Danish boundary layer (parts per million).

Element	(1) Abundance in whole rock/ abundance of residue*		(2) Abundance in residue*		Element	(1) Abundance in whole rock/ abundance of residue		(2) Abundance in residue	
	Enhanced elements†					*Depleted elements*			
V	391	± 27	330	± 31	Mn	102.0	± 1.3	21.3	± 0.5
Cr	371	± 13	358	± 9	Rb	27	± 7	35	± 4
Co	141.6	± 1.8	57.2	± 0.7	Y‡	79	± 6	6.3	± 1.8
Ni	1137	± 31	479	± 14	Zr‡	144	± 11	125	± 6
Cu‡	167	± 14	93	± 6	Nb‡	8	± 4	6.1	± 1.8
Zn‡	1027	± 49	378	± 18	Cs	1.87	± 0.19	1.51	± 0.14
As	96	± 8	68	± 4	La	61.1	± 1.6	6.8	± 0.4
Se§	46.5	± 0.6	12.1	± 0.3	Ce	57.0	± 1.2	9.7	± 0.6
Mo	29.0	± 2.5	20.3	± 1.4	Nd	63.4	± 2.7	5.4	± 0.6
Ag§	2.6	± 0.9	3.5	± 0.7	Sm	11.93	± 0.08	0.781	± 0.008
In§	0.245	± 0.022	0.086	± 0.019	Eu	2.76	± 0.11	0.121	± 0.010
Sb	8.0	± 0.4	6.7	± 0.4	Tb	1.84	± 0.04	0.148	± 0.014
Ba	1175	± 16	747	± 11	Dy	11.24	± 0.12	0.908	± 0.033
Ir	0.0643	± 0.0029	0.0416	± 0.0018	Yb	5.02	± 0.09	0.56	± 0.05
Pb‡	64	± 14	28	± 7	Lu	0.553	± 0.031	0.083	± 0.004
					Hf	4.34	± 0.16	3.88	± 0.07
	Other elements†				Ta	0.508	± 0.011	0.500	± 0.005
Sc	20.74	± 0.16	14.30	± 0.14	Th	7.1	± 0.4	1.28	± 0.06
Ga‡	30	± 6	19.8	± 3.0	U	8.63	± 0.09	0.918	± 0.024
Sr‡	1465	± 72	48.1	± 2.4					
Au	< 0.12		0.027	± 0.007					

*Column 1 minus column 2 is the amount of an element that dissolved in the acid or was lost in the firing; abundance of residue = 44.5 percent. †Elements V, Ag, and In are at least 20 percent and all other "enhanced elements" are at least a factor of 3 more abundant in the boundary residue than in the other residues. All "depleted elements" are at least 20 percent less abundant in the boundary residue than in the other residues. "Other elements" do not show a consistent pattern of boundary residue abundances relative to the others. ‡Measured by hard XRF (43). §Flux monitors were used in the NAA measurements of these elements. The indicated errors are applicable for comparing the two entries for a given element, but calibration uncertainties of possibly 10 to 20 percent must be considered when the values are used for other purposes.

extraterrestrial material at the time of the extinctions, or whether it was formed by the normal, slow accumulation of meteoritic material (19), followed by concentration in the boundary rocks by some identifiable mechanism.

There is prima facie evidence for an abnormal influx in the observations that the excess iridium occurs exactly at the time of one of the extinctions; that the extinctions were extraordinary events, which may well indicate an extraordinary cause; that the extinctions were clearly worldwide; and that the iridium anomaly is now known from two different areas in western Europe and in New Zealand. Furthermore, we will show in a later section that impact of a 10-km earth-crossing asteroid, an event that probably occurs with about the same frequency as major extinctions, may have produced the observed physical and biological effects. Nevertheless, one can invent two other scenarios that might lead

to concentration of normal background iridium at the boundary. These appear to be much less likely than the sudden-influx model, but we cannot definitely rule out either one at present.

The first scenario requires a physical or chemical change in the ocean waters at the time of the extinctions, leading to extraction of iridium resident in the seawater. This would require iridium concentrations in seawater that are higher than those presently observed. In addition, it suggests that the positive iridium anomaly should be accompanied by a compensating negative anomaly immediately above, but this is not seen.

The second scenario postulates a reduction in the deposition rate of all components of the pelagic sediment except for the meteoritic dust that carries the concentrated iridium. This scenario requires removal of clay but not of iridium-bearing particles, perhaps by currents of exactly the right velocity. These currents

must have affected both the Italian and Danish areas at exactly the time of the C-T extinctions, but at none of the other times represented by our samples. We feel that this scenario is too contrived, a conclusion that is justified in more detail elsewhere (23).

In summary, we conclude that the anomalous iridium concentration at the C-T boundary is best interpreted as indicating an abnormal influx of extraterrestrial material.

Negative Results of Tests for the Supernova Hypothesis

Considerable attention has been given to the hypothesis that the C-T extinctions were the result of a nearby supernova (11). A rough calculation of the distance from the assumed supernova to the solar system, using the measured surface density of iridium in the Gubbio

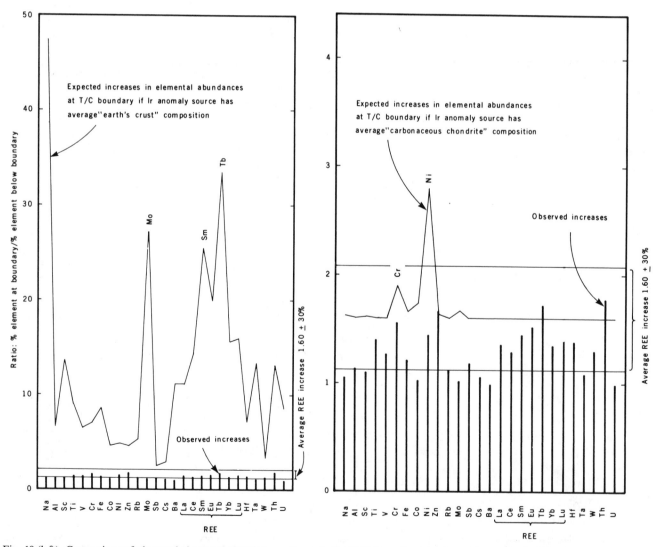

Fig. 10 (left). Comparison of observed elemental abundance patterns in the Gubbio section samples with average patterns expected for crustal material (46). Fig. 11 (right). Comparison of observed elemental abundance patterns in the Gubbio section samples with patterns expected for carbonaceous chondrites (46).

boundary layer and the amount of iridium expected to be blown off in the supernova explosion, gives about 0.1 light-year. The probability is about 10^{-9} (49) that, during the last 100 million years, a supernova occurred within this distance from the sun. Any mechanism with such a low a priori probability is obviously a one-time-only theory. Nevertheless, because the theory could be subjected to direct experimental tests, it was treated as a real possibility until we obtained two other independent pieces of evidence that forced us to reject it.

Elements heavier than nickel can be produced in stars only by neutron capture followed by beta decay. The most intense source of neutrons so far postulated is that produced by the gravitational collapse of the core of a star that leads immediately to a supernova explosion. In this environment the rapid capture of neutrons ("r process") leads to the formation of the heaviest known isotopes. The slower capture of neutrons by heavy isotopes in highly evolved stars ("s process") leads to a different mix of isotopes (50).

One heavy isotope in particular offered the possibility of testing the supernova hypothesis; this is ^{244}Pu, with a half-life of 80.5×10^6 years. The explosion of a supernova should send out an expanding shell of newly created heavy elements, with a ratio of Ir atoms to ^{244}Pu atoms equal to about 10^3. This value is inferred from the existence of an anomaly in the meteoritic abundance of heavy xenon isotopes that is interpreted as being due to the fission of ^{244}Pu (51). Any ^{244}Pu incorporated in the earth at the time of the creation of the solar system, about 4.7 billion years ago, would have decayed by 58 half-lives, or by a factor of 10^{17}, which would make it quite undetectable in the Gubbio section by the most sensitive techniques available. If the C-T extinctions were due to a supernova, and if this were the source of the anomalous Ir, each Ir atom should have been accompanied by about 10^{-3} ^{244}Pu atom, and this ^{244}Pu would have decayed by only a factor of 2.

Plutonium-244 is easily detected both by mass spectrometry and by NAA. The former is more sensitive, but the latter was immediately available. In NAA, which we utilized, ^{244}Pu is converted to ^{245}Pu, which has a half-life of 10 hours and emits many characteristic gamma rays and x-rays. Plutonium was chemically separated from 25- and 50-g batches of boundary clay and from a 50-g batch of bedding clay from below the C-T boundary, and nearly "mass-free"

samples were obtained—no carriers were added. Chemical separations were also performed on the plutonium fraction after the neutron irradiation. No significant gamma radiation was observed, other than that associated with the plutonium isotopes. In order to measure our chemical yields, Gubbio acid-soluble and acid-insoluble residues were spiked with small amounts of ^{238}Pu tracer. This plutonium isotope is easily detectable through its alpha decay, as its half-life is only 87.7 years. In addition, one of the samples was spiked with ^{244}Pu. Figure 12a shows the gamma-ray spectrum of the

sample spiked with about 20 picograms of ^{244}Pu; it indicates both the sensitivity of NAA for the detection of ^{244}Pu and the freedom of the purified sample from other elements that might interfere with the detection of ^{244}Pu. The plutonium isotopic ratios in this sample and in the tracer were also measured with a single-direction-focusing mass spectrometer 5 feet in radius.

No ^{244}Pu was detected in the Gubbio samples (Fig. 12b), with a detection limit of less than 10 percent of the amount that would be expected to accompany the measured iridium if a supernova were re-

Fig. 9. Some of the element abundances measured in acid-insoluble residues of Cretaceous, boundary layer, and Tertiary rocks near Gubbio. Data include all samples from that area measured within 19 m of the boundary. There were four samples from each of the three layers; the crosshatched areas are the standard deviations. The abundance patterns for samples from ~ 27 km north of Gubbio are similar to those shown.

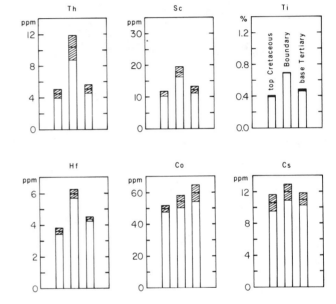

Table 3. Whole-rock composition of the Gubbio and Danish boundary layers (percent).

Element* or mineral	Abundance in boundary layer†		
	Gubbio (Contessa) measured	Denmark	
		Measured	Normalized
SiO_2	27.7 ± 0.6	29.0 ± 0.6	
Al_2O_3	12.19 ± 0.15	8.01 ± 0.17	
FeO‡	4.53 ± 0.05	4.35 ± 0.04	
MgO	1.10 ± 0.07	3.07 ± 0.10	
CaO	22.6 ± 0.4	23.1 ± 0.4	
Na_2O	0.1806 ± 0.0036	0.0888 ± 0.0018	
K_2O	2.46 ± 0.20	0.38 ± 0.04	
TiO_2	0.521 ± 0.022	0.324 ± 0.016	
S^{2-}	Not detected	~ 1.1	
PO_4^{3-}	Not detected	0.92 ± 0.09	
CO_2§	17.7 ± 0.3	18.4 ± 0.3	
Σ Trace elements	~ 0.2	~ 0.3	
Sum	89.2 ± 0.8	90.3 ± 1.0	
Difference‖	10.8 ± 0.8	9.7 ± 1.0	
Calcite		~ 90	41.5 (norm)
Quartz		5–7	~ 3
Pyrite		~ 5	~ 2
Illite		2–3	~ 1

*Abundance values are for element expressed as form shown. †Elements Si, Ca, Mg, S, P, and Gubbio Ti were measured by soft XRF (42). Some S may be lost in this sample preparation procedure. The Denmark Ti was measured by hard XRF (43). All other measurements were by NAA. Mineral analyses were done by M. Ghiorso and I. S. E. Carmichael by x-ray diffraction. ‡Total Fe expressed as FeO. §The CO_2 abundance was calculated from the Ca abundance by assuming all Ca was present as the carbonate. ‖The difference is mainly water and organic material.

sponsible for the latter. The ocean, however, can produce chemical and physical changes in depositing materials as well as diagenetic alterations in the deposited sediments, so the absence of measurable ^{244}Pu is not an absolutely conclusive argument.

The second method that was used to test whether a supernova was responsible for the iridium anomaly involved a measurement of the isotopic ratio of iridium in the boundary material. Iridium has two stable isotopes, 191 and 193, which would be expected to occur in about the same relative abundances, 37.3 to 62.7 percent, in all solar system material because of mixing the protosolar gas cloud. However, different supernovas should produce iridium with different isotopic ratios because of differences in the contributions of the r and s processes occasioned by variations in neutron fluxes, reaction times, and so on, from one supernova to the next. According to this generally accepted picture, solar system iridium is a mixture of that element produced by all the supernovas that ejected material into the gaseous nebula that eventually condensed to form the sun and its planets. A particular supernova would produce Ir with an isotopic ratio that might differ from that of solar system material by as much as a factor of 2 (*52*).

We therefore compared the isotopic ratio of Ir from the C-T boundary clay with that of ordinary Ir, using NAA. This is a new technique (*23*), which we developed because of the extreme difficulty of determining Ir isotope ratios by mass spectrometry. In our earlier analytical work we used only the 74-day ^{192}Ir, made from ^{191}Ir by neutron capture. But in this new work we also measured the 18-hour ^{194}Ir made from the heavier Ir isotope, and extensive chemical separations before and after the neutron irradiations were necessary. Figure 13 contrasts a typical gamma-ray spectrum of the kind used in the isotopic ratio measurement with one used in an Ir abundance determination. This comparison demonstrates the need for chemical purification of the iridium fraction as well as the lack of major interfering radiations.

The final result is that the isotopic ratio of the boundary Ir differs by only 0.03 ± 0.65 percent (mean + 1 standard deviation) from that of the standard. From this, we conclude that the ^{191}Ir/^{193}Ir ratio in the boundary layer and the standard do not differ significantly by more than 1.5 percent. Therefore the anomalous Ir is very likely of solar system origin, and did not come from a supernova or other source outside the solar system (*53*)—for example, during passage of the earth through the galactic arms. [In a very recent paper, Napier and Clube suggest that catastrophic events could arise from the latter (*54*).]

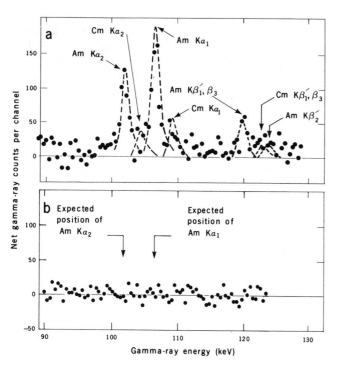

Fig. 12. Gamma-ray spectra of Pu fractions from acid-insoluble residues of irradiated boundary layer clay samples from Gubbio. (a) Sample had been spiked with ^{244}Pu and ^{238}Pu containing relatively small amounts of ^{239}Pu, ^{240}Pu, and ^{242}Pu. Dashed lines show expected energies and abundances of ^{245}Pu and equilibrated daughter radiations normalized to the 327.2-keV gamma ray of ^{245}Pu (not shown). (b) Sample had been spiked with ^{238}Pu containing relatively small amounts of ^{239}Pu, ^{240}Pu, and ^{242}Pu. No ^{244}Pu was detected.

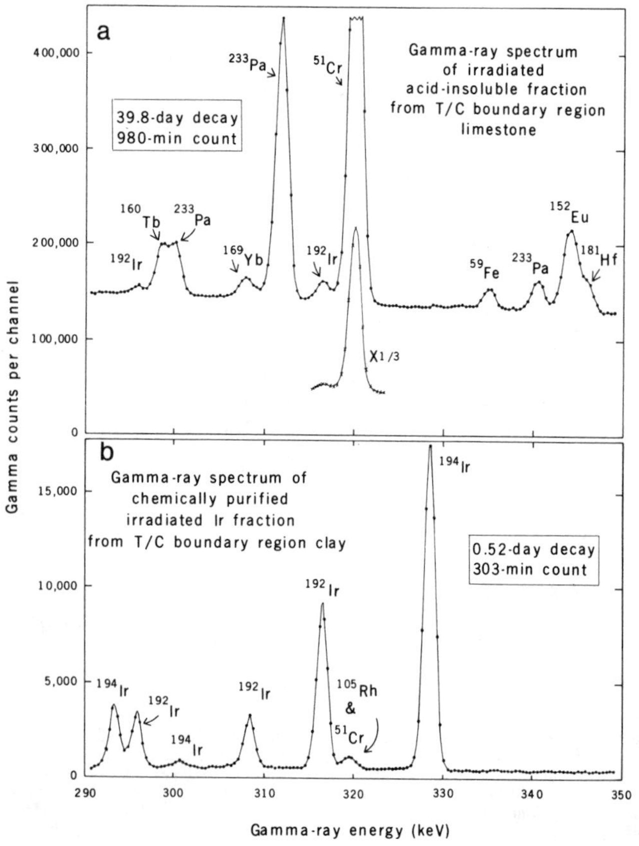

Fig. 13. (a) Gamma-ray spectrum of irradiated acid-insoluble residue from boundary layer clay at Gubbio without chemistry before or after irradiation. (b) Same as above with chemistry before and after irradiation used in isotopic ratio determinations. Counting periods, decay periods, and chemical yields are different for the two spectra.

The Asteroid Impact Hypothesis

After obtaining negative results in our tests of the supernova hypothesis, we were left with the question of what extraterrestrial source within the solar system could supply the observed iridium and also cause the extinctions. We considered and rejected a number of hypotheses (23); finally, we found that an extension of the meteorite impact hypothesis (55, 56) provided a scenario that explains most or all of the biological and physical evidence. In brief, our hypothesis suggests that an asteroid struck the earth, formed an impact crater, and some of the dust-sized material ejected from the crater reached the stratosphere and was spread around the globe. This dust effectively prevented sunlight from reaching the surface for a period of several years, until the dust settled to earth. Loss of sunlight suppressed photosynthesis, and as a result most food chains collapsed and the extinctions resulted. Several lines of evidence support this hypothesis, as discussed in the next few sections. The size of the impacting object can be calculated from four independent sets of observations, with good agreement among the four different diameter estimates.

Earth-Crossing Asteroids and Earth Craters

Two quite different data bases show that for the last billion years the earth has been bombarded by a nearly constant flux of asteroids that cross the earth's orbit. One data base comes from astronomical observations of such asteroids and a tabulation of their orbital parameters and their distribution of diameters (57). Öpik (58) computed that the mean time to collision with the earth for a given earth-crossing asteroid is about 200 million years. To a first approximation, the number of these objects with diameters greater than d drops roughly as the inverse square of d. E. M. Shoemaker [cited in (59, 60)] and Wetherill (60) independently estimated that there are at present about 700 earth-crossing asteroids with diameters greater than 1 km (Apollo objects), so there should be about seven with diameters greater than 10 km. This assumes that the power law with exponent -2 extends from the accessible 1-km-diameter range into the 10-km-diameter range. If one accepts the numbers given above, the mean time to collision for an earth-crossing asteroid with a diameter of 10 km or more would be 200 million years divided by 7, or ~ 30 million years. In a more sophisticated calculation, Shoemaker (61) estimates that a mean collision time of 100 million years is consistent with a diameter of 10 km, which is the value we will adopt. A discussion of cratering data, which leads to similar estimates, is given

in Grieve and Robertson's review article (62) on the size and age distribution of large impact craters on the earth. Rather than present our lengthy justification (23) for the estimates based on the cratering data, we will simply report the evaluation of Grieve (63), who wrote: "I can find nothing in your data that is at odds with your premise." Grieve also estimates that the diameter of the crater formed by the impact of a 10-km asteroid would be about 200 km (63). This section of our article has thus been greatly condensed now that we have heard from experienced students of the two data bases involved.

Krakatoa

The largest well-studied terrestrial explosion in historical times was that of the island volcano, Krakatoa, in the Sunda Strait, between Java and Sumatra (64). Since this event provides the best available data on injection of dust into the stratosphere, we give here a brief summary of relevant information.

On 26 and 27 August 1883, Krakatoa underwent volcanic eruptions that shot an estimated 18 km³ of material into the atmosphere, of which about 4 km³ ended up in the stratosphere, where it stayed for 2 to 2.5 years. Dust from the explosion circled the globe, quickly giving rise to brilliant sunsets seen worldwide. Recent measurements of the ^{14}C injected into the atmosphere by nuclear bomb tests confirm the rapid mixing (about 1 year) between hemispheres (65). If we take the estimated dust mass in the stratosphere (4 km³ times the assumed low density of 2 g/cm³) and spread it uniformly over the globe, it amounts to 1.6×10^{-3} g/cm². This layer did not absorb much of the incident radiation on a "straight-through" basis. However, if it were increased by a factor of about 10^3 (a rough prediction of our theory), it is most probable that the sunlight would be attenuated to a high degree.

Since the time for the colored sunsets to disappear after Krakatoa is frequently given as 2 to 2.5 years, we have assumed that the asteroid impact material in the stratosphere settled in a few years. Thus, 65 million years ago, day could have been turned into night for a period of several years, after which time the atmosphere would return relatively quickly to its normal transparent state.

What happened during the Krakatoa explosions can be expected to happen to a much greater extent during the impact of a large asteroid. An interesting dif-

ference is that extreme atmospheric turbulence would follow the impact. The asteroid would enter the atmosphere at roughly 25 km/sec and would "punch a hole" in the atmosphere about 10 km across. The kinetic energy of the asteroid is approximately equivalent to that of 10^8 megatons of TNT.

Size of the Impacting Object

If we are correct in our hypothesis that the C-T extinctions were due to the impact of an earth-crossing asteroid, there are four independent ways to calculate the size of the object. The four ways and the results obtained are outlined below.

1) The postulated size of the incoming asteroid was first computed from the iridium measurements in the Italian sections, the tabulated Ir abundances (66) in type I carbonaceous chondrites (CI), which are considered to be typical solar system material, and the fraction of erupted material estimated to end up in the stratosphere. If we neglect the latter fraction for the moment, the asteroid mass is given by $M = sA/f$, where s is the surface density of Ir (measured at Gubbio to be 8×10^{-9} g/cm²), A is the surface area of the earth, and f is the Ir mass fractional abundance in CI meteorites (0.5×10^{-6}). This preliminary value of the asteroid mass, 7.4×10^{16} g, is then divided by the estimated fraction staying in the stratosphere, 0.22, to give $M = 3.4 \times 10^{17}$ g. The "Krakatoa fraction," 0.22, is used simply because it is the only relevant number available. It could differ seriously from the correct value, however, as the two explosions are of quite different character. At a density of 2.2 g/cm³ (67), the diameter of the asteroid would be 6.6 km.

2) The second estimate comes from data on earth-crossing asteroids and the craters they have made on the earth's surface. In a sense, the second estimate comes from two quite different data bases—one from geology and the other from astronomy. Calculations of the asteroid diameter can be made from both data bases, but they will not really be independent since the two data bases are known to be consistent with each other. As shown in an earlier section, the most believable calculation of the mean time between collisions of the earth and asteroids equal to or larger than 10 km in diameter is about 100 million years. The smaller the diameter the more frequent are the collisions, so our desire to fit not only the C-T extinction, but earlier ones as well, sets the mean time between ex-

tinctions at about 100 million years and the diameter at about 10 km.

3) The third method of estimating the size of the asteroid comes from the possibility that the 1-cm boundary layer at Gubbio and Copenhagen is composed of material that fell out of the stratosphere, and is not related to the clay that is mixed in with the limestone above and below it. This is quite a surprising prediction of the hypothesis, since the most obvious explanation for the origin of the clay is that it had the same source as the clay impurity in the rest of the Cretaceous and Tertiary limestone, and that it is nearly free of primary $CaCO_3$ because the extinction temporarily destroyed the calcite-producing plankton for about 5000 years. But as discussed earlier, the material in the boundary layer is of a different character from the clay above and below it, whereas the latter two clays are very similar. To estimate the diameter of the asteroid, one can use the surface density of the boundary layer (about 2.5 g/cm^2), together with an estimate of the fraction of that material which is of asteroidal origin. The asteroid diameter is then calculated to be 7.5 km. The numbers used in this calculation are the following: clay fraction in the boundary layer, 0.5; density of the asteroid, 2.2 g/cm^3; mass of crustal material thrown up per unit mass of asteroid, ~ 60 (63); fraction of excavated material delivered to the stratosphere, 0.22 (from the Krakatoa measurements). If one uses different numbers, the diameter changes only by the cube root of the ratio of input values.

The first and the third methods are independent, even though they both depend on measurements made on the boundary material. This can best be appreciated by noting that if the Ir abundance were about the same in the earth's crust as it is in meteorites, the iridium anomaly seen in Fig. 5 would not exist. Therefore, method 1 would not exist either. The fact that method 3 could still be used is the indicator of the relative independence of the two methods.

4) The fourth method is not yet able to set close limits on the mass of the incoming asteroid, but it leads to consistent results. This method derives from the need to make the sky much more opaque than it was in the years following the Krakatoa explosion. If it is assumed that the Krakatoa dust cloud attenuated the vertically incident sunlight by about 3 percent, then an explosion involving 33 times as much material would reduce the light intensity to $1/e$. The stratospheric mass due to an explosion of the magnitude calculated in the three earlier meth-

ods—about 1000 times that of Krakatoa—would then be expected to reduce the sunlight to $\exp(-30) = 10^{-13}$. This is, of course, much more light attenuation than is needed to stop photosynthesis. But the model used in this simplistic calculation assumes that the dust is a perfect absorber of the incident light. A reasonable albedo coupled with a slight reduction in the mass of dust can raise the light intensity under the assumed "optical depth" to 10^{-7} of normal sunlight, corresponding to 10 percent of full moonlight.

Although it is impossible to make an accurate estimate of the asteroid's size from the Krakatoa extrapolation, it would have been necessary to abandon the hypothesis had a serious discrepancy been apparent. In the absence of good measurements of the solar constant in the 1880's, it can only be said that the fourth method leads to asteroid sizes that are consistent with the other three.

Until we understand the reasons for the factor of 10 difference in Ir content of the boundary clay between Denmark and Italy, we will be faced with different values for the asteroid diameter based on the first method. The "Danish diameter" is then 6.6 km $\times$ $10^{1/3}$ = 14 km. The second and third estimates are unchanged; the second does not involve measurements made on the boundary layer, and the third uses the thickness of the clay, which is only slightly greater in Denmark than in Italy. The fourth method is based on such an uncertain attenuation value, from Krakatoa, that it is not worth recalculating. We conclude that the data are consistent with an impacting asteroid with a diameter of about 10 $\pm$ 4 km.

Biological Effects

A temporary absence of sunlight would effectively shut off photosynthesis and thus attack food chains at their origins. In a general way the effects to be expected from such an event are what one sees in the paleontological record of the extinction.

The food chain in the open ocean is based on microscopic floating plants, such as the coccolith-producing algae, which show a nearly complete extinction. The animals at successively higher levels in this food chain were also very strongly affected, with nearly total extinction of the foraminifera and complete disappearance of the belemnites, ammonites, and marine reptiles.

A second food chain is based on land plants. Among these plants, existing individuals would die, or at least stop pro-

ducing new growth, during an interval of darkness, but after light returned they would regenerate from seeds, spores, and existing root systems. However, the large herbivorous and carnivorous animals that were directly or indirectly dependent on this vegetation would become extinct. Russell (2) states that "no terrestrial vertebrate heavier than about 25 kg is known to have survived the extinctions." Many smaller terrestrial vertebrates did survive, including the ancestral mammals, and they may have been able to do this by feeding on insects and decaying vegetation.

The situation among shallow marine bottom-dwelling invertebrates is less clear; some groups became extinct and others survived. A possible base for a temporary food chain in this environment is nutrients originating from decaying land plants and animals and brought by rivers to the shallow marine waters.

We will not go further into this matter, but we refer the reader to the proceedings of the 1976 Ottawa meeting on the C-T extinctions. This volume reproduces an extensive discussion among the participants of what would happen if the sunlight were temporarily "turned off" (5, pp. 144–149). Those involved in the discussion seemed to agree that many aspects of the extinction pattern could be explained by this mechanism, although a number of puzzles remained.

We must note, finally, an aspect of the biological record that does not appear to be in accord with the asteroid impact hypothesis or with any sudden, violent mechanism. Extinction of the foraminifera and nannoplankton occurs within reversed geomagnetic polarity zone Gubbio G− in the Gubbio section (30). Butler and co-workers (68, 69) have studied the nonmarine sequence of the San Juan Basin of New Mexico and have found a polarity sequence that appears to be correlated with the reversal sequence at Gubbio. In the San Juan Basin, the highest dinosaur fossils are found in the normal polarity zone (anomaly 29) that follows what is identified as the Gubbio G− zone. It would thus appear that the dinosaur and foram-nannoplankton extinctions were not synchronous. (Extinctions occurring in the same polarity zone in distant sections would not establish either synchroneity or diachroneity.) Three comments on the San Juan Basin work have been published (70) calling attention to the possibility of an unconformity at the boundary, in which case the correlation of the magnetic polarity zones could be in error and the extinctions might still be synchronous. Lindsay et al. (69) argue strongly against a major hiatus, but admit that "the case is not completely closed." Russell (71) has noted stratigraphic evidence against a diachronous extinction in the continental and marine realm.

Resolution of the question of whether the extinctions could have been synchronous will depend on further paleomagnetic studies. In the meantime we can state that the asteroid impact hypothesis predicts that the apparently diachronous timing of the foram-nannofossil and dinosaur extinctions will eventually be shown to be incorrect.

Problems in Boundary Clay Composition

One would expect from the simplest considerations of our hypothesis that the boundary layer resulted from crustal material (enriched in certain elements by the asteroidal matter) that was distributed worldwide in the stratosphere and then fell into the ocean. This material would be subjected to chemical and physical processes in the atmosphere and then in the ocean, which would alter the composition. The enhancements of metals having water-insoluble sulfides in the Danish C-T boundary compared to the Italian might be related to an anaerobic environment during deposition of the former and an aerobic one for the latter. Hydrogen sulfide can be produced by bacteria in oxygen-deficient waters, and this would precipitate those metals if they were available. This would not, however, explain the striking depletion of some trace elements in the Danish boundary or its very high Ir abundance. If chondritic Ir with an abundance of ~ 500 ppb were diluted 60-fold with crustal material, the Ir abundance should be ~ 8 ppb rather than the 65 ppb observed. Possible solutions to these difficulties may arise when better estimates of the extent of mixing of asteroidal and terrestrial material in the atmosphere are made, and when the boundary layer chemistry is studied at additional locations and a better understanding of the marine chemistry is achieved.

Implications

Among the many implications of the asteroid impact hypothesis, if it is correct, two stand out prominently. First, if the C-T extinctions were caused by an impact event, the same could be true of the earlier major extinctions as well. There have been five such extinctions since the end of the Precambrian, 570 million years ago, which matches well

the probable interval of about 100 million years between collisions with 10-km-diameter objects. Discussions of these extinction events generally list the organisms affected according to taxonomic groupings; it would be more useful to have this information given in terms of interpreted ecological or food-chain groupings. It will also be important to carry out iridium analyses in complete stratigraphic sections across these other boundaries. However, E. Shoemaker (private communication) predicts that if some of the extinctions were caused by the collision of a "fresh" comet (mostly ice), the Ir anomaly would not be seen even though the extinction mechanism was via the same dust cloud of crustal material, so the absence of a higher Ir concentration at, for example, the Permian-Triassic boundary would not invalidate our hypothesis. According to Shoemaker, cometary collisions in this size range could be twice as frequent as asteroidal collisions.

Second, we would like to find the crater produced by the impacting object. Only three craters 100 km or more in diameter are known (62). Two of these (Sudbury and Vredefort) are of Precambrian age. For the other, Popigay Crater in Siberia, a stratigraphic age of Late Cretaceous to Quaternary and a potassium-argon date of 28.8 million years (no further details given) have been reported (72, 73). Thus, Popigay Crater is probably too young, and at 100-km-diameter probably also too small, to be the C-T impact site. There is about a 2/3 probability that the object fell in the ocean. Since the probable diameter of the object, 10 km, is twice the typical oceanic depth, a crater would be produced on the ocean bottom and pulverized rock could be ejected. However, in this event we are unlikely to find the crater, since bathymetric information is not sufficiently detailed and since a substantial portion of the pre-Tertiary ocean has been subducted.

References and Notes

1. D. A. Russell, *Geol. Assoc. Can. Spec. Rep. 13* (1975), p. 119.
2. ———, in (5), p. 11.
3. M. B. Cita and I. Premoli Silva, *Riv. Ital. Paleontol. Stratigr. Mem. 14* (1974), p. 193.
4. D. A. Russell, *Annu. Rev. Earth Planet. Sci. 7*, 163 (1979).
5. K-TEC group (P. Béland *et al.*), *Cretaceous-Tertiary Extinctions and Possible Terrestrial and Extraterrestrial Causes* (Proceedings of Workshop, National Museum of Natural Sciences, Ottawa, 16 and 17 Nov. 1976).
6. T. Birkelund and R. G. Bromley, Eds., *Cretaceous-Tertiary Boundary Events*, vol. 1, *The Maastrichtian and Danian of Denmark* (Symposium, University of Copenhagen, Copenhagen, 1979); W. K. Christiansen and T. Birkelund, Eds., *ibid.*, vol. 2, *Proceedings*.
7. H. Tappan, *Palaeogeogr. Palaeoclimatol. Palaeoecol. 4*, 187 (1968); T. R. Worsley, *Nature (London) 230*, 318 (1971); W. T. Holser, *ibid. 267*, 403 (1977); D. M. McLean, *Science*

200, 1060 (1978); *ibid.* 201, 401 (1978); S. Gartner and J. Keany, *Geology 6*, 708 (1978).
8. E. G. Kauffman, in (6), vol. 2, p. 29.
9. A. G. Fischer, in (6), vol. 2, p. 11; ——— and M. A. Arthur, *Soc. Econ. Paleontol. Mineral. Spec. Publ. 25* (1977), p. 19.
10. J. F. Simpson, *Geol. Soc. Am. Bull. 77*, 197 (1966); J. D. Hays, *ibid.* 82, 2433 (1971); C. G. A. Harrison and J. M. Prospero, *Nature (London) 250*, 563 (1974).
11. O. H. Schindewolf, *Neues Jahrb. Geol. Palaeontol. Monatsh.* 1954, 451 (1954); *ibid.* 1958, 270 (1958); A. R. Leoblich, Jr., and H. Tappan, *Geol. Soc. Am. Bull. 75*, 367 (1964); V. I. Krasovski and I. S. Shklovsky, *Dokl. Akad. Nauk SSSR 116*, 197 (1957); K. D. Terry and W. H. Tucker, *Science 159*, 421 (1968); H. Laster, *ibid.* 160, 1138 (1968); W. H. Tucker and K. D. Terry, *ibid.*, p. 1138; D. Russell and W. H. Tucker, *Nature (London) 229*, 553 (1971); M. A. Ruderman, *Science 184*, 1079 (1974); R. C. Whitten, J. Cuzzi, W. J. Borucki, J. H. Wolfe, *Nature (London) 263*, 398 (1976).
12. S. Gartner and J. P. McGuirk, *Science 206*, 1272 (1979).
13. A. Boersma and N. Schackleton, in (6), vol. 2, p. 50; B. Buchardt and N. O. Jorgensen, in (6), vol. 2, p. 54.
14. L. Christensen, S. Fregerslev, A. Simonsen, J. Thiede, *Bull. Geol. Soc. Den. 22*, 193 (1973).
15. N. O. Jorgensen, in (6), vol. 1, p. 33, vol. 2, p. 62; M. Renard, in (6), vol. 2, p. 70.
16. H. P. Luterbacher and I. Premoli Silva, *Riv. Ital. Paleontol. Stratigr. 70*, 67 (1964).
17. H. Pettersson and H. Rotschi, *Geochim. Cosmochim. Acta 2*, 81 (1952).
18. V. M. Goldschmidt, *Geochemistry* (Oxford Univ. Press, New York, 1954).
19. J. L. Barker, Jr., and E. Anders, *Geochim. Cosmochim. Acta 32*, 627 (1968).
20. R. Ganapathy, D. E. Brownlee, P. W. Hodge, *Science 201*, 1119 (1978).
21. A. M. Sarna-Wojcicki, H. R. Bowman, D. Marchand, E. Helley, private communication.
22. J. H. Crocket and H. Y. Kuo, *Geochim. Cosmochim. Acta 43*, 831 (1979).
23. These are briefly discussed in L. W. Alvarez, W. Alvarez, F. Asaro, H. V. Michel, *Univ. Calif. Lawrence Berkeley Lab. Rep. LBL-9666* (1979).
24. A description of the NAA techniques is given in Alvarez *et al.* (23), appendix II; I. Perlman and F. Asaro, in *Science and Archaeology*, R. H. Brill, Ed. (MIT Press, Cambridge, Mass., 1971), p. 182.
25. These limestones belong to the Umbrian sequence, of Jurassic to Miocene age, which has been described in V. Bortolotti, P. Passerini, M. Sagri, G. Sestini, *Sediment. Geol. 4*, 341 (1970); A. Jacobacci, E. Centamore, M. Chiocchini, N. Malferrari, G. Martelli, A. Micarelli, *Note Esplicative Carta Geologica d'Italia (1:50,000), Foglio 190: "Cagli"* (Rome, 1974).
26. H. P. Luterbacher and I. Premoli Silva, *Riv. Ital. Paleontol. Stratigr. 33*, 162 (1962); I. Premoli Silva, L. Paggi, S. Monechi, *Mem. Soc. Geol. Ital. 15*, 21 (1976).
27. S. Monechi, in (6), vol. 2, p. 164.
28. D. V. Kent, *Geology 5*, 769 (1977); M. A. Arthur, thesis, Princeton University (1979).
29. O. Renz, *Eclogae Geol. Helv. 29*, 1 (1936); *Serv. Geol. Ital. Mem. Descr. Carta Geol. Ital. 29*, 1 (1936).
30. M. A. Arthur and A. G. Fischer, *Geol. Soc. Am. Bull. 88*, 367 (1977); I. Premoli Silva, *ibid.*, p. 371; W. Lowrie and W. Alvarez, *ibid.*, p. 374; W. M. Roggenthen and G. Napoleone, *ibid.*, p. 378; W. Alvarez, M. A. Arthur, A. G. Fischer, W. Lowrie, G. Napoleone, I. Premoli Silva, W. M. Roggenthen, *ibid.*, p. 383; W. Lowrie and W. Alvarez, *Geophys. J. R. Astron. Soc. 51*, 561 (1977); W. Alvarez and W. Lowrie, *ibid. 55*, 1 (1978).
31. Locations of the sections studied are: (i) Bottaccione Gorge at Gubbio: 43°21.9'N, 12°35.0'E (0°7.9' east of Rome); (ii) Contessa Valley, 3 km northwest of Gubbio: 43°22.6'N, 12°33.7'E (0°6.6' east of Rome); (iii) Petriccio suspension bridge, 2.3 km west-southwest of Acqualagna: 43°36.7'N, 12°38.7'E (0°11.6' east of Rome); (iv) Acqualagna, road cut 0.8 km southeast of town: 43°36.7'N, 12°40.8'E (0°13.7' east of Rome); and (v) Gorgo a Cerbara: 43°36.1'N, 12°33.6'E (0°6.5' east of Rome). We thank E. Sannipoli, W. S. Leith, and S. Marshak for help in sampling these sections.
32. J. H. Crocket, J. D. McDougall, R. C. Harriss, *Geochim. Cosmochim. Acta 37*, 2547 (1973).
33. Location: 55°16.7'N, 12°26.5'E. We thank I. Bank and S. Gregersen for taking W.A. to this locality.
34. A. Rosenkrantz and H. W. Rasmussen, *Guide to Excursions A42 and C37* (21st International

Geological Congress, Copenhagen, 1960), part 1, pp. 1-17.

35. F. Surlyk, in (6), vol. 1, p. 164.
36. C. Heinberg, in (6), vol. 1, p. 58.
37. H. W. Rasmussen, in (6), vol. 1, p. 65; P. Gravesen, in (6), vol. 1, p. 72; U. Asgaard, in (6), vol. 1, p. 74.
38. E. Hakansson and E. Thomsen, in (6), vol. 1, p. 78.
39. S. Floris, in (6), vol. 1, p. 92.
40. I. Bang, in (6), vol. 1, p. 108.
41. K. Perch-Nielsen, in (6), vol. 1, p. 115.
42. Soft x-ray fluorescence measurements for major element determinations were made by S. Flexser and M. Sturz of Lawrence Berkeley Laboratory.
43. Hard x-ray fluorescence measurements for trace element determinations were made by R. D. Giauque of Lawrence Berkeley Laboratory.
44. F. G. Walton Smith, Ed. *CRC Handbook of Marine Science* (CRC Press. Cleveland, 1974), vol. 1, p. 11.
45. H. A. Wollenberg *et al.*, *Univ. Calif. Lawrence Berkeley Lab. Rep. LBL-7092*, revised 1980.
46. *Encyclopaedia Britannica* (Benton, Chicago, ed. 15, 1974), vol. 6, p. 702.
47. K. K. Turekian, *Oceans* (Prentice-Hall, Englewood Cliffs, N.J., 1976), p. 122.
48. J. H. Crocket, *Can. Mineral.* **17**, 391 (1979); J. R. Ross and R. Keays, *ibid.*, p. 417.
49. I. S. Shklovsky, *Supernovae* (Wiley, New York, 1968), p. 377.
50. D. D. Clayton, *Principles of Stellar Evolution and Nucleosynthesis* (McGraw-Hill, New York, 1968), pp. 546-606.
51. D. N. Schramm, *Annu. Rev. Astron. Astrophys.* **12**, 389 (1974).
52. C. F. McKee, personal communication.
53. These observations were reported at the American Geophysical Union meeting in May 1979 and at the Geological Society of America meet-

ing in November 1979 [W. Alvarez, L. W. Alvarez, F. Asaro, H. V. Michel, *Eos* **60**, 734 (1979); *Geol. Soc. Am. Abstr. Programs* **11**, 350 (1979)].
54. W. M. Napier and S. V. M. Clube, *Nature (London)* **282**, 455 (1979).
55. H. C. Urey, *ibid.* **242**, 32 (1973).
56. E. J. Öpik, *Ir. Astron. J.* **5** (No. 1), 34 (1958).
57. E. M. Shoemaker, J. G. Williams, E. F. Helin, R. F. Wolfe, in *Asteroids*, T. Gehrels, Ed. (Univ. of Arizona Press, Tucson, 1979), pp. 253-282.
58. E. J. Öpik, *Adv. Astron. Astrophys.* **2**, 220 (1963); *ibid.* **4**, 302 (1964); *ibid.* **8**, 108 (1971). These review articles give references to Öpik's extensive bibliography on meteorites, Apollo objects, and asteroids.
59. C. R. Chapman, J. G. Williams, W. K. Hartmann, *Annu. Rev. Astron. Astrophys.* **16**, 33 (1978).
60. G. W. Wetherill, *Sci. Am.* **240** (No. 3), 54 (1979).
61. E. M. Shoemaker, personal communication.
62. R. A. F. Grieve and P. B. Robertson, *Icarus* **38**, 212 (1979).
63. R. A. F. Grieve, personal communication.
64. G. J. Symons, Ed., *The Eruption of Krakatoa and Subsequent Phenomena* (Report of the Krakatoa Committee of the Royal Society, Harrison, London, 1888).
65. I. U. Olson and I. Karlen, *Am. J. Sci. Radiocarbon Suppl.* **7** (1965), p. 331; T. A. Rafter and B. J. O'Brien, *Proc. 8th Int. Conf. Radiocarbon Dating* **1**, 241 (1972).
66. U. Krähenbühl, *Geochim. Cosmochim. Acta* **37**, 1353 (1973).
67. B. Mason, *Space Sci. Rev.* **1**, 621 (1962-1963).
68. R. F. Butler, E. H. Lindsay, L. L. Jacobs, N. M. Johnson, *Nature (London)* **267**, 318 (1977); E. H. Lindsay, L. L. Jacobs, R. F. Butler, *Geology* **6**, 425 (1978).
69. E. H. Lindsay, R. F. Butler, N. M. Johnson, in preparation.

70. W. Alvarez and D. W. Vann, *Geology* **7**, 66 (1979); J. E. Fassett, *ibid.*, p. 69; S. G. Lucas and J. K. Rigby, Jr., *ibid.*, p. 323.
71. D. A. Russell, *Episodes 1979 No. 4* (1979), p. 21.
72. V. L. Masaytis, M. V. Mikhaylov, T. V. Selivanovskaya, *Sov. Geol. No. 6* (1971), pp. 143-147; translated in *Geol. Rev.* **14**, 327 (1972).
73. V. L. Masaytis, *Sov. Geol. No. 11* (1975), pp. 52-64; translated in *Int. Geol. Rev.* **18**, 1249 (1976).
74. It will be obvious to anyone reading this article that we have benefited enormously from conversations and correspondence with many friends and colleagues throughout the scientific community. We would particularly like to acknowledge the help we have received from E. Anders, J. R. Arnold, M. A. Arthur, A. Buffington, I. S. E. Carmichael, G. Curtis, P. Eberhard, S. Gartner, R. L. Garwin, R. A. F. Grieve, E. K. Hyde, W. Lowrie, C. McKee, M. C. Michel (who was responsible for the mass spectrometric measurements), J. Neil, B. M. Oliver, C. Orth, B. Pardoe, I. Perlman, D. A. Russell, A. M. Sessler, and E. Shoemaker. One of us (W.A.) thanks the National Science Foundation for support, the other three authors thank the Department of Energy for support, and one of us (L.W.A.) thanks the National Aeronautics and Space Administration for support. The x-ray fluorescence measurements of trace elements Fe and Ti by R. D. Giauque and of major elements by S. Flexser and M. Sturz were most appreciated. We appreciate the assistance of D. Jackson and C. Nguyen in the sample preparation procedures. We are grateful to T. Lim and the staff of the Berkeley Research Reactor for many neutron irradiations used in this work. We also appreciate the efforts of G. Pefley and the staff of the Livermore Pool Type Reactor for the irradiations used for the Ir isotopic ratio measurements.

Appendix A
Biography

Luis W. Alvarez was born in San Francisco, California, on June 13, 1911. From the University of Chicago he received his Bachelor of Science degree in 1932, Master of Science in 1934, and Doctor of Philosophy in 1936. Alvarez joined the University of California Radiation Laboratory as a Research Fellow in 1936. He is now Professor Emeritus of Physics and Senior Research Scientist at the Lawrence Berkeley National Laboratory. He was on leave at the Massachusetts Institute of Technology Radiation Laboratory from 1940 to 1943, the University of Chicago Metallurgical Laboratory in 1943, and the Los Alamos Laboratory from 1944 to 1945.

Early in his scientific career Alvarez worked concurrently in the fields of optics and cosmic rays. He is codiscoverer of the "East-West effect" in cosmic rays. At Berkeley he concentrated his work in the field of nuclear physics. In 1937 he gave the first experimental demonstration of the existence of the phenomenon of K-electron capture by nuclei. Another early development was a method for producing beams of very slow neutrons. This method subsequently led to a fundamental investigation of neutron scattering in ortho- and parahydrogen with Pitzer and to the first measurement of the magnetic moment of the neutron with Bloch. With Wiens he produced the first mercury-198 lamp, a device developed by the National Bureau of Standards into what was for fifteen years the universal standard of length. Alvarez and Cornog discovered the radioactivity of hydrogen-3 (tritium) and showed that helium-3 was stable. Tritium is best known as a source of thermonuclear energy, while helium-3 has become of importance in low-temperature research.

During World War II Alvarez was responsible at MIT for inventing and developing three important radar systems. The best known of these, Ground Control Approach, is a blind landing system of civilian as well as military value. He also worked with Fermi at the first nuclear reactor and later at Los Alamos on various problems related to the atomic bomb. He flew as a scientific observer at the first two nuclear explosions at Alamogordo and Hiroshima.

Dr. Alvarez was responsible for the design and construction of the Berkeley 40 foot proton linear accelerator, which was completed in 1947. In 1951 he published the first suggestion for charge exchange acceleration that quickly led to the development of the "tandem Van de Graaff accelerator." For the next fifteen years he was engaged in high-energy physics using the six-billion electron volt Bevatron at the University of California Radiation Laboratory. His main efforts were concentrated on the development and use of large liquid hydrogen bubble chambers and on the development of high-speed devices to measure and analyze the millions of photographs produced each year by the bubble chamber complex. The net result of this work was the discovery by Dr. Alvarez's research group of a large number of previously unknown "fundamental particle resonances."

In 1967 Alvarez initiated a program to study primary cosmic rays using balloons and superconducting magnets. In 1965 he proposed an idea to "X-ray" the Second Pyramid at Giza using cosmic-ray muons. This led to the Joint UAR–USA Pyramid Project, which proved that the Second Pyramid is unique among the large pyramids, in that it has only one chamber. In 1979 he and his son Walter, with two collaborators, showed that 65 million years ago Earth was hit by a 10 km diameter object from the solar system (asteroid or comet) that led to the death of more than half the existing species, among them the dinosaurs. They postulated that the killing mechanism was starvation resulting from the cessation of photosynthesis, due to the darkness caused by the cloud of dust thrown into the stratosphere by the vaporization of the object and additional crustal and mantle material.

In the field of optics, his invention of the variable-focus thin lens has led to the introduction of new products in the eye-care field, and his invention of methods of stabilizing the image in hand-held cameras and viewing devices has also given rise to a number of new and useful products.

In 1946 Alvarez was awarded the Collier Trophy by the National Aeronautical Association for the development of Ground Control Approach. In 1947 he was awarded the United States Medal for Merit and in 1953 the John Scott Medal and Prize by the City of Philadelphia for the same work. In 1960 he was named California Scientist of the Year for his research in high-energy physics. In 1961 he was awarded the Einstein Medal for his contribution to the physical sciences. In 1963 he was awarded the Pioneer Award by the American Institute of Electrical and Electronic Engineers. For contributions to high-energy physics he was awarded in 1964 the National Medal of Science, in 1965 the Michelson Award, and in 1968 the Nobel Prize in Physics. In 1978 he was inducted into the National Inventors Hall of Fame and received the University of Chicago Alumni Medal. In 1981 he received the Wright Prize from Harvey Mudd College.

Alvarez was a member of the President's Science Advisory Committee in 1972–1973 and Associate Director-at-Large, Lawrence Berkeley Laboratory in 1954–1959 and 1975–1978. He served as a Director of the Hewlett-Packard Corporation in 1957–1984; and Chairman of the Board of Humphrey Instruments in 1965–1978 and Schwem Technology since 1979.

Family

Married:
    Janet Lucile Landis, December 28, 1958
    Children: Donald Luis, October 15, 1965
            Helen Landis, November 9, 1967

Former Marriage:
    Geraldine Smithwick, 1936
    Children: Walter Alvarez, October 3, 1940
            Jean Alvarez, October 3, 1944

Father:
    Walter C. Alvarez (deceased)
Mother:
    Harriet Smyth Alvarez (deceased)
Sisters:
    Gladys Alvarez Mead, Berkeley CA
    Bernice Alvarez Brownson, San Mateo CA

Brother:
    Robert S. Alvarez, South San Francisco CA

Education:
    1918 to 1924 Madison School, San Francisco CA
    1924 to 1926 Polytechnic High School, San Francisco CA
    1926 to 1928 Rochester High School, Rochester MN
    1928 to 1936 University of Chicago, Chicago IL

Organizational Membership
    Aircraft Owners and Pilots Association (1967)
    American Academy of Arts and Sciences (Fellow 1958)
    American Association for the Advancement of Science (Fellow 1978)
    American Philosophical Society (Elected 1953)
    American Physical Society (Fellow 1939 and President 1969)
    Bohemian Club, San Francisco (1951)
    Faculty Club, Berkeley (1938)
    Institut d'Egypte (Elected Associate Member 1971)
    Mira Vista Country Club, El Cerrito CA (1952)
    National Academy of Engineering (Elected 1969)
    National Academy of Sciences (Elected 1947)
    Optical Society of America (Fellow 1976)

Honorary Degrees
    Sc.D., University of Chicago (1967)
    Sc.D., Carnegie-Mellon University (1968)
    Sc.D., Kenyon College (1969)
    Sc.D., University of Notre Dame (1976)
    Sc.D., Ains Shams University, Cairo (1979)
    Sc.D., Pennsylvania College of Optometry (1981)

Appendix B
Publications

1932 "A Simplified Method for the Determination of the Wavelength of Light," School Science and Math. **32**, 89–91; and Educational Focus, **3** (2), 11–21.

1933 "A Positively Charged Component of Cosmic Rays," with A. H. Compton, Phys. Rev. **43**, 835–836.

1934 "Scientific Work in the Century of Progress Stratosphere Balloon," with A. H. Compton, C. L. Fordney, H. G. Gale, G. F. Smith, F. C. Meier, R. A. Millikan, G. S. Monk, A. Piccard, J. Piccard, T. G. W. Settle, and R. J. Stephenson, Proc. Nat. Acad. Sci. **20**, 79–81.

"On the Interior Magnetic Field in Iron," Phys. Rev. **45**, 225–226.

1935 "Artificial Radioactivity Induced by Cosmic Rays," Phys. Rev. **47**, 320–321.

"The Energies of X-Ray Photoelectrons," Phys. Rev. **47**, 636.

"A Thin-Walled Geiger-Counter," with J. S. Allen, Rev. Sci. Instrum. **6**, 329.

1936 "The Diffraction Grating at Grazing Incidence," J. Opt. Soc. Am. **26**, 343–346.

1937 "Removal of the Ion Beam of the Cyclotron from the Magnetic Field," with E. M. McMillan and A. H. Snell, Phys. Rev. **51**, 148–149.

"Nuclear K Electron Capture," Phys. Rev. **52**, 134–135.

1938 "Cloud Chamber Studies in the Cyclotron Magnetic Field," with W. M. Brobeck, Phys. Rev. **53**, 213.

"Search for Short-lived Radioelements," Phys. Rev. **53**, 215.

"Neutron Yields from Deuteron Reactions at High Energy," Phys. Rev. **53**, 326.

"Electron Capture and Internal Conversion in Ga^{67}," Phys. Rev. **53**, 606.

"Collimated, Variable Energy Beam of Pure Thermal Neutrons," Phys. Rev. **54**, 235.

"Capture of Orbital Electrons by Nuclei," Phys. Rev. **54**, 486–497.

"Production of Collimated Beams of Monochromatic Neutrons in the Temperature Range 300°–10° K," Phys. Rev. **54**, 609–617.

1939 "Heavily Ionizing Particles from Uranium," with G. K. Green, Phys. Rev. **55**, 417.

"Scattering of Ultra-slow Neutrons in Ortho- and Para Hydrogen," with K. S. Pitzer, Phys. Rev. **55**, 596.

"Initial Performance of the 60-Inch Cyclotron of the William H. Crocker Radiation Laboratory, University of California," with E. O. Lawrence, W. M. Brobeck, D. Cooksey, D. R. Corson, E. M. McMillan, W. W. Salisbury, and R. L. Thornton, Phys. Rev. **56**, 124.

"He^3 in Helium," with R. Cornog, Phys. Rev. **56**, 379.

"Helium and Hydrogen of Mass 3," with R. Cornog, Phys. Rev. **56**, 613.

1940 "A Quantitative Determination of the Neutron Moment in Absolute Nuclear Magnetons," with F. Bloch, Phys. Rev. **57**, 111–122.

"Radioactive Hydrogen," with R. Cornog, Phys. Rev. **57**, 248.

"Magnetic Moment of the Neutron," with F. Bloch, Phys. Rev. **57**, 352.

"Isomeric Silver and the Weizsacker Theory," with A. C. Helmholz and E. Nelson, Phys. Rev. **57**, 660–661.

"High Energy Carbon Nuclei," Phys. Rev. **58**, 192–193.

"Radioactive Hydrogen-A Correction," with R. A. Cornog, Phys. Rev. **58**, 197.

"Scattering of 20° Neutrons in Ortho- and Para- Hydrogen," with K. S. Pitzer, Phys. Rev. **58**, 1003–1004.

"Spectroscopically Pure Mercury (198)," with J. Wiens, Phys. Rev. **58**, 1005.

1941 "Recoil from K Capture," with A. C. Helmholz and B. T. Wright, Phys. Rev. **60**, 160.

1942 "Microwave Linear Radiators," MIT Rad. Lab. Report 366, June.

1946 "The Design of a Proton Linear Accelerator," Bull. Am. Phys. Soc. **21**(5), 21.

1947 "The Measurement of Short Time Intervals," Phys. Rev. **72**, 741.

1948 "Properties of the 4-second Neutron Emitter of Low Z," Phys. Rev. **74**, 1217.

1949 "N^{17}, a Delayed Neutron Emitter, Phys. Rev. **75**, 1127–1132.

"Nitrogen 12," Phys. Rev. **75**, 1815–1818.

"A Proposed Experimental Test of the Neutrino Theory," UCRL-328 April.

1950 "Mode Problems in Long Cavities," Engineering Note 200-10, M-24 (Classified) and UCID-947 October.

"Three New Delayed Alpha Emitters of Low Mass," Phys. Rev. **80**, 519–523.

"Relative Densities of Sun and Moon," Am. J. Phys. **18**, 468.

"Electrical Detection of Artificially Produced Mesons," with A. Longacre, V. G. Ogren, and R. E. Thomas, Phys. Rev. **77**, 752.

1951 "Energy Doubling in dc Accelerators," Rev. Sci. Instrum. **22**, 705–706.

1952 "Successive Diffractions by a Concave Grating," with F. A. Jenkins, J. Opt. Soc. Am. **42**, 699–705.

1954 "Neon-18," with J. D. Gow, Phys. Rev. **94**, 365–367.

1955 "Berkeley Proton Linear Accelerator," H. Bradner, J. V. Franck, H. Gordon, J. D. Gow, L. C. Marshall, F. Oppenheimer, W. K. H. Panofsky, C. Richman, and J. R. Woodyard, Rev. Sci. Instrum. **26**, 111–133.

"The Lifetime of the τ Meson," with S. Goldhaber, Nuovo Cimento **X2**, 344–345.

"The Bubble Chamber Program at UCRL," AEC Proposal for 72-inch bubble chamber, April, p. 19.

"Elastic Scattering of 1.6-Mev Gamma Rays from Carbon," with F. S. Crawford and M. L. Stevenson, Phys. Rev. **98**, 280.

1956 "Lifetime of K Mesons," with F. Crawford, M. L. Good, and M. L. Stevenson, Phys. Rev. **101**, 503–505.

"Radiation Hazards," Research Reviews (2), 25–32.

1957 "Catalysis of Nuclear Reactions by Mu Mesons," H. Bradner, F. S. Crawford, Jr., J. A. Crawford, P. Falk-Vairant, M. L. Good, J. D.

Gow, A. H. Rosenfeld, F. Solmitz, M. L. Stevenson, H. K. Ticho, and R. D. Tripp, Phys. Rev. **105**, 1127–1128.

"Interactions of K⁻ Mesons With Protons," with H. Bradner, P. Falk-Vairant, J. D. Gow, A. H. Rosenfeld, F. T. Solmitz, and R. D. Tripp, Nuovo Cimento **X5**, 1026–1046.

"Excerpts from a Russian Diary," Phys. Today **10** (5), 244.

"Further Excerpts from a Russian Diary," Phys. Today **10** (6), 22.

1958 "High-Energy Physics With Hydrogen Bubble Chambers," in *Proceedings of the Second U.N. International Conference on Peaceful Uses of Atomic Energy,* (United Nations, Geneva, 1958), **30**, 164–165.

"Liquid Hydrogen Bubble Chambers," in *Proceedings of the CERN Symposium on High Energy Accelerators and Pion Physics,* edited by A. Citron, G. von Dardel, B. d'Espagnat, Y. Goldschmidt-Clermont, and C. Peyrou, (CERN, Geneva, 1956), *2,* 13–15.

"Elastic Scattering of 1.6 Mev Gamma Rays from H, Li, C and Al Nuclei," with F. S. Crawford, Jr. and M. L. Stevenson, Phys. Rev. **112**, 1267–1273.

1959 "Neutral Cascade Hyperon Event," with P. Eberhard, M. L. Good, W. Graziano, H. K. Ticho, and S. G. Wojcicki, Phys. Rev. Lett. **2**, 215–219.

1960 "Hodoscope Design to Minimize Photomultiplier Use," Rev. Sci. Instrum. **31**, 76.

"Polarized Nucleons as Targets in Hydrogen Chambers," Physics Note 151 and UCID-1122 February.

"The Interactions of Strange Particles," UCRL-9354 August (1959) and in the proceedings of the *Ninth International Conference on High Energy Physics,* (USSR Academy, Moscow, 1960), pp. 471–526.

"A Proposed Device for the Rapid Measurement of Bubble Chamber Film," Physics Note 223, October.

"An Analogue Device for Erasing Beam Tracks from Bubble Chamber Photographs," Physics Note 231, November.

"Resonance in the $\Lambda\pi$ System," with M. Alston, P. Eberhard, M. L. Good, W. Graziano, H. K. Ticho, and S. G. Wojcicki, Phys. Rev. Lett. **5**, 520–524.

1961 "Resonance in the K-π System," with M. Alston, P. Eberhard, M. L. Good, W. Graziano, H. K. Ticho, and S. G. Wojcicki, Phys. Rev. Lett. **6**, 300–302.

"K^- Interactions in Hydrogen at 1.15 Bev/c," with M. Alston, P. Eberhard, M. L. Good, W. Graziano, H. K. Ticho, and S. G. Wojcicki, UCRL-9551 February, and Bull. Am. Phys. Soc. **6**, 292.

"The Σ/Λ Branching Ratio Y^*_1," with M. Alston, P. Eberhard, M. L. Good, W. Graziano, H. K. Ticho, and S. G. Wojcicki, UCRL-9675 February.

"Evidence for a $T=0$ Resonance in the $\Sigma\pi$ System," with M. Alston, P. Eberhard, M. L. Good, W. Graziano, H. K. Ticho, and S. G. Wojcicki, UCRL-9686 April.

"Study of Resonances of the Σ-π System," with M. H. Alston, P. Eberhard, M. L. Good, W. Graziano, H. K. Ticho, and S. G. Wojcicki, Phys. Rev. Lett. **6**, 698–702.

"Liquid Hydrogen Bubble Chambers," in *Experimental Cryophysics,* edited by F. E. Hoare, L. C. Jackson, and N. Kurti (Butterworths, London), pp. 258–264.

"Evidence for a $T=0$ Three-Pion Resonance," with B. C. Maglić, A. H. Rosenfeld, and M. L. Stevenson, Phys. Rev. Lett. **7**, 178–182.

1962 "Ξ^- Hyperons in the Reaction $K^- + p \rightarrow \Xi^- K^+$," with J. P. Berge, R. Kalbfleisch, J. Button-Shafer, F. T. Solmitz, M. L. Stevenson, and H. K. Ticho, UCRL-10236 June and in the *Proceedings of the 1962 International Conference on High-Energy Physics at CERN,* edited by J. Prentki, (CERN, Geneva, 1962), pp. 433–437.

"SMP-1 Scanning and Measuring Projector," with P. Davey, R. Hulsizer, J. Snyder, A. J. Schwemin, and R. Zane, UCRL-10109 April.

"Study of Resonances in the Σ-π System," with M. H. Alston, M. Ferro-Luzzi, A. H. Rosenfeld, H. K. Ticho, and S. G. Wojcicki, UCRL-10233 June and in the *Proceedings of the 1962 International Conference on High-Energy Physics at CERN,* edited by J. Prentki (CERN, Geneva, 1962), pp. 311–314.

"Spin and Parity of the ω Meson," with B. C. Maglić, A. H. Rosenfeld, and M. L. Stevenson, Phys. Rev. **125**, 687–690.

"Adventures in Nuclear Physics," UCRL-10476 October.

1963 "The 1660 MeV Y_1^* Hyperon," M. H. Alston, M. Ferro-Luzzi, D. O. Huwe, G. R. Kalbfleisch, D. H. Miller, J. J. Murray, A. H. Rosenfeld, J. B. Shafer, F. T. Solmitz, and S. G. Wojcicki, UCRL-10569 and Phys. Rev. Lett. **10**, 184–188.

"Design of an Electromagnetic Detector for Dirac Monopoles," Physics Note 470, September.

"A Possible Explanation of High Energy Cosmic Ray Phenomena in Terms of Dirac Magnetic Monopoles," Physics Note 479, November.

1964 "Observation of a Nonstrange Meson of Mass 959 MeV," with G. R. Kalbfleisch, A. Barbaro-Galtieri, O. I. Dahl, P. E. Eberhard, W. E. Humphrey, J. S. Lindsey, D. W. Merrill, J. J. Murray, A. Rittenberg, R. R. Ross, J. B. Shafer, F. T. Shively, D. M. Siegal, G. A. Smith, and R. D. Tripp, UCRL-11358 and Phys. Rev. Lett. **12**, 527–530.

"Electrostatic Potential of the Earth and Moon," with R. Golden and D. Judd, Physics Note 502, April.

"Production of $S=0,-1$ Resonant States in K^-p Interactions at 2.45 GeV/c," with R. R. Ross, J. H. Friedman, D. M. Siegel, S. Flatte, A. Barbaro-Galtieri, J. Button-Shafer, O. I. Dahl, P. Eberhard, W. E. Humphrey, G. R. Kalbfleisch, J. S. Lindsey, D. W. Merrill, J. J. Murray, A. Rittenberg, F. T. Shively, G. A. Smith, and R. D. Tripp. UCRL-11424 July; and in the *Proceedings of the XIIth International Conference on High Energy Physics,* edited by Ya.A. Smorodinsky, (ATOMIZDAT, Moscow, 1966), Vol. **1**, p. 642–646.

"Study of High Energy Interactions, Using a 'Beam' of Primary Cosmic Ray Protons," Physics Note 503, May.

"Proposal for High Altitude Particle Physics Experiment," with W. E. Humphrey, UCBSSL-192 September.

"LRL 25-inch Bubble Chamber," with J. D. Gow, F. Barrera, G. Eckman, J. Shand, R. Watt, D. Norgren, and H. P. Hernandez, UCRL-11521 July and in the *Proceedings of the XIIth International Conference on High Energy Physics,* edited by Ya.A. Smorodinsky, (ATOMIZDAT, Moscow, 1966), Vol. **2**, p. 483–486.

1965 "A Proposal to "X-Ray" the Egyptian Pyramids to Search for Presently Unknown Chambers," Physics Note 544, March.

"A Pseudo Experience in Parapsychology," Science **148**, 1541.

"Parapsychology and Spontaneous Cases," Science **150**, 436.

"A Scientist's Debt to Michelson," (Case Inst. Tech., Cleveland, 1965).

1966 "Round-Table Discussion on Bubble Chambers," UCRL-17096 September; and in *Proceedings of the 1966 International Conference on Instrumentation for High Energy Physics,* (SLAC, Stanford, 1966), pp. 271–295.

1968 "The Use of Liquid Noble Gases in Particle Detectors with 1) High Spatial Resolution over a Large Area and 2) High Energy Resolution as Total Absorption Counters," Physics Note 672, November.

1969 "Recent Developments in Particle Physics," UCRL-18696 and Science **165**, 1071–1091.

"Putting the Moon Where It Belongs," Nature (London) **223**, 1082–1083.

"Cosmic Ray Studies With a Superconducting Magnet in a Space Station Facility," with J. A. Anderson and A. Buffington, UCBSSL unnumbered July.

"The Prospect of High Spatial Resolution for Counter Experiments: A New Particle Detector Using Electron Multiplication in Liquid Argon," with S. E. Derenzo, R. A. Muller, and R. G. Smits, UCRL-19254 October; and in *1969 Summer Study,* edited by A. Roberts, NAL SS-154, **3**, 79–102.

1970 "Ernest Orlando Lawrence," UCRL-17359 January; and Nat. Acd. Sci. Biographical Memoirs **41**, 251–294.

"Dedication for Sulamith Goldhaber," in *Advances in Particle Physics,* edited by R. E. Marshak and R. Cool, (New York: Wiley), 2, vii–ix.

"Search for Magnetic Monopoles in the Lunar Samples of Apollo 11," with P. H. Eberhard, R. R. Ross, and R. D. Watt, UCRL-19440 (1969); Science **167**, 701–703 (1970); and in *Proceedings of Apollo 11 Lunar Science Conference 3,* edited by A. A. Levinson (New York, Pergamon 1970), Vol. **3**, pp. 1953–1957.

"Search for Hidden Chambers in the Pyramids," with J. A. Anderson, F. El Bedwei, J. Burkhard, A. Fakhry, A. Girgis, A. Goneid, F. Hassan, D. Iverson, G. Lynch, Z. Miligy, A. H. Moussa, Mohammed-Sharkawi, and L. Yazolino, Science **167**, 832–839.

"Tune Without a Piper," Nature (London) **227**, 534.

"Superconducting Magnetic Spectrometer Experiment for HEAO," Technical Proposal, with A. Buffington, C. L. Deney, J. P. Dooher, A. J. Favale, R. Golden, R. Kirz, R. Madey, R. Muller, E. Pickup, E. S. Schneid, L. H. Smith, P. Sub, H. J. Vroom, and M. A. Wahlig, UCBSSL-374 May.

"Particle Detectors Based on Noble Liquids," with R. A. Muller, S. E. Derenzo, R. G. Smits, and H. Zaklad, UCRL-20135 September; and in *Proceedings of the International Conference on Instrumentation for High Energy Physics,* edited by V. P. Dzhelepov, (USSR Academy, Dubna, 1970), pp. 301–304.

"Recent Developments in High-Resolution Noble Liquid Counters," with S. E. Derenzo, D. B. Smith, R. G. Smits, H. Zaklad, and R. A. Muller, UCRL-20118 September; and in *National Accelerator Laboratory 1970 Summer Study,* NAL SS-181, pp. 45–74.

"Proof of J. J. Thomson's Famous Theorem Regarding Electric and Magnetic Charges," Physics Note 722, December.

1971 "Search for Magnetic Monopoles in Lunar Material," with P. H. Eberhard, R. R. Ross and R. D. Watt, UCRL-20835 and Phys. Rev. D**4**, 3260–3272.

"Liquid-Filled Proportional Counter," with R. A. Muller, S. E. Derenzo, G. Smadja, D. B. Smith, R. G. Smits, and H. Zaklad, UCRL-20811 June and Phys. Rev. Lett. **27**, 532–535.

"A Magnetic Monopole Detector Utilizing Superconducting Elements," with M. Antuna Jr., R. A. Byrns, P. H. Eberhard, R. E. Gilmer, E. H. Hoyer, R. R. Ross, H. H. Stellrecht, J. D. Taylor, and R. D. Watt, UCRL-19756 April (1970) and Rev. Sci. Instrum. **42**, 326–330.

"Measurement of the Primary Cosmic Ray Nuclear Rigidity Spectra for Individual Elements of Charge $Z>2$," with L. Smith, A. Buffington, G. Smoot, and M. Wahlig, UCBSSL HEAO Note 179, December.

"High Precision Charged Particle Detector using Noble Liquids," with S. E. Derenzo, G. Smadja, R. G. Smits, H. Zaklad, and R. A. Muller, Nature (London) **233**, 617.

"A Search for Antimatter in Primary Cosmic Rays," with A. Buffington, L. H. Smith, G. F. Smoot, and M. A. Wahlig, Particle Physics Note 176, November (1971); and Nature (London) **236**, 335–338 (1972).

1972 "Transverse/Optics I. A Thin Lens With Variable Spherical Power," J. Opt. Soc. Am. **62**, 727.

"Forward," in *Adventures in Experimental Physics,* edited by B. Maglich (Princeton: World Science), vol. α, pp. i–iv.

"Story of the Search for Chambers in the Pyramids," in *Adventures in Experimental Physics,* edited by B. Maglich, (Princeton: World Science), vol. α, pp. 157–188.

"A Liquid Xenon Radioisotope Camera," with H. Zaklad, S. E. Derenzo, R. A. Muller, G. Smadja, and R. G. Smits, LBL-338 March and in *1972 Thirteenth Scintillation and Semiconductor Detectors,* edited by R. L. Chase, Nucl. Sci. **NS-19**, 206–213.

"A Measurement of Cosmic Ray Rigidity Spectra Above 5 GeV/*c* of Elements from Hydrogen to Iron," with L. H. Smith, A. Buffington, G. F. Smoot, and W. A. Wahlig, UCBSSL unnumbered.

"Search for Antimatter in Primary Cosmic Rays," with A. Buffington, L. H. Smith, G. F. Smoot, and M. A. Wahlig, Nature (London) **236**, 335–338.

"Liquid-Xenon-Filled Wire Chambers," S. E. Derenzo, R. Flagg, S. G. Louie, F. G. Mariam, T. S. Mast, and A. J. Schwemin, with R. G. Smits and H. Zaklad, LBL-1321 October; and in *Proceedings of the XVI International Conference on High Energy Physics,* edited by J. D. Jackson and A. Roberts (NAL, Batavia, 1972), **2**, 388–402.

"Reminiscences—Early History of the Search for Solar Neutrinos," in *Proceedings of Solar Neutrino Conference,* edited by F. Reines and V. Trimble, (Irvine: Univ. of California), pp. 1–3.

1973 "Search for Magnetic Monopoles in Lunar Material Using an Electromagnetic Detector," with R. R. Ross, P. H. Eberhard, and R. D. Watt, LBL-1730 May and Phys. Rev. D**8**, 698–702.

"A Signal Generator for Ray Davis' Neutrino Detector," Physics Note 767, March.

"High-Resolution Liquid-Filled Multi-Wire Chambers for Use in High-Energy Beams," with S. E. Derenzo, A. Schwemin, R. G. Smits and H. Zaklad, LBL-1791 April and in the *Proceedings of the 1973 International Conference on Instrumentation for High Energy Physics,* edited by S. Stipcich (INFN, Frascati, 1973), pp. 305–312.

"Liquid Xenon Filled Wire Chambers for Medical Imaging Applications," with S. E. Derenzo, T. F. Budinger, R. G. Smits, and H. Zaklad, LBL-2092 May and in *Proceedings of the Symposium on Advanced Technology Arising from Particle Physics Research,* edited by R. C. Arnold, G. H. Thomas, and B. W. Wicklund (ANL, Argonne, 1973), pp. 11.1–11.16.

"Voice of a Physicist, an Invited Commentary," in *Adventures in Experimental Physics,* edited by B. Maglich, (Princeton: World Science), vol. γ, pp. v-vi.

1974 "Berkeley: A Lab Like No Other," Bull. Atomic Scientists **30** (4), 18–23.

" 'Aether Drift' and the Isotropy of the Universe. A Proposed Search for Anisotropies in the Primordial Black Body Radiation," with A. Buffington, P. Dauber, T. Mast, R. A. Muller, C. Orth, and G. Smoot, UCBSSL unnumbered, January.

"Liquid Xenon Multiwire Proportional Chambers for Nuclear Medicine Applications," with H. Zaklad, S. E. Derenzo and T. F. Budinger, LBL-3000 May.

"A New γ-Ray Imaging Camera," Physics Note 748, May.

"A Doppler-Shifted Black Body Spectrum in Small β Approximation," Physics Note 782, May.

1975 "A Mechanical Analog of the Synchrotron Illustrating Phase Stability and Two-Dimensional Focusing," with R. Smits and G. Senecal, LBL-2466 December (1973) and Am. J. Phys. **43**, 293–296 (1975).

"Evidence at the 10^{-18} Probability Level against the Production of Magnetic Monopoles in Proton Interactions at 300 GeV/c," with P. H. Eberhard, R. R. Ross, J. D. Taylor, and H. Oberlack, LBL-3680 February and Phys. Rev. **D11**, 3099–3104.

"Alfred Loomis. Obituary," Phys. Today 28(11), 84–85. (See also under 1980).

"Analysis of a Reported Magnetic Monopole," LBL-4260 September and in *Proceedings 1975 International Symposium on Lepton and Photon Interactions at High Energies,* edited by W. T. Kirk, (SLAC, Stanford), pp. 967–979.

1976 "A Proposed Search for Anti-iron Cosmic Rays," Astrophysics Note 328, September.

"A Physicist Examines the Kennedy Assassination Film," LBL-3884 July (1975) and Am. J. Phys. **44**, 813–827.

"Search for Hidden Chambers in the Pyramids," Nat. Geographic Soc. Research Reports—1968 Projects.

1977 "Mass Spectroscopy With the Berkeley 88-inch Cyclotron," with E. J. Stephenson, D. J. Clark, R. A. Gough, W. R. Holley, A. Jain, and R. Muller, Bull. Am. Phys. Soc. **22**, 579.

"Quarks With Unit Charge: A Search for Anomalous Hydrogen," with R. A. Muller, W. R. Holley, and E. J. Stephenson, Science **196**, 521–523.

1978 "Development of Variable-Focus Lenses and a New Refractor," J. Am. Optometric Assn. **49**, 24–29.

1979 "Anomalous Iridium Levels at the Cretaceous-Tertiary at Gubbio, Italy," with W. Alvarez, F. Asaro and H. V. Michel, in *Proceedings of the Symposium on Cretaceous-Tertiary Boundary Events on,* edited by W. K. Christensen and T. Birkelund, (Univ. Copenhagen, Copenhagen), **2**, 69.

"Experimental Evidence in Support of an Extra-Terrestrial Trigger for the Cretaceous-Tertiary Extinctions," with W. Alvarez, F. Asaro, and H. V. Michel, Eos **60**, 734.

"Anomalous Iridium Levels at the Cretaceous-Tertiary Boundary at Gubbio, Italy: Negative Results of Tests for a Supernova Origin," with W. Alvarez, F. Asaro, and H. V. Michel, Geol. Soc. Am. Abstr. **11**, 378.

1980 "Alfred Lee Loomis (1887–1975)," Nat. Acd. Sci. Biographical Memoirs **51**, 309–341.

"Extraterrestrial Cause for the Cretaceous-Tertiary Extinction," with W. Alvarez, F. Asaro, and H. Michel, LBL-9666 November (1979) and Science **208**, 1095–1108.

"Anomalous Iridium at the Cretaceous-Tertiary Boundary: The Impact Hypothesis," with W. Alvarez, F. Asaro, and H. V. Michel, EOS **61**, 259.

"Results of a Dating Attempt—Chemical and Physical Measurements Relevant to the Cause of the Cretaceous-Tertiary Extinctions," with F. Asaro, H. V. Michel, and W. Alvarez, LBL-11613 September and in *Proceedings of the Symposium on Nuclear and Chemical Dating Techniques,* edited by L. A. Currie, (Amer. Chem. Soc., Houston) Series 176, 401–409 (1982).

1981 "Asteroid Extinction Hypothesis," with D. V. Kent, G. C. Reid, C. Brown, E. Randall, W. Alvarez, F. Asaro, and H. Michel, Science **211**, 654–656.

"Personal Recollections," in *Professor Ryokichi Sagane* (Tokyo, 1981), pp. 415–422.

"The Early Days of Accelerator Mass Spectrometry," LBL-12846 May; *Proceedings of Conference on Accelerator Mass Spectrometry* (ANL, Argonne, 1981); and Phys. Today **35** (1), 25–32 (1982).

"Distribution of Iridium and Other Elements Near the Cretaceous-Tertiary Boundary in Hole 465A: Preliminary Results," with H. V. Michel, F. Asaro, and W. Alvarez, (Rpts. Deep Sea Drilling Proj., edited by L. N. Stout et al., 847–849.

"Asteroids and Dinosaurs," in *Proceedings of the Celebration of, the 50th Anniversary of the Lawrence Berkeley Laboratory—Symposium and Banquet Speeches,* LBL-13613 October, pp. 3–26.

1982 "Geochemical Anomalies Near the Eocene-Oligocene and Permian-Triassic Boundaries," with F. Asaro, W. Alvarez, and H. V. Michel, in *Proceedings of the Conference on Large Body Impacts and Terrestrial Evolution: Geological, Climatological and Biological Implications,* edited by L. T. Silver and P. H. Schultz. Geo. Soc. of Am. Special Paper **190**, 517–528.

"Iridium and Other Geochemical Profiles Near the Cretaceous-Tertiary Boundary in a Brazos River Section in Texas," with F. Asaro,

H. V. Michel, W. Alvarez, R. F. Maddocks, F. Rosalie, and T. Bunch, in *Guidebook of Excursions and Related Papers for the Eighth International Symposium on Ostracoda,* edited by R. F. Maddocks, (Houston: Univ. Houston), pp. 238–241.

"Iridium Anomaly Approximately Synchronous with Terminal Eocene Extinctions," with W. Alvarez, F. Asaro, and H. Michel, Science **216**, 886–888.

"Micropaleontological, Mineralogical and Chemical Analysis of K-T Boundary Clay in the Northern Apennines," with A. Montanari, W. Alvarez, F. Asaro, and H. Michel, Geol. Soc. Am. Abstr. **14**, 569.

"New Data on the Cretaceous-Tertiary Extinction," with W. Alvarez, F. Asaro, H. V. Michel, M. A. Arthur, W. E. Dean, D. A. Johnson, M. Kastner, F. Maurasse, R. R. Revelle, and D. A. Russell, AAAS Pub. **82-2**, 47.

"Current Status of the Impact Theory for the Terminal Cretaceous Extinction," with W. Alvarez, F. Asaro, and H. V. Michel, Geol. Soc. Am. Special Paper **190**, 305–315.

"Reexamination of Acoustic Evidence in the Kennedy Assassination," Committee on Ballistic Acoustics, National Research Council Commission on Physical Sciences, Mathematics, and Resources, and in Science **218**, 127–133.

"Major Impacts and Their Geological Consequences," with W. Alvarez, F. Asaro, and H. V. Michel, Geol. Soc. Am. **14** (7), 431–432.

1983 "Abundance Profiles of Iridium and Other Elements Near the Cretaceous-Tertiary Boundary in Hole 516F of Deep Sea Drilling Project Leg 72," with H. V. Michel, F. Asaro, W. Alvarez and D. A. Johnson, in *Initial Rep. Deep Sea Drilling Proj.* **72**, 931–936, edited by E. Whalen.

"Experimental Evidence that an Asteroid Impact Led to the Extinction of Many Species 65 Million Years Ago," Proc. of the Nat. Acad. of Sci. **80**, 627–642.

"Alfred Lee Loomis—Last Great Amateur of Science," LBL-6700, July (1977) and Phys. Today **36** (1), 25–34.

"Development of Radar," Letter to the Editor and Response, Phys. Today **36** (6), 101–104.

1984 "Spheroids at the Cretaceous-Tertiary Boundary are Altered Impact Droplets of Basaltic Composition," with B. Bohor, A. Montanari, R. L. Hay, W. Alvarez, F. Asaro, H. V. Michel, and J. Smit, Geol. **11**, 695–696.

"Impact Theory of Mass Extinctions and the Invertebrate Fossil Record," with W. Alvarez, E. G. Kauffman, F. Surlyk, F. Asaro, and H. V. Michel, Science **223**, 1135–1141.

"The Precursor of the K-T Boundary Clays at Stevns Klint, Denmark and DSDP Hole 465A," with M. Kastner, F. Asaro, H. V. Michel, and W. Alvarez, Science **226**, 137–143.

"Element Profile of Ir and Other Elements Near the K-T Boundary in Hole 577B," with H. V. Michel, F. Asaro, and W. Alvarez, Science **223**, 1183–1186.

"The End of the Cretaceous: Sharp Boundary or Gradual Extinction?" with W. Alvarez, F. Asaro, and H. V. Michel, Science **223**, 1135–1141.

1987 "Mass Extinctions Caused by Large Bolide Impacts," LBL 22786 December and Phys. Today **40**, 24–33.

Appendix C
U.S. Patents

"Communication System," No. 2,479,195; August 16, 1949. GCA Indicator for airplanes.

"Radio Distance and Direction Indicator," No. 2,480,208; August 30, 1949. EAGLE blind bombing.

"Airway Monitoring and Control Systems," No. 2,514,436; July 11, 1950. TRICON, Part II, traffic control.

"Radio Beacon and Discrimination Circuit Therefor," No. 2,527,474; October 24, 1950. Minor circuit.

"Radio-Echo Detection and Location Apparatus for Approaching Hostile Craft," No. 2,530,418; November 21, 1950. VIXEN.

"Linear Accelerator," No. 2,545,595; March 20, 1951. The first proton linear accelerator.

"Aircraft Control System," No. 2,555,101; May 29, 1951. with L. H. Johnston. GCA.

"Radio Beacon and System Utilizing It," No. 2,568,265; September 18, 1951. The basic microwave transponder.

"Antenna System with Variable Directional Characteristics, No. 2,605,413; July 29, 1952. Scannable microwave linear arrays used in GCA, EAGLE and MEW.

"Pulse Discriminating Circuit," No. 2,626,352; January 20, 1953. Circuit for method of delayed coincidences.

"Radio Navigation System," No. 2,668,287; February 2, 1954. TRICON, Part I, navigational features.

"Thermionically Emissive Element," No. 2,672,567; March 16, 1954. For satellite use but made obsolete by the transistor.

"Radioactivity Measurement," No. 2,685,027; July 27, 1954. Magnetic memory for short-lived radioactivities.

"Vertical Determination Device," No. 2,706,793; April 19, 1955. with B. Rossi and F. C. Chromey. Uses cosmic rays to sense the effect of gravity on the atmosphere.

"Coded Pulse Generating Circuit," No. 2,717,960; September 13, 1955. Coding circuit for microwave beacon transmitter.

"Electrode Construction for Cathode-Ray Tubes," No. 2,777,083; January 8, 1957. For Lawrence color TV tubes.

"Suppressed Side-Lobe Radio Receiving System," No. 2,804,614; August 27, 1957. Eliminates side lobes in radar antennas.

"Golf Training Device," No. 2,825,569; March 4, 1958. Stroboscopic device.

"Color Television Display Screen," No. 2,878,411; March 17, 1959. Low-Z coating to suppress electron scattering.

"Electronuclear Reactor," No. 2,933,442; April 19, 1960. with E. O. Lawrence and E. M. McMillan. The MTA accelerator.

"X-Ray Spectroscope Comprising Plural Sources, Filters, Fluorescent Radiators and Comparative Detectors," No. 3,114,832; December 17, 1963. X-ray detector for explosives in airplane baggage.

"Optical Range Finder with Variable-Angle Exponential Prism," No. 3,299,768; January 24, 1967. Stadimetric range finder, nautical.

"Two-Element Variable Power Spherical Lens," No. 3,305,294; February 21, 1967. Variable focus lens, elements move transverse to optic axis.

"Scanning Apparatus for Aiding in the Determination of Point Coordinates of Paths of Charged Particles as Recorded on Photographic Film," No. 3,366,794; January 30, 1968. The Scanning and Measuring Projector, SMP.

"Gyroscopically Controlled Accidental Motion Compensator for Optical Instruments," No. 3,378,326; April 16, 1968. The first inertially stabilized optical system. Assigned to Bell and Howell.

"Gyroscopic Lens," No. 3,434,771; March 25, 1969. EYEBALL, a version of the above patent. Assigned to Bell and Howell.

"Stabilized Zoom Optical Device," No. 3,468,596; September 23, 1969. A stabilized mono-binocular.

"Variable-Power Lens and System," No. 3,507,565; April 21, 1970. with W. E. Humphrey. Improved variable focus lens.

"Subatomic Particle Detector with Liquid Electron Multiplication Medium," No. 3,659,105; April 25, 1972. with S. E. Derenzo, R. A. Muller, R. G. Smits and H. Zaklad. Liquid xenon counters.

"Aircraft Warning System," No. 3,563,651; February 16, 1971. with W. E. Humphrey. Uses ultraviolet light in the "solar blind" region.

"Method of Making Fresnelled Optical Element Matrix," No. 3,739,455; June 19, 1973.

"Method of Forming an Optical Element of Reduced Thickness," No. 3,829,536; August 13, 1974.

"Deuterium Tagged Articles such as Explosives and Method for Detection Thereof," No. 4,251,726; February 17, 1981. Uses (γ,n)-reaction on deuterium with neutron detection.

"Color Television Viewer," No. 4,301,468; November 17, 1981. Hand-held color TV system with large image.

"Stabilized Zoom Binocular," No. 4,316,649; February 23, 1982. with A. J. Schwemin. Impracticable system, evolved into stabilized zoom mono-binoculars.

"Stand Alone Collision Avoidance System," No. 4,317,119; February 23, 1982. For use on large commercial aircraft.

"Television Viewer," No. 4,399,455; April 16, 1983. Field sequential projection system for use with computers.

"Stabilized Zoom Device," No. 4,417,788; November 29, 1983. with A. J. Schwemin.

"Dead Reckoning Range-Finding Device for Cart," No. 4,480,310; October 30, 1984. For golf cart with computer output showing distance to the green.

"Optically Stabilized Camera Lens System," No. 4,615,590; October 7, 1986. With A. J. Schwemin.

Commentators

Frank Asaro, Senior Staff Scientist, Chemistry Division, Lawrence Berkeley National Laboratory, Berkeley, California

Hans A. Bethe, Professor Emeritus, Physics Department, Cornell University, Ithaca, New York

Andrew Buffington, Senior Staff Scientist, Astronomy Department, University of California, San Diego, California

Robert Cornog, President, Cornog Enterprises, Los Angeles, California

Philippe H. Eberhard, Senior Staff Scientist, Physics Division, Lawrence Berkeley Laboratory, Berkeley, California

Richard L. Garwin, Fellow, IBM Thomas J. Watson Research Center, Yorktown Heights, New York

William Humphrey, Executive Vice President, Humphrey Instruments, San Leandro, California

J. David Jackson, Professor, Physics Department, University of California, Berkeley, California

Lawrence Johnston, Professor, Physics Department, University of Idaho, Moscow, Idaho

Franklin Miller, Jr., Professor Emeritus, Physics Department, Kenyon College, Gambier, Ohio

Richard A. Muller, Professor, Physics Department, University of California, Berkeley, California

H. Victor Neher, Professor Emeritus, Physics Department, California Institute of Technology, Pasadena, California

W. K. H. Panofsky, Director Emeritus and Professor, Stanford Linear Accelerator Center, Stanford University, Stanford, California

Kenneth S. Pitzer, Professor, Chemistry Department, University of California, Berkeley, California

Norman F. Ramsey, Professor Emeritus, Harvard University, Cambridge, Massachusetts.

Peter H. Rose, President IBSD, Eaton Corporation, Beverly, Massachusetts

Emilio Segrè, Professor Emeritus, Physics Department, University of California, Berkeley, California

M. Lynn Stevenson, Professor, Physics Department, University of California, Berkeley, California

Cornelius A. Tobias, Professor, Biophysics Department, University of California, Berkeley, California

Robert D. Watt, Senior Staff Scientist Emeritus, Stanford Linear Accelerator Center, Stanford University, Stanford, California

Jacob H. Wiens, Professor Emeritus, San Mateo College, San Mateo, California (deceased)

Stanley G. Wojcicki, Professor, Physics Department, Stanford University, Stanford, California